Climate, Planetary and Evolutionary Sciences

Guido Visconti

Editor

Climate, Planetary and Evolutionary Sciences

A Machine-Generated Literature Overview

 Springer

Editor
Guido Visconti
Dipartimento di Scienze Fisiche e Chimiche
Università Degli Studi dell'Aquila
L'Aquila, Italy

ISBN 978-3-030-74712-1   ISBN 978-3-030-74713-8   (eBook)
https://doi.org/10.1007/978-3-030-74713-8

This Springer imprint is published by the registered company Springer Nature Switzerland AG
The registered company address is: Gewerbestrasse 11, 6330 Cham, Switzerland

Foreword

This book presents the result of an innovative challenge, to create a systematic literature overview driven by machine-generated content. The topics were inspired in conversation and collaboration with Prof. Guido Visconti, who devised a series of questions and related keywords for the machine to query, discover, collate and structure by artificial intelligence clustering. The AI-based approach seemed especially suitable to provide an innovative perspective as the topics are indeed both complex and interdisciplinary, for example, climate, planetary and evolution sciences, so it presented quite the task! Springer Nature has published much on these topics in its journals over the years, so the exercise was for the machine to identify the most relevant content and present it in a structured way that the reader would find useful.

The automatically generated literature summaries in this book are intended as a springboard to further discoverability on a specific topic. They are particularly useful to readers with limited time, looking to learn more about the subjects quickly, especially if they are new to the topics. They can help to identify the most relevant Springer Nature research within this field and therefore be highly useful, e.g., for developing class reading lists or discovering content for human-written literature reviews. Springer Nature seeks to support anyone who needs a fast and efficient start in their content discovery journey, from the undergraduate student exploring interdisciplinary content, to master or Ph.D. thesis researchers developing research questions, to the practitioner searching for support materials, this book can serve as an inspiration, to name a few examples.

So why did we take on this experimental approach? We want to continue our innovation journey that started with the first machine-generated research book on Lithium-Ion Batteries published in 2019. The amazing number of downloads clearly demonstrated that there is a strong interest to further explore the role of artificial intelligence in research publishing. Since then, our efforts continued to experiment further with discoverability and how to find new ways to identify highly relevant content within the myriad of publications available, structuring and summarizing them in a way useful and time saving for readers.

From a publishing perspective, the user, which can be the author or the reader, is in the focus of all our activities. We have conducted surveys and user research to identify solutions best serving their needs. As mentioned from the beginning, this

AI-driven approach should be considered as a joint journey of author, publisher and machine. It is important to us as a publisher to make the advances in technology easily understandable and accessible to our authors and find new ways of AI-based author services that allow human–machine interaction to generate readable, usable, collated, research content.

The human is the start and end of every project. The author can pick the topic query and/or question to be answered and define the criteria for the data set that will be summarized by the machine. The algorithm then identifies and clusters the data set. Our user interface allows the author to dive into the content, interact with the AI and finally refine the clustering. The machine-generated structure can then be used by the author to inform themselves, identify individual blind spots and publications they are unaware of, as well as pinpoint articles in a speedy manner or to generate a summary of that research for incorporation into a publication, courses or further reading, as in this book.

Transparency is one of the cornerstones of our efforts. As with all experiments, it will not present perfection. We are on a journey based on machine-generated content and would like to continue to share this journey with the research community, at the various milestones we reach together with our authors. Looking forward, we hope the reader will be inspired by this concept and embrace this innovation for their own enquiries for machine-generated content. As we continue to think of new ideas, product development, further human–machine interactions, we would be pleased to hear your feedback. Simply send an e-mail to stephanie.preuss@springer.com with any comments or ideas. If you are interested in editing your own AI-supported book, please also contact us with your ideas.

Much appreciation goes to Prof. Guido Visconti, who took on this challenge with an open mind. His professional curation and insightful introductions set the scene to each chapter and enable the readers to start their journey of content discovery.

Robert Doe
Stephanie Preuss
Springer Nature

Preface

Robert Doe wrote me an e-mail some time ago with a new idea for a book. His suggestion was to use artificial intelligence to do some kind of mining of the vast scientific literature accumulated in the Springer publications. He asked me to suggest the topics to be searched and then write some comments on the contents of the resulting papers. I accepted the invitation right away and prepared a list that reflected what I think are the most interesting and promising topics in the field of climate, atmospheric and planetary evolution, geobiology and some esoteric ideas about the Fermi paradox, Gaia stuff, exoplanets and the implications of their discoveries on our future.

My idea of science (at least at my age) is that, on one side, the money spent on it should be always justified while on the other side it must imply some fun on the part of the person who is involved. The result is that we have very serious stuff, like the stochastic climate models or the response theory, and then have some relaxation with Gaia or the discussion on life in the universe. The topics given here are quite arbitrary and may be considered as a test whose results could be applied to any other field.

The "machine," after finding the papers, would make a summary of the content, and for each relevant section it would add an acknowledgment. The different chapters however are not quite independent and can even be grouped. The topics climate and atmosphere are included in the first five chapters and relate to different aspects of the topic so that climate studies are treated in their most recent development (response theory and stochastic weather and climate modeling) but also in their most useful application (downscaling). The maximum entropy principle (MEP) can be also considered to be part of climate studies, although the most promising application seems to be in the field of ecology. The Gaia theory could also be considered as part of the climate studies, but of course its implications are more important than that. At the present time, it is still very controversial with heavy and not negligible objections but at the same time apparently a revival could be based on the application of Markov blankets. The chapter on geobiology is strictly related to climate studies because we do include papers on the biogeochemical cycles which determine the concentration of the greenhouse gases in the atmosphere.

The most entertaining content of this exercise is concentrated in three chapters on astrobiology, exoplanets and the Fermi paradox. However, if the papers are read carefully we could find some useful insights about the future of our planet and the implications to discover even elementary forms of life on other planets. As of today, we have experienced no contacts with extraterrestrial intelligence and some begin to have doubts about the existence of life, at the least in our galaxy. One of the explanations for that is the existence of the so-called Great Filter, an episode (natural or artificial) in the history of civilizations that annihilate any further development. If we should find any form of life on any other planets, this would mean that life is not a rare event and we on Earth should expect the Great Filter in our future. These ideas are those of Adam Frank that encourage and justify the study of astrobiology.

What about the performance of AI? In the science fiction novel by Robert Sheckley, "Ask a foolish question" (1953), a supercomputer is built that should be able to answer the ultimate question. His name is "The Answerer" but was limited in its answers by the context of the questions posed. The moral of the story is that "…In order to ask a question you must already know most of the answer." In our case, we have the same problem, and when we asked the machine about "dynamical systems and climate" or "response theory and climate change," we got all kinds of answers so we decided to introduce a chapter on downscaling and one on progress in climate modeling. For the rest, the AI worked just fine considering that this is limited to papers appearing in journals and not to books or material from meetings, etc.

We have noticed that the scientific debate in the papers is rather lively but especially in progress in climate modeling we noticed a tendency to use quite sophisticated new mathematical tools and a growing distance from the classical practitioners of modeling. This fact could be in response to the invocation of the authorities of the trade (like Tim Palmer), to be more and more involved in the climate studies of mathematicians and physicists. However, we must keep in mind that each scientific endeavor has its tools like biology and physics but they are both respectable sciences.

The Preface usually concludes with acknowledgments, and I have to thank all the people that permitted the use of the "machine."

L'Aquila, Italy   Guido Visconti

About the Machine

This book was created using the Dimensions Auto-Summarization platform, which was developed and customized by Digital Science in collaboration with Springer Nature in 2020. The platform builds on a repository of over 8 million Springer Nature publications and supports book and journal editors with a systematic production pipeline and user interface that combines many aspects of state-of-the-art Natural Language Processing (NLP) and Artificial Intelligence (AI): multi-level document clustering and ordering based on datasets retrieved from the Dimensions search engine, single document extractive auto-summarization as well as bibliography and manuscript management.

Contents

About the Editor

Guido Visconti is Professor Emeritus at Università dell'Aquila, Italy and a member of the National Academy of Lincei. He has held a Fulbright Fellowship (1968–69), University of Maryland (USA). NATO Fellowship, Department of Meteorology, MIT, USA, (1976, 1977). NATO Senior Fellowship, NCAR, USA (1986–1987). He has served as committee member of the Intergovernmental Panel for Climatic Change (IPCC) and Member of the International Ozone Commission (WMO). In his long career he was the Principal Investigator UARS Correlative Measurements Program (NASA) and Atmospheric Effects of Supersonic Airplane (AESA, NASA). He is the author of the following books; *Fundamentals of Physics and Chemistry of the Atmosphere*, Second Edition, (2016), *Problems, Philosophy and Politics of Climate Science*, (2018) and *Fluid dynamics* (2020).

Chapter 1
Origin and Evolution of Atmospheres

Introduction by Guido Visconti

The evolution of the Earth's atmosphere is not limited to its chemical composition but also involves other physical characteristics like pressure and temperature. It is obvious the impact of such variables on the climate of the planet. On the other hand, there is a clear influence of the geological history of the planet on the atmospheric composition. Here the term geological refers mainly to the solid part of the planet and its interactions with the ocean and the atmosphere.

Life is another important factor which affects the evolution of the planet's atmosphere. Traces of life existed quite early in the Earth's history (around 3.8 billion years ago) and remained "dormient" (or simply they were exploring most efficient evolutionary ways) for more than a couple of billion years until it appeared to have a quite rapid surge after the end of the "snow ball Earth" half a billion years ago. Before that time dominant chemical reducing atmospheric environment was changed through the action of micro-bacteria that among other things provided hazes that contributed to compensate for the lower luminosity of the Sun at that early time.

There is a basic misunderstanding about the role of the photosynthesis as the sources of atmospheric oxygen. We know that photosynthesis (through which oxygen is produced) and respiration (through which oxygen is consumed) do not balance on a global scale because a tiny part of the photosynthesized material is subtracted to the respiration process (for example due to flooding, burial, etc.). This imbalance constitutes a source of the atmospheric oxygen but elementary calculations show that the consumption of all organic compounds would deplete the atmospheric oxygen less than 1%. The attention should be directed in other directions and plate tectonics is the most important. Oxygen is then produced during the formation of reduced sedimentary material and consumed when the same sediments are brought up to the surface. The role of microbiota is important also in this case.

A possible proof of this hypothesis is the explanation of the Great Oxidation Event when about 2.5 billion year ago a sudden increase in the abundance of oxygen was observed. The rising of atmospheric oxygen has been a fundamental factor for the

© The Author(s), under exclusive license to Springer Nature Switzerland AG 2021
G. Visconti (ed.), *Climate, Planetary and Evolutionary Sciences*,
https://doi.org/10.1007/978-3-030-74713-8_1

development of the major forms of life as we know today. One of the hypotheses is that during that time the removal of the carbon dioxide from the atmosphere changed from a sea floor process to a continental process.

The atmospheric composition influences the climate of our planet. A nice proof of these effects are the so-called "faint young Sun paradox" that was discovered by Carl Sagan and George Mullen in 1972. The Sun is a star which changes its luminosity of about 5% every billion years. This means that in the past the Earth should have experienced very cold periods which do not exist in the geologic records. Most of the explanation of this paradox invokes changes in the atmospheric composition.

The study of the evolution of atmospheric composition has obvious implications for astrobiology. First of all, hopefully, we could observe exoplanets with their atmosphere undergoing different phases of evolution. The existence of forms of life at different stages of development could indicate clues of the possible outcome of the influence of an industrialized society on the environment.

Machine-Generated Summaries

Keywords: *atmosphere, atmospheric, model, organic, value, evolution, cycle, isotope, water, rate, temperature, gas, archean, life, component*

Origin and Evolution of the Atmospheres of Early Venus, Earth and Mars

https://doi.org/10.1007/s00159-018-0108-y

Abstract-Summary

We review the origin and evolution of the atmospheres of Earth, Venus and Mars from the time when their accreting bodies were released from the protoplanetary disk a few million years after the origin of the Sun.

The evolution scenario of early Earth is then compared with the atmospheric evolution of planets where no active plate tectonics emerged like on Venus and Mars.

This review concludes with a discussion on the implications of understanding Earth's geophysical and related atmospheric evolution in relation to the discovery of potential habitable terrestrial exoplanets.

Extended

The present review, therefore, aims to piece together the latest findings in astrophysics, planetary formation, geophysics and atmosphere evolution, to understand why early Earth evolved to a Class I habitat and why Venus and Mars did not.

Introduction

There are arguments that point to the fact that significant amounts of noble gases have been trapped from the protoplanetary disk, and left in solar composition in the interiors of early Venus and Earth (e.g. Mizuno et al. [1]; Sasaki and Nakazawa [2]; Porcelli and Pepin [3]; Dixon et al. [4]; Porcelli et al. [5]; Becker et al. [6]; Halliday [7]; Yokochi and Marty [8]).

Understanding the origin and evolution of the atmospheres of Venus, Earth and Mars and why Earth's atmosphere evolved differently compared to those of its neighboring planets, so that life could arise and evolve, is crucial for the search of life and atmospheric bio-markers on potentially habitable terrestrial exoplanets (e.g. Fridlund et al. [9]).

The present review, therefore, aims to piece together the latest findings in astrophysics, planetary formation, geophysics and atmosphere evolution, to understand why early Earth evolved to a Class I habitat and why Venus and Mars did not.

The Young Sun

While the long-term evolution of the Sun's bolometric radiation is quite well understood, following from calculations of the Sun's internal structure and nuclear reactions (e.g. Sackmann and Boothroyd [10]), the evolution of the high-energy radiation is less clear.

How the long-term evolution of the magnetically induced high-energy radiation is related to solar/stellar rotation started being clarified in the late sixties.

Cluster samples show a wide dispersion in rotation periods for ages up to a few hundred Myr, after which they gradually converge to a unique, stellar mass-dependent value (Soderblom et al. [11]).

An evolutionary decay law for high-energy radiation, therefore, needs to account for the dispersion of rotation periods.

One further parameter that is important for atmospheric modelling, that is, the hardness of high-energy radiation.

The X-ray hardness (the relative amount of "harder" to "softer" radiation) decreases with decreasing X-ray surface flux, because more active stars are dominated by hotter coronal plasma (Johnstone et al. [12]).

The First 100 Myr: From Planetary Embryos to Protoplanets

Important implications of the briefly described truncation mechanism are the scattering of primitive planetesimals onto planet-crossing orbits during the early formation of the planets, and a much faster accretion time of proto-Venus and Earth (e.g. Walsh et al. [13]).

For planetary embryos and planets that accreted after the first 2–3 Myr of the solar system, magma oceans can form as a result of the conversion of gravitational potential energy into heat upon metal–silicate separation and core formation (e.g. Rubie et al. [14]) and of the dissipation of the kinetic energy of highly energetic accretionary impacts (e.g. Tonks and Melosh [15]; Canup [16]).

In agreement with planetary dynamics model results related to the "Grand Tack" hypothesis (Walsh et al. [13]; O'Brien et al. [17]) and the recent Ru isotope analysis of primitive meteoritic material (Fischer-Gödde and Kleine [18]), it is now expected that volatile-rich carbonaceous chondritic bodies were scattered from the outer into the inner Solar System early in Earth's, Venus' and Mars' accretion phase.

CO_2 Atmosphere Loss to the Sub-surface and Space

A large amount of water (Pearson et al. [19]; Schmandt et al. [20]; Plümper et al. [21]) interacted with Earth's mantle environment during the early Hadean, possibly leading to conditions favorable for a proto-plate tectonics regime.

Although the main processes which drive plate tectonics are not fully understood, the minimum requirements are a sufficient mass relevant for the heat flow to drive mantle convection and a certain amount of water in the planet's interior to lubricate plate motion (e.g. Regenauer-Lieb et al. [22]; Solomatov [23]).

One should also note that at ages as studied by Way et al. [24] the EUV flux of the Sun would be too weak so that the water of an Earth-like ocean cannot escape easy to space when it is in the atmosphere.

As long as a magma ocean formed, the remaining oxygen and atmospheric nitrogen could have been incorporated into Venus' hot magmatic crust, where the oxygen oxidized the upper mantle material (e.g. Gillmann et al. [25]; Lichtenegger et al. [26]; Wordsworth [27]).

Origin and Evolution of Atmospheric N_2

The dissolution of nitrogen in early magma oceans or hot magmatic surface environments could have led to a significant incorporation of nitrogen into the mantle from very early in Earth's history (Marty [28]).

Some fixed nitrogen remains buried in the sediments, to be released during metamorphism (e.g. Haendel et al. [29]) or subducted.

Subduction of N-bearing sediments and altered oceanic lithosphere constitutes the primary flux of surficial nitrogen into the deep Earth (Busigny et al. [30]; Halama et al. [31]).

The above discussion highlights the importance of biological reactions in the cycling of nitrogen through Earth's oceans, atmosphere, and sediments (e.g. Cartigny and Marty [32]; Stüeken et al. [33]; Zerkle and Mikhail [34]).

The current amount of nitrogen in Earth's crust and mantle is highly uncertain and estimated between surface partial pressures of 0.32–5.6 bar (e.g. Goldblatt et al. [35]; Johnson and Goldblatt [36]; Marty [37]; Marty and Dauphas [38]).

Response of Earth's N2 Atmosphere Against the Sun's EUV Flux During the Archean

If this expansion occurs, strong solar wind ion pick up loss rates (Lichtenegger et al. [39]; Scherf et al. [40]), in addition to the high thermal loss rates (Tian et al. [41, 42]) take place.

If this outgassing rate was responsible for the build-up to the present atmospheric partial surface pressure within about 1.0 Gyr, one has to enhance the outgassing rate by about 20 times.

This rate is about 100 times lower than the modeled mass loss rate to space (Lichtenegger et al. [39]).

Earth's Atmosphere at the Time of Life's Origin

Airapetian et al. [43] studied the prebiotic chemistry and greenhouse warming of the early Earth by including solar activity effects (i.e. superflares, frequent collisions with Coronal Mass Ejections (CME), solar wind-induced shocks, high energy particles and compressed magnetosphere) caused by the active young Sun.

According to Airapetian et al. [43], these molecules, which have been relevant for the formation of life on Earth (i.e. HCN polymerization, known to produce various amino acids, the building blocks of proteins (Miyakawy and Others 2002)), rain out into the surface water reservoirs, where complex life may have formed and will participate in higher order organic chemistry.

According to Kasting et al. [44], data of the redox state of chromite inclusions in diamonds and sulfide indicate that the mantle oxidation process was slow, so that reduced conditions could have prevailed until about 2.5–4.0 Gyr ago.

The Makings of Earth-Like Habitats

Planets in the habitable zone should have the right amount of heat producing radioactive materials (i.e. U, Th, K, etc.) and water, which set the initial conditions for the thermal evolution of the mantle and later onset of plate tectonics and magnetic dynamos during several Gyr (Murthy et al. [45]; O'Neill et al. [46, 47]; Jellinek and Jackson [48]; Van Kranendonk [49]; Noack and Breuer [50]).

The active intrinsic magnetic dynamo (e.g. Murthy et al. [45]; O'Neill et al. [46]; Jellinek and Jackson [48]), which is also linked to plate tectonics, was/is important for the protection of surface life against high energy particles and should have played a role against large nonthermal escape rates of nitrogen caused by the high EUV flux of young stars.

Conclusions

Larger planetary embryos which developed magma oceans most likely lost the main fraction of volatiles due to hydrodynamic escape, in the inner Solar System after their magma oceans solidified.

After Mars finalized its accretion after a few Myr and the planet's final magma ocean solidified, a steam atmosphere similar to early Venus and Earth was outgassed, but with a lower amount of volatiles.

For a better understanding of the origin and evolution of the atmospheres of early Venus, Earth and Mars it will be crucial to gain a clearer knowledge of the early evolution of the Sun, in particular of its EUV flux during the first Gyr.

Elemental and isotopical fractionations in the atmosphere and in the bulk composition of the Earth and to a lesser degree of Mars and Venus are known.

Acknowledgement

A machine generated summary based on the work of Lammer, Helmut; Zerkle, Aubrey L.; Gebauer, Stefanie; Tosi, Nicola; Noack, Lena; Scherf, Manuel; Pilat-Lohinger,

Elke; Güdel, Manuel; Grenfell, John Lee; Godolt, Mareike; Nikolaou, Athanasia (2018 in The Astronomy and Astrophysics Review).

The Role of Primordial Atmosphere Composition in Organic Matter Delivery to Early Earth

https://doi.org/10.1007/s12210-020-00878-x

Summary

A model of the atmospheric entry of sub-mm grains is employed to evaluate the effect of the chemical composition of the primordial Earth's atmosphere on the grain heating, in the context of organic matter delivery.

The present work shows that: the total gas budget of the atmosphere is not highly relevant as far as the determination of the heating associated with slowing to subsonic speed is concerned; accordingly, light components (which are expected to be present in a primordial atmosphere and more abundant in the upper one) may be the primary ones in the evaluation of momentum and heat transfer in such scenarios.

Strong reduced heating is obtained in the case of an upper atmosphere rich in light components, showing that the composition of the primordial Earth atmosphere may represent the key issue in the delivery of thermolabile organic matter enclosed in sub-mm extraterrestrial grains.

Extended

The present work will focus on grains in the dimension range 0.01–1 mm, which represent the peak in the distribution on extraterrestrial matter delivery to Earth (Jenniskens et al. [51]).

Future studies should discuss the possibility that energetic (in the grain frame) hydrogen or methane molecules may react with a grain surface covered by an organic layer, producing more complex species of prebiotic significance.

Introduction

Life-related molecules, which are particularly thermolabile, might reach the Earth's surface associated to solid particles; in this way, the mineral composition of these grains may provide the necessary thermal protection against the high temperatures reached during the first stages of the atmospheric entry process.

The present work will focus on grains in the dimension range 0.01–1 mm, which represent the peak in the distribution on extraterrestrial matter delivery to Earth (Jenniskens et al. [51]).

There is no reason to assume that an early Earth atmosphere (with a mixture of components of such very different molar mass, like nitrogen and hydrogen) should have uniform composition in its lower and upper regions: most probably, the early atmosphere was enriched in low mass components at high altitudes.

Sekine and others suggested that the impact of iron meteorites, around 4 Gyr ago, may represent the most important atmospheric methane sources in the early Earth (Sekine et al. [52]).

Model of the Isothermal Atmosphere
The atmospheric entry model of sub-mm size grains developed by the authors (Micca Longo and Longo [53, 54]; Micca Longo et al. [55, 56]) is applied to the heating effect study concerning different hypotheses of primordial Earth atmosphere, in the context of organic matter delivery and origin of Life.

An isothermal atmospheric model is considered, which allows a simple comparison of different test cases regarding the upper atmosphere composition.

The present model includes a 2D geometry, an isothermal atmospheric profile, power balance, evaporation, ablation, radiation losses; furthermore, it includes additional features like chemical changes, stoichiometry, chemical effects in power balance (in the case of carbonate/sulphate grain atmospheric entry).

A crucial aspect of the entry model is the study of the relation between momentum and energy transfer based on an atomic model of the atmosphere and of the grain components.

Results
With such a high atmospheric layer, the reduction of the value of the gravitational acceleration g with height d should be included in the isothermal model.

The thermal trend is quite similar to the one concerning the present-day atmosphere: the carbonate decomposition is able to keep the grain temperature quite low, until the process is complete; then, when the grain is totally converted into oxide, it experiences a peak temperature.

By virtue of the invariance principle discussed in the previous sections, an effective synoptic representation of the effect of different hypothetical compositions with respect to the primordial atmosphere can be obtained by setting the grain speed on the x-axis.

The carbonate enhances the thermal mitigation effect, due to its lower density with respect to silicates (the global, uniform reduction visible in the figure) and, in the high-speed range, to its decomposition reaction.

Discussion
Although this result needs to be verified in the future using a more advanced description of the interaction (as mentioned above), such low temperatures have been not previously reported in the simulations of these phenomena.

Conclusions
The present investigation, supported by numerical calculations, leads us to conclude that different chemical compositions of the upper primordial atmosphere may have actually produced very different outcomes, in terms of heating, dealing with grains passing through it.

The preliminary analysis based on a very simplified model of energy transfer suggests that light components, if abundant in the upper atmosphere of the primordial Earth, may have led to higher organic delivery rates, due to lower heating.

Acknowledgement
A machine generated summary based on the work of Micca Longo, Gaia; Longo, Savino (2020 in Rendiconti Lincei. Scienze Fisiche e Naturali).

Co-evolution of Primitive Methane-Cycling Ecosystems and Early Earth's Atmosphere and Climate

https://doi.org/10.1038/s41467-020-16374-7

Abstract-Summary
In spite of their low productivity, the evolution of methanogenic metabolisms strongly modifies the atmospheric composition, leading to a warmer but less resilient climate.

As the abiotic carbon cycle responds, further metabolic evolution (anaerobic methanotrophy) may feed back to the atmosphere and destabilize the climate, triggering a transient global glaciation.

Although early metabolic evolution may cause strong climatic instability, a low $CO:CH_4$ atmospheric ratio emerges as a robust signature of simple methane-cycling ecosystems on a globally reduced planet such as the late Hadean/early Archean Earth.

Extended
In spite of extremely low productivity, metabolic evolutionary innovation in primitive methane-based biospheres is predicted to cause distinctive shifts in atmospheric composition, such as a decreasing $CO:CH_4$ ratio as greater metabolic complexity evolves.

Introduction
Previous studies [57–59] have addressed the productivity of primitive, chemolithotrophic ecosystems and their influence on the young Earth's equilibrium atmospheric conditions.

Such studies relied on equilibrium analyses of the planetary ecosystem; they made strongly simplifying assumptions on the function of chemolithotrophic microbial metabolisms, and did not close the feedback loop linking biological activity, atmospheric composition, and climate.

Although these studies showed that primitive biospheres may have had a significant impact on the planet's early atmosphere and climate, their ability to quantify this impact and estimate the underlying biomass productivity was limited.

Theory based on equilibrium analyses could not address the coupled dynamics of metabolic evolution and planetary surface conditions, whereby evolutionary changes might trigger significant atmospheric and climatic events and lead to novel steady states.

Advancing existing models is needed to generate hypotheses on the history of atmospheric and climatic conditions that metabolic evolutionary innovation may have driven on the early Earth.

Results

We first consider the direct effects on the early Earth's atmosphere and climate, on a relatively short timescale of ~10^6 years, of the transition from the initially cool, lifeless state to a planet populated by one of three methanogenic biospheres, MG, AG + AT, or MG + AG + AT.

Under these specific conditions, the formation of organic hazes may overwhelm the warming effect of methanogenic ecosystems and leave the planet in a globally glaciated state.

In the MG + AG + AT ecosystem the two methanogenic pathways interact synergistically, leading to a nonlinear, multiplicative increase in biomass production at low and high temperature.

With a deposition rate corresponding to our lowest value of 10^7 molecules cm^{-2} s^{-1} (Ref. [60]), and a hydrothermal removal rate of $[H_2SO_4]_{oc.} \times 7.2 \times 10^{12}$ L y^{-1} (ref [61]), we obtain an abiotic oceanic concentration of 0.4 mM. Such a stock is sufficient for methanotrophs to consume most of the atmospheric CH_4, leading to the global cooling described above.

Discussion

Our results confirm the contention that the late Hadean/early Archean planet was most likely habitable to methane-cycling chemolithotrophic biospheres and that under the assumption of high enough H_2 supply, these biospheres were key factors of the climatic and atmospheric evolution of the planet [57–59, 62, 63].

On short time scales (10^5–10^6 years) the evolution of methanogenic biospheres may have considerably warmed the climate and influenced its resilience, in spite of a very low ecosystem productivity.

A general result is that MG and AG + AT ecosystems are characterized by extremely low biomass production relative to their planetary impact on the atmosphere and climate.

By performing equilibrium analyses of the planetary system on longer time-scales (10^7–10^8 years), on which the carbon cycle responds to ecosystem function and sequential metabolic diversification of the biosphere, we find that the climate regulation of the planet by the abiotic carbon cycle largely buffers the influence of early methanogenic activity on climate.

Methods

Catabolism produces energy used by anabolism for biomass production, which determines cell growth and division; a fraction of energy produced by catabolism is used for cell maintenance.

Both metabolic rate and maintenance cost increase with cell size, but not as fast as structural biomass.

The biomass specific rates of metabolism and energy consumption for maintenance decrease with cell size.

Grid, we derived the following simple parametrization of the mean surface temperature as a function of pCO_2 and pCH_4 (expressed in bar): In the coupled biological-planetary model, we assume that the climate is always at equilibrium, meaning that the timescale of climate convergence is shorter than biological and geochemical timescales.

The reaction rates of (R1) and (R2) can be parameterized respectively as (in molecules $cm^{-2}\,s^{-1}$):where u (between 0.5 and 1) is given by To simulate the evolution of pCO_2 with time and feedbacks between the microbial community, climate, and the carbon cycle, we use a simple carbon cycle model based on Ref. [64].

Acknowledgement
A machine generated summary based on the work of Sauterey, Boris; Charnay, Benjamin; Affholder, Antonin; Mazevet, Stéphane; Ferrière, Régis (2020 in Nature Communications).

The Origin and Degassing History of the Earth's Atmosphere Revealed by Archean Xenon

https://doi.org/10.1038/ncomms15455

Abstract-Summary
The initial isotopic composition of atmospheric Xe remains unknown, as do the mechanisms involved in its depletion and isotopic fractionation compared with other reservoirs in the solar system.

High precision analyses of noble gases trapped in fluid inclusions of Archean quartz (Barberton, South Africa) that reveal the isotopic composition of the paleo-atmosphere at $\approx$3.3 Ga. The Archean atmospheric Xe is mass-dependently fractionated by $12.9 \pm 2.4\ \permil\,u^{-1}$ ($\pm 2\sigma$, s.d.) relative to the modern atmosphere.

The primordial Xe component delivered to the Earth's atmosphere is distinct from Solar or Chondritic Xe but similar to a theoretical component called U-Xe.

Introduction
When corrected for mass-dependent isotope fractionation, atmospheric Xe is depleted in its heavy isotopes (^{134}Xe and ^{136}Xe) relative to Solar or Chondritic Xe, and cannot be related to any known cosmochemical component [65, 66].

Recent studies of Archean barite and quartz samples from North Pole, Pilbara (NW Australia) demonstrated that, 3.5 to 3.0 Ga ago, atmospheric Xe had an isotopic composition less isotopically fractionated than the modern atmospheric Xe relative to any of the potential primordial components [67–70].

Xenon escape processes could have also led to mass-independent isotope fractionation, in addition to the mass-dependent one, that could account for the unique isotope composition of modern atmospheric Xe.

Depletion in radiogenic ^{129}Xe relative to the modern atmosphere allows us to compute a degassing rate from the Earth's mantle to the atmosphere over the last 3.3 Ga. Furthermore, Archean Xe originates from a primordial component different from all other known reservoirs of Xe in the solar system and similar to the theoretical U-Xe.

Results

The initial ^{40}Ar/^{36}Ar is 458 ± 4 ($\pm 2\sigma$, s.d.) for sample BMGA3-9, which is higher than the modern atmospheric ratio of 298.6 (Ref. [71]), this may be explained by the presence of some ^{40}Ar excess uncorrelated with the chlorine content.

This value is a minimum value for the Archean atmospheric ^{40}Ar/^{36}Ar ratio but is in broad agreement with previous estimates and prediction (143 ± 21, 3.5 Ga ago, (ref 70)) and models invoking a peak in crustal extraction between 3.8 and 2.5 Ga (Refs. [70, 72]).

Xenon in Barberton quartz thus has an Archean isotopic composition that differs from the modern atmosphere.

The isotopic fractionation of xenon in Barberton quartz relative to the isotopic composition of the modern atmosphere was computed using the light stable, non-fissiogenic, non-radiogenic isotopes of Xe (^{126,128,130}Xe) plus ^{131}Xe, for which production by the fission of ^{238}U is small [73].

Discussion

The atmospheric increase in ^{129}Xe(I) excess between 3.3 Ga and the present day corresponds to an integrated ^{129}Xe(I) degassing rate of 8 ± 4 mol a^{-1} ($\pm 1\sigma$, s.d.) (Methods).

Estimates for the mantle ^{130}Xe/^{3}He ratio range from 0.85×10^{-3} to 3.5×10^{-3} (Refs. [74–76]) and values between 1.29 and 1.7 for the ^{129}Xe(I)/^{130}Xe (Ref. [77]), this leads to a modern degassing rate of 1.37 ± 0.88 ($\pm 1\sigma$, s.d.) mol a^{-1} of ^{129}Xe(I).

Depletions in ^{134}Xe and ^{136}Xe for the primordial component similar to U-Xe as recorded by Barberton quartz may reflect either a mass-independent isotope fractionation process, not yet identified, or the presence of a nucleosynthetic anomaly (for example, r-process deficit) in the early atmosphere compared with other major components of the solar system (Chondritic or Solar Xe).

The intense ^{129}Xe(I) degassing rate of 8 ± 4 ($\pm 1\sigma$, s.d.) mol a^{-1} integrated over 3.3 Ga probably reflects degassing of the whole mantle in the active early Earth.

Methods

The irradiation parameter J is determined from the measured ^{40}Ar/^{39}Ar ratio in the Hb3gr hornblende standards that were irradiated in the same tubes as the samples (Eq. 1): where t_m is the age of Hb3gr of 1074.9 ± 3.5 Ma (Ref. [78]) and λ is the total decay constant (5.531×10^{-10} a^{-1}, Ref [79]).

In units of moles cm^{-3} STP: Following irradiation, samples were analysed in two successive steps: (1) step-crushing to release fluids trapped in fluid inclusions; (2) step-heating up to 1,700 °C to release K, Cl and Ar trapped in small inclusions, and present in the quartz lattice.

The fitting method led to a $^{40}Ar/K$ value of 6.63×10^{-5} ($\pm 8 \times 10^{-6}$, $\pm 2\sigma$, s.e.m.) for B that formally corresponds to an age of 3.3 (± 0.1) Ga (2σ, s.e.m.). ($^{40}Ar/^{36}Ar)_0$, representative of the initial $^{40}Ar/^{36}Ar$ trapped in Barberton quartz, is 458 ± 4 (2σ, s.e.m.) higher than the present day atmospheric value of 298.6.

Additional Information

Acknowledgement
A machine generated summary based on the work of Avice, Guillaume; Marty, Bernard; Burgess, Ray (2017 in Nature Communications).

Timescales of Oxygenation Following the Evolution of Oxygenic Photosynthesis

https://doi.org/10.1007/s11084-015-9460-3

Abstract-Summary
It is widely accepted that the invention of oxygenic photosynthesis ultimately resulted in the rise of oxygen by ca 2.35 Gya, but it is debated whether this occurred more or less immediately as a proximal result of the evolution of oxygenic Cyanobacteria or whether they originated several hundred million to more than one billion years earlier in Earth history.

Calculations illustrate that oxygenation would have overwhelmed redox buffers within ~100 kyr following the emergence of oxygenic photosynthesis, a geologically short amount of time unless rates of primary production were far lower than commonly expected.

This result arises because of the multiscale nature of the carbon and oxygen cycles: rates of gross primary production are orders of magnitude too fast for oxygen to be masked by Earth's geological buffers, and can only be effectively matched by respiration at non-negligible O_2 concentrations.

These results suggest that oxygenic photosynthesis arose shortly before the rise of oxygen, not hundreds of millions of years before it.

Extended
It is widely appreciated that burial and weathering (plus volcanic) fluxes are in balance on million year timescales for an atmosphere of ~20% O_2 by volume. (e.g., Lasaga and Ohmoto [80]).

Introduction

These arguments often rely on redox buffers, geologically-sourced reduced compounds in the atmosphere and/or oceans (like Fe^{2+} or CH_4), which reacted with molecular oxygen to prevent its environmental accumulation (Schidlowski [81]; Gaillard and Others [82]; Kump and Barley [83]).

The depletion of these redox buffers over geological time due either to changes in source fluxes or reaction with oxygen is thought to eventually allow oxygen to rise (e.g., Holland 2009).

This tight coupling of primary production and aerobic respiration has maintained a stably oxygenated atmosphere over Phanerozoic time (Glasspool and Scott [84]), but this balance may not have applied soon after oxygenic photosynthesis evolved, as efficient mechanisms for aerobic respiration may not have evolved until after the evolution of oxygenic photosynthesis (Gribaldo and Others [85]).

Until oxygen concentrations were reached that made aerobic respiration an efficient O_2 sink, O_2 sourced from oxygenic phototrophs would largely titrate reduced compounds such as methane and ferrous iron in seawater and the atmosphere—redox buffers that might prevent accumulation of oxygen in the atmosphere depending on their relative abundances and reaction kinetics (Lyons and Others [86]).

Basic Accounting

While this serves as no more than a rough approximation, it highlights that despite the overall reduced state of the Archean Earth, absolute abundances of redox buffer compounds are small compared to the anticipated fluxes of oxygen produced through oxygenic photosynthesis.

Model Summary

If the Faint Young Sun is instead counteracted by, for instance, high concentrations of CO_2 as has been elsewhere proposed (Owan and Others [87]), these concentrations may be substantially lower and would lower existing reduced pools and accelerate atmospheric oxidation; we therefore utilize these upper concentration ranges as a conservative estimate.

Photolysis of methane, and hydrogen escape to space are modeled as first order reactions with time constants based on rates from Catling and Others [88], resulting in a net loss of reducing power from the Earth system over time without consumption of O_2.

This form of equation is used for reduced compounds with the exceptions of methane and hydrogen, which are taken as functions following the modified form: $X(t) = X(t-1) + X_f * t * e^{-c*t}$, where c is a constant based on estimated lifetimes for these compounds in the Archean atmosphere (Catling and Others [88]).

Results

Goldblatt and Colleagues [89] showed that the balance between photosynthetic oxygen production and geologic flux of reduced compounds is fundamental to determining steady state oxygen concentrations and that oxygenation of the fluid Earth would proceed rapidly in less than 150 kyr, a conclusion similar to the model results here.

As most organic carbon fixed in the modern ocean is quickly respired, the balance of O_2 in the atmosphere can be quickly perturbed if burial rates are altered, and uncertainty in burial efficiency introduces a large degree of uncertainty into our model results.

Although hydrothermal activity, volcanic outgassing, and crustal production rates are minor components of this model, if their rates are far higher as suggested by some authors (e.g., Kump and Barley [83]) this could delay oxygenation somewhat (though a doubling of either factor affects the time to oxygenation by no more than a few thousand years).

Conclusions

From the perspective of the calculations presented here, the characteristic timescales of environmental oxygenation [certainly the timescales required to impact sensitive redox proxies like redox-sensitive detrital grains (Johnson and Others [90]) and mass independent S isotope fractionation (Pavlov and Kasting [91]) following the evolution of oxygenic photosynthesis are geologically rapid—unless rates of GPP are far lower than typically thought.

It has been shown that extant Cyanobacteria experience a defect in growth rate when exposed to high dissolved iron concentrations—an effect termed anaerobic iron toxicity—suggesting that perhaps iron toxicity may have delayed oxygenation following the evolution of oxygenic photosynthesis (Swanner and Others [92]).

It is reasonable to expect that Cyanobacteria rapidly adapted and became specialized within a variety of environments following the evolution of oxygenic photosynthesis.

This radiation likely allowed the Cyanobacteria to quickly adapt to an oxygenic photosynthetic lifestyle and occupy new niches and environments, with a dominant role in photosynthetic ecosystems.

Acknowledgement

A machine generated summary based on the work of Ward, Lewis M.; Kirschvink, Joseph L.; Fischer, Woodward W. (2015 in Origins of Life and Evolution of Biospheres).

Emergence of Life: Physical Chemistry Changes the Paradigm

https://doi.org/10.1186/s13062-015-0060-y

Abstract-Summary

To re-energize the research and define a new experimental paradigm, we advance four premises to better understand the physicochemical complexities of life's emergence: (1) Chemical and Darwinian (biological) evolutions are distinct, but become continuous with the appearance of heredity.

We discuss these premises in relation to current 'constructive' (non-evolutionary) paradigm of origins research—the process of complexification of chemical matter 'from the simple to the complex'.

This paradigm artificially avoids planetary chemical complexity and the natural tendency of molecular compositions toward maximum disorder embodied in the second law of thermodynamics.

Introduction: The Problem

Our premises identify physicochemical conditions for the emergence of life, and suggest new experiments with gradients of electromagnetic radiation, temperature and water activity that keep complex chemical mixtures in cyclic disequilibria, i.e., 'repeatedly stoked with energy', and hence in continuous physicochemical evolution [93].

The diurnal disequilibria arise naturally when Earth's rotation converts 'constant' solar radiation into cyclic energy gradients that drive chemical reactions at Earth's oceanic and rocky surfaces; hydrothermal vents release metallic ions and other compounds into the ocean, enriching the ocean's molecular complexity.

At the biological end of the chemical evolutionary continuum, microbial cells can also be viewed as (self-constructing) chemically evolving open thermodynamic systems that exchange materials and energy with their environments as they grow and divide.

Cellular chemical complexity has been characterized by the term 'crowding' resulting from the high total volume fraction of all molecules [94], some of which may be unknown or individually at low concentrations, which leads to our last premise.

Discussion

The evolutionary continuity between chemistry and biology, our first premise, is uncontroversial: complex non-equilibrium chemical matter inevitably evolves such that under some conditions the emergence of living states becomes imperative, eliminating the discontinuous ('miraculous') mechanism of 'life being breathed' into inanimate matter by external experimenters.

Our three remaining premises address conditions for physicochemical evolution toward cellular life: planetary energies driving chemical evolution (premise 2) under complex—multicomponent, multiphase, crowded, and non-equilibrium molecular conditions (premise 3), thereby enabling the evolution of molecular recognition and cellular self-organization (premise 4).

According to this interpretation of chemical complexity as a nominally extensive property of evolving matter, the universal (cosmic) emergence of life progresses from the large and complex—gravitational and nuclear evolution of stars providing chemical elements, molecules and macromolecules in planetary disequilibria—to the small and complex, the evolutionary chemistry of phase-separated micron-sized microspaces.

Conclusions

These premises lead to a new kind of non-equilibrium chemistry, i.e., evolutionary chemistry, which deals with supramacromolecular ('cellular') evolution of inanimate open thermodynamic systems, including their boundaries ('membranes'), involving thousands of different molecules, driven by cyclic gradients of temperature, electromagnetic radiation and chemical potentials of environmental chemicals.

These experiments define a paradigm that addresses the compositions and processes that are critical for the conversion of complex non-equilibrium chemical states into living states.

Current experimental paradigms for origins research based on 'complexification of matter' are untenable because they artificially eliminate chemical complexity and the thermal disordering effects of the second law of thermodynamics.

Reviewers' Comments

To me, the key points are (i) the continuity of chemical and biological evolution phases and (ii) the importance of complex chemical (micro) environments and molecular crowding that in all likelihood were required to hatch life.

The authors repeatedly refer to the second law of thermodynamics and its "violations" by evolving complex systems, in some cases even making disingenuous statements such as "The paradigms of molecular replicators and self-organizing metabolic networks violate the second law of thermodynamics..." Certainly, they are well aware that the second law cannot be violated within its domain of applicability, i.e., in systems at equilibrium.

This is not the type of thermodynamics that is important for understanding origin and evolution of life and complexity.

We thank the reviewer for the positive comments regarding the necessity to understand origin of life problems in terms of molecular forces.

Acknowledgement

A machine generated summary based on the work of Spitzer, Jan; Pielak, Gary J.; Poolman, Bert (2015 in Biology Direct).

Prebiotic Chemistry in Neutral/Reduced-Alkaline Gas–Liquid Interfaces

https://doi.org/10.1038/s41598-018-36579-7

Abstract-Summary

We experimentally demonstrate the uniqueness of alkaline aerosols as prebiotic reactors that produce an undifferentiated accumulation of a variety of multi-carbon biomolecules resulting from high-energy processes (in our case, electrical discharges).

Using simulation experiments, we demonstrate that the detection of important biomolecules in tholins increases when plausible and particular local planetary environmental conditions are simulated.

A greater diversity in amino acids, carboxylic acids, N-heterocycles, and ketoacids, such as glyoxylic and pyruvic acid, was identified in tholins synthetized from reduced and neutral atmospheres in the presence of alkaline aqueous aerosols than that from the same atmospheres but using neutral or acidic aqueous aerosols.

Extended

Plausible alkaline environmental conditions and the analytical findings checked for the tholins synthetized under such conditions lead to proposing a vision about the origin of life such as that previously suggested by Eschemnoser via "an iconoclastic attitude among genetic, metabolism, and compartmentalization towards one of openness to horizontal transfer of ideas and insights" in the field of the prebiotic chemistry [95].

Introduction

Since this successful experiment, most of the recent Miller-type experiments simulating earth conditions used water in vapour form (e.g. [96]) or liquid water (e.g. [97–99]).

These water surfaces can correspond to oceans, internal seas, lakes or rives and so on, which can in turn present different compositions in terms of salinity and pH. The introduction of these parameters together with others that are more widely studied, such as the compositions of the gas mixtures, might lead to finding new clues about the puzzling trouble of the origin of life.

We investigate the role of the pH in the presence of aqueous aerosols in Miller-type experiments to understand what conditions are more favourable to the accumulation of organics, which may be the main characters in a plausible emergence of a primitive biology.

A wide screening for polar organic molecules was performed via GC–MS, whereas in the case of Johnson and Others [100], HPLC was used as analytical technique, and amino acids were mainly reported, although in both cases, the role of the aerosols was revealed.

Materials and Methods

For the identification of polar organic molecules in all freeze-dried fractions: (i) The samples were hydrolysed with 6 M HCl at 110 °C for 24 h and then freeze dried to remove water, HCl and any volatile organics; (ii) two milligrams of each hydrolysed sample in 75 μL of BSTFA with 1% TMCS [N,O-bis(trimethylsilyl)trifluoroacetamide with trimethylchlorosilane, from Thermo Scientific] was heated at 70 °C for 19 h to obtain the respective TMS derivatives; and (iii) the derivatized samples were analysed by GC–MS using the following GC oven program: 60 °C (initial temperature) with a hold time of 1.5 min, heating to 130 °C at 5 °C/min with a hold time of 11 min, heating to 180 °C at 10 °C/min with a hold time of 10 min and heating to 220 °C/min at 20 °C/min with a final hold time of 15 min.

Results

The triplot showed a positive correlation between experiments performed with salts and aerosols and the total number of organic compounds identified, number of carboxylic acids and number of amino acids.

The high pH experiments showed a positive correlation with the number of polyols, number of N-heterocycles and milligrams of hydrophilic tholin.

The NH_3-based atmosphere showed a positive correlation with the number of polyols, milligrams of hydrophilic tholin, milligrams of hydrophobic tholin and high final pH. The experiments were plotted on different areas of the diagram depending on their experimental characteristics.

The experiments performed under extreme conditions (lowest pH and upper pH) presented a high correlation with the total number of organic compounds identified, especially if they were performed in the presence of aerosols and salts (experiments 4, 8 and 9).

Discussion

The identification and formation of carboxylic acids under possible prebiotic conditions is a key component of the autotrophic hypothesis about the origin of life and the emergence of a primitive metabolism [101].

In the abovementioned "glyoxylate scenario", glyoxylate and its formal dimer, dihydroxyfumarate, are suggested to be the key starting materials of the chemical constitution of a possible metabolism, serving as a source of the main biomonomers, such as sugars, amino acids, pyrimidines and the constituents of the rTCA cycle [102].

Prebiotic synthesis of cyanuric acid (h13) is demonstrated, indicating that its production is possible using alkaline aqueous aerosols, a reductive atmosphere and spark discharges.

Plausible alkaline environmental conditions and the analytical findings checked for the tholins synthetized under such conditions lead to proposing a vision about the origin of life such as that previously suggested by Eschemnoser via "an iconoclastic attitude among genetic, metabolism, and compartmentalization towards one of openness to horizontal transfer of ideas and insights" in the field of the prebiotic chemistry [95].

Conclusions

Our experiments demonstrate both the formation of key elements of the rTCA cycle and of some sugar precursors.

This is the first time that an α-ketoacid that forms part of the rTCA cycle has been detected in tholins from spark discharge experiments.

The yields increase with the increase of the initial pH for the experiments of the $CH_4 + NH_3 + H_2$ series, and the yield for the experiments at an initial pH 12 is greater with NH_3.

Contrasting our analysis with the control experiments demonstrates that the presence of aerosols, salts and alkaline pH values leads to a greater diversity of organic compounds.

Acknowledgement

A machine generated summary based on the work of Mompeán, Cristina; Marín-Yaseli, Margarita R.; Espigares, Patricia; González-Toril, Elena; Zorzano, María-Paz; Ruiz-Bermejo, Marta (2019 in Scientific Reports).

Constraining the Rise of Oxygen with Oxygen Isotopes

https://doi.org/10.1038/s41467-019-12883-2

Abstract-Summary

After permanent atmospheric oxygenation, anomalous sulfur isotope compositions were lost from sedimentary rocks, demonstrating that atmospheric chemistry ceded its control of Earth's surficial sulfur cycle to weathering.

Mixed signals of anoxia and oxygenation in the sulfur isotope record between 2.5 and 2.3 billion years (Ga) ago require independent clarification, for example via oxygen isotopes in sulfate.

We show <2.31 Ga sedimentary barium sulfates (barites) from the Turee Creek Basin, W. Australia with positive sulfur isotope anomalies of $\Delta^{33}S$ up to $+1.55‰$ and low $\delta^{18}O$ down to $-19.5‰$. The unequivocal origin of this combination of signals is sulfide oxidation in meteoric water.

Introduction

Within marine and terrestrial settings, microbial sulfate reduction (MSR) processes exert the most important controls on sulfur isotopic fractionation of SO_4^{2-}.

It is generally accepted that S-MIF results from atmospheric photochemical reactions operating under low O_2 of <0.001% of the present atmospheric level (PAL) of oxygen, causing surface sulfur fluxes of insoluble S_0 with $\Delta^{33}S > 0‰$ and soluble SO_4^{2-} with $\Delta^{33}S < 0‰$ that were not homogenised during their transfer into sedimentary rocks [91, 103].

The slow disappearance of $\Delta^{33}S$ signals from the rock record after 2.45 Ga may be attributable to the increasingly important oxidative weathering of an older, S-MIF-bearing, continental sulfide reservoir whose anomalous isotope compositions (i.e., $\Delta^{33}S > 0.4‰$) would be transferred to sulfate until this source was either exhausted or negligible as compared to contemporaneous, non-anomalous, sulfur sources [104].

Results and Discussion

Palaeoproterozoic seawater appears to be faithfully recorded in Kazput carbonate that is, as mentioned previously, on the order of 10‰ lower than modern carbonates but comparable to carbonates from around 2.3 Ga. Therefore, the Kazput barites recording the lowest $\delta^{18}O$ indicate their precursor sulfate was oxidised in the most evolved meteoric waters, placing this water–oxygen source for this sulfate firmly on land, as compared to a seawater $\delta^{18}O$ that was likely around $-10\%o$. Although sulfate-water oxygen isotope fractionation during sulfide oxidation requires further study, we take a median sulfide-derived sulfate-water $\delta^{18}O$ fractionation value of $+10\%o$ [105], as compared to the barite, to roughly estimate that the water sources for Kazput sulfate may have ranged between -30 and $-8\%o$. Considering a possible seawater composition around $-10\%o$, this range of source water $\delta^{18}O$ is appropriate for meteoric sources.

Methods

Barite extractions were performed on ~100 g of powdered drill core rock sample using a modified barite purification technique (the DDARP method) from Bao 2006 [106] that was originally developed to purify sulfate samples for triple oxygen isotope measurements.

Sample powders were decarbonated in HCl-acidified solution, rinsed in distilled water, and then treated for 3 days with constant stirring in a 0.05 M Diethylen-etriaminepentaacetic acid (DTPA) and 1 M NaOH solution to dissolve barium sulfate.

The sample was once again redissolved in 0.05 M DTPA and 1 M NaOH solution by agitating overnight then re-precipitated by acidification with HCl (to pH < 2) and addition of $BaCl_2$ solution.

Purified barite samples required additional wet chemistry for quadruple sulfur isotope analysis, all done at the Institut de Physique du Globe de Paris in Paris, France.

The sample Ag_2S precipitate was washed in triplicate in distilled water, oven-dried overnight, then ready for sulfur isotope analysis.

Acknowledgement

A machine generated summary based on the work of Killingsworth, B. A.; Sansjofre, P.; Philippot, P.; Cartigny, P.; Thomazo, C.; Lalonde, S. V. (2019 in Nature Communications).

Book Reading List

Atmospheres of Earth and the Planets
by McCormac, B. (Ed) *(1975).*

This book contains the lectures presented at the Summer Advanced Study Institute, 'Physics and Chemistry of Atmospheres' which was held at the University of Liege, Belgium, during the period July 29-August 9, 1974. One-hundred nineteen persons from eleven different countries attended the Institute. The authors and publisher have made a special effort for rapid publication of an up-to-date status of the physics and chemistry of the atmospheres of Earth and the plan etc., which is an ever-changing area. Special thanks are due to the lecturers for their diligent preparation and excellent presentations.

Please see https://www.springer.com/gp/book/9789027705754 for original source.

Astrophysics of Exoplanetary Atmospheres
by Bozza, V. (Ed), Mancini, L. (Ed), Sozzetti, A. (Ed) *(2018)*.

In this book, renowned scientists describe the complexity of exoplanetary atmospheres and all of the observational techniques that are employed to probe them. Readers will also find a panoramic description of the atmospheres of the planets within the Solar System, with explanation of considerations especially relevant to exoplanets.

Please see https://www.springer.com/gp/book/9783319897004 for original source.

Biosphere Origin and Evolution
by Dobretsov, N. (Ed), Kolchanov, N. (Ed), Rozanov, A. (Ed), Zavarzin, G. (Ed) *(2008)*.

Modern natural science shows that the infancy of life on Earth experienced prebiotic evolution and included the emergence of primitive self-reproducing biologic forms and their systems. The subsequent coevolution of inorganic environment and biologic systems resulted in global propagation of life over the Earth and its enormous diversification. Diverse living organisms colonized the land, water, and atmosphere, as well as upper layers of the lithosphere, thereby forming the biosphere. The book covers notions by scientists of various branches on the evolutionary relationship between the biosphere and geosphere, evolution features at various levels of living matter organization, and problems of prebiotic evolution and life origin.

Please see https://www.springer.com/gp/book/9780387686554 for original source.

Chemical Evolution: Physics of the Origin and Evolution of Life
by Chela-Flores, J. (Ed), Raulin, F. (Ed) *(1996)*.

Leading researchers in the area of the origin and evolution of life in the universe contributed to *Chemical Evolution: Physics of the Origin and Evolution of Life*. This volume provides a review of this interdisciplinary field. In 35 chapters many aspects of the origin of life are discussed by 90 authors, with particular emphasis on the early paleontological record: physical, chemical, biological, and informational aspects of life's origin, instrumentation in exobiology and system exploration; the search for habitable planets and extraterrestrial intelligent radio signals.

Please see https://www.springer.com/gp/book/9780792341116 for original source.

Chemical Evolution and the Origin of Life
by Rauchfuss, H. *(2008)*.

Up to now, we do not have a generally accepted theory about the origin of life and about the process of development of life, we only have a great number of—to some extent even contradictory—hypotheses. Meanwhile there came up some scientific findings beyond thought only a few years ago.

Horst Rauchfuss is comparing the different theories from the view of the latest results and is giving an exciting and easy understandable insight into the present state of research.

Please see https://www.springer.com/gp/book/9783540788225 for original source.

Earth's Early Atmosphere and Oceans, and The Origin of Life
by Shaw, G. H. *(2016).*

This book provides a comprehensive treatment of the chemical nature of the Earth's early surface environment and how that led to the origin of life. This includes a detailed discussion of the likely process by which life emerged using as much quantitative information as possible. The emergence of life and the prior surface conditions of the Earth have implications for the evolution of Earth's surface environment over the following 2–2.5 billion years. The last part of the book discusses how these changes took place and the evidence from the geologic record that supports this particular version of early and evolving conditions.

Please see https://www.springer.com/gp/book/9783319219714 for original source.

Evolutionary Biology: Genome Evolution, Speciation, Coevolution and Origin of Life
by Pontarotti, P. (Ed) *(2014).*

This book includes the most essential contributions presented at the 17th Evolutionary Biology Meeting in Marseille, which took place in September 2013. It consists of 18 chapters organized according to the following categories: Molecular and Genome Evolution, Phylogeography of Speciation and Coevolution, Exobiology and Origin of Life.

Please see https://www.springer.com/gp/book/9783319076225 for original source.

Exobiology: Matter, Energy, and Information in the Origin and Evolution of Life in the Universe
by Chela-Flores, J. (Ed), Raulin, F. (Ed) *(1998).*

Leading researchers in the area of the origin, evolution and distribution of life in the universe contributed to *Exobiology: Matter, Energy, and Information in the Origin and Evolution of Life in the Universe.* This volume provides a review of this interdisciplinary field. In 50 chapters many aspects that contribute to exobiology are reviewed by 90 authors. These include: historical perspective of biological evolution; cultural aspects of exobiology, cosmic, chemical and biological evolution, molecular biology, geochronology, biogeochemistry, biogeology, and planetology.

Please see https://www.springer.com/gp/book/9780792351726 for original source.

Evolution, Origin of Life, Concepts and Methods
by Pontarotti, P. (Ed) (*2019*).

This book presents 15 selected contributions to the 22nd Evolutionary Biology Meeting, which took place in September 2018 in Marseille. They are grouped under the following major themes: Origin of Life and Concepts and Methods.

Please see https://www.springer.com/gp/book/9783030303624 for original source.

Origin and Evolution of Planetary Atmospheres
by Lammer, H. (*2013*).

Based on the author's own work and results obtained by international teams he coordinated, this SpringerBrief offers a concise discussion of the origin and early evolution of atmospheres of terrestrial planets during the active phase of their host stars, as well as of the environmental conditions which are necessary in order for planets like the Earth to obtain N_2-rich atmospheres. Possible thermal and non-thermal atmospheric escape processes are discussed in a comparative way between the planets in the Solar System and exoplanets. Lastly, a hypothesis for how to test and study the discussed atmosphere evolution theories using future UV transit observations of terrestrial exoplanets within the orbits of dwarf stars is presented.

Please see https://www.springer.com/gp/book/9783642320866 for original source.

Origin and Evolution of Biodiversity
by Pontarotti, P. (Ed) (*2018*).

The book includes 19 selected contributions presented at the 21st Evolutionary Biology Meeting, which took place in Marseille in September 2017. The chapters are grouped into the following five categories: Genome/Phenotype Evolution, Self/Nonself Evolution, Origin of Biodiversity, Origin of Life and Concepts.

Please see https://www.springer.com/gp/book/9783319959535 for original source.

Philosophy and the Origin and Evolution of the Universe
by Agazzi, E. (Ed), Cordero, A. (Ed) (*1991*).

It has often been noted that a kind of double dynamics char- terizes the development of science. On the one hand the progress in every discipline appears as the consequence of an increasing specialization, implying the restriction of the inquiry to very partial fields or aspects of a given domain. On the other hand, an opposite (but one might better say a complementary) trend points towards the construction of theoretical frameworks of great ge- rality, the aim of which seems to correspond not so much to the need of providing «explanations» for the details accumulated through partial investigation, as to the desire of attaining an - rizon of global comprehension of the whole field.

Please see https://www.springer.com/gp/book/9789401055956 for original source.

Planetary Atmospheres
Sagan, C. (Ed), Owen, T. (Ed), Smith, H. (Ed) *(1971)*.

Please see https://www.springer.com/gp/book/9789027701657 for original source.

The Early Evolution of the Atmospheres of Terrestrial Planets
by Trigo-Rodriguez, J. (Ed), Raulin, F. (Ed), Muller, C. (Ed), Nixon, C. (Ed) *(2013)*.

"The Early Evolution of the Atmospheres of Terrestrial Planets" presents the main processes participating in the atmospheric evolution of terrestrial planets. A group of experts in the different fields provide an update of our current knowledge on this topic.

Please see https://www.springer.com/gp/book/9781461451907 for original source.

Towards Revealing the Origin of Life
by Ikehara, K. *(2021)*.

The origin of life has been investigated by many researchers from various research fields, such as Geology, Geochemistry, Physics, Chemistry, Molecular Biology, Astronomy and so on. Nevertheless, the origin of life remains unsolved. One of the reasons for this could be attributed to the different approaches that researchers have used to understand the events that happened on the primitive Earth. The origins of the main three members of the fundamental life system, as gene, genetic code and protein, could be only separately understood with these approaches.

Please see https://www.springer.com/gp/book/9783030710866 for original source.

The Evolving Universe and the Origin of Life
Teerikorpi, P., Valtonen, M., Lehto, K., Lehto, H., Byrd, G., Chernin, A. *(2009)*.

Sir Isaac Newton famously said, regarding his discoveries, "If I have seen further it is by standing upon the shoulders of giants."

The Evolving Universe and the Origin of Life describes, complete with fascinating biographical details of the thinkers involved, the ascent to the metaphorical shoulders accomplished by the greatest minds in history. For the first time, a single book can take the reader on a journey through the history of the universe as interpreted by the expanding body of knowledge of humankind. From subatomic particles to the protein chains that form life, and expanding in scale to the entire universe, this book covers the science that explains how we came to be.

Please see https://www.springer.com/gp/book/9780387095349 for original source.

References

1. Mizuno H, Nakazawa K, Hayashi C (1980) Dissolution of the primordial rare gases into the molten Earth's material. Earth Planet Sci Lett 50:202–210
2. Sasaki S, Nakazawa K (1989) did a primary solar-type atmosphere exist around the proto-Earth? Icarus 85:21–42
3. Porcelli D, Pepin RO (2000) Rare gas constraints on early Earth history. In: Canup RM, Righter K (eds) Origin of the Earth and Moon. University of Arizona Press, Tucson, pp 435–458
4. Dixon ET, Honda M, McDougall I, Campbell IH, Sigridsson I (2000) Preservation of neon-solar neon isotopic ratios in Icelandic basalds. Earth Planet Lett 180:309–324
5. Porcelli D, Cassen P, Woolum D (2001) Deep Earth rare gases: initial inventories, capture from the solar nebula and losses during Moon formation. Earth Planet Sci Lett 193:237–251
6. Becker RH, Clayton RN, Galimov EM, Lammer H, Marty B, Pepin RO, Wieler R (2003) Isotopic signatures of volatiles in terrestrial planets. Space Sci Rev 106:377–410
7. Halliday AN (2014) The origin and earliest history of the Earth. In: Davis AM (ed) Planets, asteroids, comets and the solar system. Treatise on geochemistry, pp 149–211
8. Yokochi R, Marty B (2004) A determination of the neon isotopic composition of the deep mantle. Earth Planet Sci Lett 225:77–88
9. Fridlund M, Eiroa C, Henning T, Herbst T, Kaltenegger L, Léger A, Liseau R, Lammer H, Selsis F, Beichman C, Danchi W, Lunine J, Paresce F, Penny A, Quirrenbach A, Röttgering H, Schneider J, Stam D, Tinetti G, White GJ (2010) A roadmap for the detection and characterization of other Earths. Astrobiology 10:113–119
10. Sackmann IJ, Boothroyd AI (2003) Our Sun. V. A bright young Sun consistent with helio-seismology and warm temperatures on ancient Earth and Mars. Astrophys J 583:1024–1039
11. Soderblom DR, Stauffer JR, MacGregor KB, Jones BF (1993) The evolution of angular momentum among zero-age main-sequence solar-type stars. Astrophs J 409:624–634
12. Johnstone CP, Güdel M, Stökl A, Lammer H, Tu L, Kislyakova KG, Lüftinger T, Odert P, Erkaev NV, Dorfi EA (2015) The evolution of stellar rotation and the hydrogen atmospheres of habitable-zone terrestrial planets. Astrophys J Lett 815:A12
13. Walsh KJ, Morbidelli A, Raymond SN, O'Brien DP, Mandell AM (2011) A low mass for Mars from Jupiter's early gas-driven migration. Nature 475:206–209
14. Rubie DC, Nimmo F, Melosh HJ (2015) Formation of the Earth's core. In: Schubert G (ed) Treatise on geophysics, vol 9, pp 43–79
15. Tonks WB, Melosh HJ (1993) Magma ocean formation due to giant impacts. J Geophys Rev 98:5319–5333
16. Canup RM (2004) Origin of terrestrial planets and the Earth-Moon system. Phys Today 57:56–62
17. O'Brien DP, Walsh KJ, Morbidelli A, Raymond SN (2014) Water delivery and giant impacts in the 'Grand Tack' scenario. Icarus 239:74–84
18. Fischer-Gödde M, Kleine T (2017) Ruthenium isotopic evidence for an inner solar system origin of the later vaneer. Nature 541:525–527
19. Pearson DG, Brenker FE, Nestola F, McNeill J, Nasdala L, Hutchinson MT, Matveev S, Mather K, Silversmit G, Schmitz S, Vekemans B, Vincze L (2014) Hydrous mantle transition zone indicated by ringwoodite included within diamond. Nature 507:221–224
20. Schmandt B, Jacobsen SD, Becker TW, Liu Z, Dueker KG (2014) Dehydration melting at the top of the lower mantle. Science 344:165–1268
21. Plümper O, John T, Podladchikov YuY, Vrijmoed JC, Scambelluri M (2017) Fluid escape from subduction zones controlled by channel-forming reactive porosity. Nat Geosci 10:150–156
22. Regenauer-Lieb K, Yuen DA, Branlund J (2001) The initiation of subduction: criticality by addition of water? Science 294:578–581
23. Solomatov VS (2004) Initiation of subduction by small-scale convection. J Geophys Res 109:B01412

24. Way MJ, Del Genio AD, Kiang NY, Sohl LE, Grinspoon DH, Aleinov I, Kelley M, Clune T (2016) Was Venus the first habitable world of our solar system? Geophys Res Lett 43:8376–8383

25. Gillmann C, Chasseﬁére E, Lognonné P (2009) A consistent picture of early hydrodynamic escape of Venus atmosphere explaining present Ne and Ar isotopic ratios and low oxygen atmospheric content. Earth Planet Sci Lett 286:503–513

26. Lichtenegger HIM, Kislyakova KG, Odert P, Erkaev NV, Lammer H, Gröller H, Johnstone CP, Elkins-Tanton L, Tu L, Güdel M, Holmström M (2016) Solar XUV and ENA-driven water loss from early Venus' steam atmosphere. J Geophys Res 121:4718–4732

27. Wordsworth RD (2016) Atmospheric nitrogen evolution on Earth and Venus. Earth Planet Sci Lett 447:103–111

28. Marty B (2012) The origins and concentrations of water, carbon, nitrogen and noble gases on Earth. Earth Planet Sci Lett 313:56–66

29. Haendel D, Mühle K, Nitzsche H-M, Stiehl G, Wand U (1986) Isotopic variations of the fixed nitrogen in metamorphic rocks. Geochim Cosmochim Acta 50:749–758

30. Busigny V, Cartigny P, Philippot P (2011) Nitrogen isotopes in ophiolitic metagabbros: a re-evaluation of modern nitrogen fluxes in subduction zones and implication for the early Earth atmosphere. Geochim Cosmochim Acta 75:7502–7521

31. Halama R, Bebout GE, John T, Scambelluri M (2014) Nitrogen recycling in subducted mantle rocks and implications for the global nitrogen cycle. Int J Earth Sci 103:2081–2099

32. Cartigny P, Marty B (2013) Nitrogen isotopes and mantle geodynamics: the emergence of life and the atmosphere–crust–mantle connection. Elements 9:359–366

33. Stüeken EE, Kipp MA, Koehler MC, Schwieterman EW, Johnson B, Buick R (2016) Modeling N_2 pN_2 through geological time: Implications for planetary climates and atmospheric biosignatures. Astrobiology 16:949–963

34. Zerkle AL, Mikhail S (2017) The geobiological nitrogen cycle: from microbes to the mantle. Geobiology 15:343–352

35. Goldblatt C, Claire MW, Lenton TM, Matthews AJ, Watson AJ, Zahnle KJ (2009) Nitrogen-enhanced greenhouse warming on early Earth. Nat Geosci 2:891–896

36. Johnson B, Goldblatt C (2015) The nitrogen budget of Earth. Earth Sci Rev 148:150–173

37. Marty B (1995) Nitrogen content of the mantle inferred from N_2 N_2 -Ar correlation in oceanic basalts. Nature 377:326–329

38. Marty B, Dauphas N (2003) The nitrogen record of crust-mantle interaction and mantle convection from Archean to present. Earth Planet Sci Lett 206:397–410

39. Lichtenegger HIM, Lammer H, Grießmeier J-M, Kulikov YuN, von Paris P, Hausleitner W, Krauss S, Rauer H, Kulikov YuN, von Paris P, Hausleitner W, Krauss S, Rauer H (2010) Aeronomical evidence for higher CO_2 CO_2 levels during Earth's Hadean epoch. Icarus 210:1–7

40. Scherf M, Khodachenko ML, Blokhina M, Johnstone C, Alexeev I, Belenkaya E, Tarduno JA, Kulikov Yu N, Tu L, Lichtenegger HIM, Güdel M, Lammer H (2018) On the Earth's paleo-magnetosphere the late Hadean eon and possible implications for the ancient terrestrial atmosphere. Earth Planet Sci Lett (submitted)

41. Tian F, Kasting JF, Liu H-L, Roble RG (2008a) Hydrodynamic planetary thermosphere model: 1. Response of the Earth's thermosphere to extreme solar EUV conditions and the significance of adiabatic cooling. J Geophys Res 113:E05008

42. Tian F, Solomon SC, Qian L, Lei J, Roble RG (2008b) Hydrodynamic planetary thermo-sphere model: 2. Coupling og an electron transport/energy deposition model. J Geophys Res 113:E07005

43. Airapetian VS, Glocer A, Gronoff G, Hébrard E, Danchi W (2016) Prebiotic chemistry and atmospheric warming of early Earth by an active young Sun. Nat Geosci 9:452–455

44. Kasting JF, Eggler DH, Raeburn SP (1993) Mantle redox evolution and the case for a reduced Archean atmosphere. J Geol 101:245–257

45. Murthy VR, van Westrenen W, Fei Y (2003) Experimental evidence that potassium is a substantial radioactive heat source in planetary cores. Nature 423:163–165

46. O'Neill CO, Jellinek AM, Lenardic A (2007) Conditions for the inset of plate tectonics on terrestrial planets and moons. Earth Planet Sci Lett 261:20–32
47. O'Neill HSC, Palme H (2017) Collisional erosion and the non-chondritic composition of the terrestrial planets. Phil Trans R Soc a 366:4205–4238
48. Jellinek AM, Jackson MG (2015) Conenctions between the bulk composition, geodynamics and habitability of Earth. Nat Geosci 8:587–593
49. Van Kranendonk MJ (2011) Onset of plate tectonics. Science 333:413–414
50. Noack L, Breuer D (2014) Plate tectonics on rocky exoplanets: influence of initial conditions and rheology. Planet Space Sci 98:41–49
51. Jenniskens P, Wilson MA, Packan D, Laux CO, Kruger CH, Boyd ID, Popova O, Fonda M (2000) Meteors: a delivery mechanism of organic matter to the early Earth. In: Leonid storm research, 57–70. Springer
52. Sekine Y, Sugita S, Kadono T, Matsui T (2003) Methane production by large iron meteorite impacts on early Earth. J Geophys Res 108:5070
53. Micca Longo G, Longo S (2017) Thermal decomposition of $MgCO_3$ $MgCO_3$ during the atmospheric entry of micrometeoroids. Int J Astrobiol 16(4):368–378
54. Micca Longo G, Longo S (2018) Theoretical analysis of the atmospheric entry of sub-mm meteoroids of $Mgx\ Ca(1-x)\ CO_3$ $Mgx\ Ca(1-x)CO_3$ composition. Icarus 310:194–202
55. Micca Longo G, Piccinni V, Longo S (2019) Evaluation of $CaSO_4$ $CaSO_4$ micrograins in the context of organic matter delivery: thermochemistry and atmospheric entry. Int J Astrobiol 18(4):345–352
56. Micca Longo G, D'Elia M, Fonti S, Longo S, Mancarella F, Orofino V (2019) Kinetics of white soft minerals (WSMs) decomposition under conditions of interest for astrobiology: a theoretical and experimental study. Geosciences 9(2)
57. Ozaki K et al (2018) Effects of primitive photosynthesis on Earth's early climate system. Nat Geoscis 11:55
58. Kasting JF et al (2001) A coupled ecosystem-climate model for predicting the methane concentration in the Archean atmosphere. Orig Life Evol Biosph. 31:271–285
59. Kharecha P, Kasting J, Siefert J (2005) A coupled atmosphere–ecosystem model of the early Archean Earth. Geobiology 3:53–76
60. Ono S et al (2003) New insights into Archean sulfur cycle from mass-independent sulfur isotope records from the Hamersley Basin. Australia. Earth Planet Sci Lett 213:15–30
61. Wong M et al (2017) Nitrogen oxides in early Earth's atmosphere as electron acceptors for life's emergence. Astrobiology 17:975–983
62. Kasting JF (2005) Methane and climate during the Precambrian era. Precambrian Res 137:119–129
63. Kasting JF, Ono S (2006) Palaeoclimates: the first two billion years. Philos Trans R Soc Lond B Biol Sci 361:917–929
64. Krissansen-Totton J et al (2018) Constraining the climate and ocean pH of the early Earth with a geological carbon cycle model. Proc Natl Acad Sci USA 115:4105–4110
65. Pepin RO (1991) On the origin and early evolution of terrestrial planet atmospheres and meteoritic volatiles. Icarus 92:2–79
66. Takaoka N (1972) An interpretation of general anomalies of xenon and the isotopic composition of primitive xenon. Mass Spectrom. 20:287–302
67. Srinivasan B (1976) Barites: anomalous xenon from spallation and neutron-induced reactions. Earth Planet Sci Lett 31:129–141
68. Pujol M, Marty B, Burnard P, Philippot P (2009) Xenon in Archean barite: Weak decay of 130Ba, mass-dependent isotopic fractionation and implication for barite formation. Geochim Cosmochim Acta 73:6834–6846
69. Pujol M, Marty B, Burgess R (2011) Chondritic-like xenon trapped in Archean rocks: a possible signature of the ancient atmosphere. Earth Planet Sci Lett 308:298–306
70. Pujol M, Marty B, Burgess R, Turner G, Philippot P (2013) Argon isotopic composition of Archaean atmosphere probes early Earth geodynamics. Nature 498:87–90

71. Lee J-Y et al (2006) A redetermination of the isotopic abundances of atmospheric Ar. Geochim Cosmochim Acta 70:4507–4512
72. Stuart FM, Mark DF, Gandanger P, McConville P (2016) Earth-atmosphere evolution based on new determination of Devonian atmosphere Ar isotopic composition. Earth Planet Sci Lett 446:21–26
73. Ragettli RA, Hebeda EH, Signer P, Wieler R (1994) Uranium-xenon chronology: precise determination of λsf* 136Ysf for spontaneous fission of 238 U. Earth Planet Sci Lett 128:653–670
74. Moreira M, Kunz J, Allègre C (1998) Rare gas systematics in popping rock: isotopic and elemental compositions in the upper mantle. Science 279:1178–1181
75. Holland G, Ballentine CJ (2006) Seawater subduction controls the heavy noble gas composition of the mantle. Nature 441:186–191
76. Mukhopadhyay S (2012) Early differentiation and volatile accretion recorded in deep-mantle neon and xenon. Nature 486:101–104
77. Trieloff M, Kunz J (2005) Isotope systematics of noble gases in the Earth's mantle: possible sources of primordial isotopes and implications for mantle structure. Phys Earth Planet Inter 148:13–38
78. Schwarz WH, Trieloff M (2007) Intercalibration of 40Ar–39Ar age standards NL-25, HB3gr hornblende, GA1550, SB-3, HD-B1 biotite and BMus/2 muscovite. Chem Geol 242:218–231
79. Steiger RH, Jäger E (1977) Subcommission on geochronology: convention on the use of decay constants in geo-and cosmochronology. Earth Planet Sci Lett 36:359–362
80. Lasaga AC, Ohmoto H (2002) Geochim Cosmochim Acta 66:361–381
81. Schidlowski M (1983) Precambrian Res 20:319–335
82. Gaillard F, Scaillet B, Arndt NT (2011) Nature 479:229–232
83. Kump LR, Barley ME (2007) Nature 448:1033–1036
84. Glasspool IJ, Scott AC (2010) Nat Geosci 3(9):627–630
85. Gribaldo S, Talla E, Brochier-Armanet C (2009) Trends Biochem Sci 34:375–381
86. Lyons TW, Reinhard CT, Planavsky NJ (2014) Nature 506:307–315
87. Owan T, Cess RD, Ramanathan V (1979) Nature 277:640–642
88. Catling DC, Zahnle K, McKay C (2001) Science 293:839–843
89. Goldblatt C, Lenton TM, Watson AJ (2006) Nature 443:683–686
90. Johnson JE, Gerpheide A, Lamb M et al (2014) Geol Soc Am Bull 126:813–830
91. Pavlov AA, Kasting JF (2002) Astrobiology 2(1):27–41
92. Swanner ED, Mloszewska v, Cirpka O et al (2015) Nat Geosci 8:126–12130
93. Spitzer J, Poolman B (2009) The role of biomacromolecular crowding, ionic strength and physicochemical gradients in the complexities of life's emergence. Microbiol Mol Biol Revs 73:371–388
94. Zhou HX, Rivas G, Minton AP (2008) Macromolecular crowding and confinement: biochemical, biophysical, and potential physiological consequences. Annu Rev Biophys 37:375–397
95. Eschenmoser A (2007) On a Hypothetical generational relationship between HCN and constituents of the reductive citric acid cycle. Origin Life Evol Biosph 43:554–573
96. Utsumi Y, Hattori T (2002) Synthesis of ammonium and organic compounds from N_2, H_2O vapour, and CO_2 gas mixture by synchrontron radiation induced photochemical reactions at atmospheric pressure and at room temperature. Rev Sci Instrum 73:1387–1389
97. Takahashi J et al (2005) Photochemical abiotic synthesis of amino-acid precursors from simulated planetary atmospheres by vacuum ultraviolet light. J App Phys 98:024097
98. Plankensteiner K, Reiner H, Schranz B, Rode BM (2004) Prebiotic formation of amino acids in a neutral atmosphere by electric discharge. Angew Chem Int Ed 43:1886–1895
99. Takano Y, Ohashi A, Kaneko T, Kobayashi K (2003) Abiotic synthesis of high-molecular-weight organics from an inorganic gas mixture of carbon monoxide, ammonia, and water by 3 MeV proton irradiation. Appl Phys Lett 84:1410–1412
100. Johnson AP et al (2008) The Miller volcanic spark discharge experiment. Science 322:404
101. Morowitz HJ, Kostelnik JD, Yang J, Cody GD (2000) The origin of intermediary metabolism. Proc Natl Acad Sci USA 97:7704–7708

102. ter Braak CJF, Šmilauer P (2002) CANOCO reference manual and CanoDraw for Window's user's guide: software for canonical community ordination (version 4.5). Ithaca: Microcomputer Power, 500 p
103. Halevy I (2013) Production, preservation, and biological processing of mass-independent sulfur isotope fractionation in the Archean surface environment. Proc Natl Acad Sci 110:17644–17649
104. Farquhar J, Wing BA (2003) Multiple sulfur isotopes and the evolution of the atmosphere. Earth Planet Sci Lett 213:1–13
105. Gomes ML, Johnston DT (2017) Oxygen and sulfur isotopes in sulfate in modern euxinic systems with implications for evaluating the extent of euxinia in ancient oceans. Geochim Cosmochim Acta 205:331–359
106. Bao H (2006) Purifying barite for oxygen isotope measurement by dissolution and reprecipitation in a chelating solution. Anal Chem 78:304–309

Chapter 2
Downscaling, Regional Models and Impacts

Introduction by Guido Visconti

One of the reasons why General Circulation Models (GCM) were invented was the necessity to predict the geographical distribution of climatic changes especially at regional level. Until a few decades ago GCM had a resolution of a few hundred kilometers that was not enough for the above purpose so that around the early eighties Regional Climate Models (RCM) were introduced. They were applied to a regional domains (like the Mediterranean area) and could then use a much smaller grid space than GCM and in theory could take into account the regional characteristics of orography, surface and so on. In order to run them they use boundary conditions produced by a GCM with a technique known as nesting. The use of RCM is a typical modeling application that does not solve basic physical problems and it is more like a magnifying glass applied to a GCM. One reason to achieve higher resolution is in fact to solve accurately small scales effects like convection dynamics. The RCM approach had great success especially in terms of published papers and soon the same models found an application to weather forecast in the form of the so called dynamical downscaling. In this case the coarse scale weather forecasts were detailed on a specific region using an RCM driven this time with the GCM forecast results. This is one of the downscaling methods the other being the statistical downscaling. In this case the statistical correlation between different scales are used that could be obtained when high-resolution weather data were available. All these processes at the end are necessary to evaluate the impact at regional level of the climatic changes or of the weather events. Of particular interest are the consequences on the hydrological cycle where the high resolution is really important if one thinks about the formation of precipitation and its effects on the ground.

In principle the use of RCM should add some knowledge to the large scale (GCM) results and as a matter of fact this constitutes what is called the "added value" problem that is the question is to understand if all the results obtained with RCM were already in the GCM. The conclusions on the long debate about added value indicate that RCM do not answer fundamental questions like those related to small scale convection but

G. Visconti (ed.), *Climate, Planetary and Evolutionary Sciences*, https://doi.org/10.1007/978-3-030-74713-8_2

show greater details over mountain regions and coastal zones while the effects are not so evident in ocean regions. Besides RCM can resolve mesoscale weather phenomena and also represent better heavy rainfall. There is an intrinsic limit to increase of resolution in RCM and this is related to the fact that below some grid dimension the hydrostatic approximation is no longer valid and new numerical schemes have to be used. This aspect could be solved with a technique known as telescoping where beside a GCM and an RCM a high resolution model (at the level of cloud convection) is used. This approach could resolve scales at the tornado level.

As mentioned before, dynamic downscaling is one of the possible applications of RCM. First of all, downscaling can be performed on spatial and temporal characteristics of climate projections. In spatial downscaling methods are used to derive finer-resolution spatial climate information from coarser-resolution GCM output at specific regions. Temporal downscaling refers to the derivation of fine-scale temporal information (like daily rainfall sequences) from coarser-scale temporal GCM output (like monthly or seasonal rainfall amounts). Statistical downscaling requires establishing some empirical relationship between large-scale data and local data (at the station level). Such relationships could be obtained using historical data. Based on the assumption that such correlation is maintained through the climatic change they are used to extrapolate the GCM projections at the local level. Statistical downscaling implies only the handling of data and does not require expensive numerical facilities to run very complex models so apparently should be quite suitable for underdeveloped countries which on the other hand may not have extensive weather records to establish the correlation. It is to notice that RCM should be adapted to the specific geographical area of interest and the quality of results depends on the quality of driving GCM. On the other hand, RCM results may be used for long term planning and especially for impact studies. The statistical method is indicated in particular for modeling and management of water and other natural resources.

Machine-Generated Summaries

Keywords: *scenario, drive, rainfall, gcms, approach, sst, rcm, area, boundary, surface, daily, summer, domain, perform, output.*

Joint Variable Spatial Downscaling

https://doi.org/10.1007/s10584-011-0167-9

Abstract-Summary
Joint Variable Spatial Downscaling (JVSD), a new statistical technique for downscaling gridded climatic variables, is developed to generate high resolution gridded datasets for regional watershed modeling and assessments.

The proposed approach differs from previous statistical downscaling methods in that multiple climatic variables are downscaled simultaneously and consistently to produce realistic climate projections.

In the bias correction step, JVSD uses a differencing process to create stationary joint cumulative frequency statistics of the variables being downscaled.

The results show that the proposed downscaling method is able to reproduce the sub-grid climatic features as well as their temporal/spatial variability in the historical periods.

Comparisons are also performed for precipitation and temperature with other statistical and dynamic downscaling methods over the southeastern US and show that JVSD performs favorably.

The downscaled sequences are used to assess the implications of GCM scenarios for the Apalachicola-Chattahoochee-Flint river basin as part of a comprehensive climate change impact assessment.

Extended
The results show that there is no significant statistical difference between dynamic downscaling (DDS) and JVSD with no bias correction.

The downscaled sequences are used to drive hydrologic watershed models which have been shown to represent sufficiently well the dynamics of runoff and other hydrologic variables.

The downscaled sequences should capture climatic mean and variability trends.

Introduction
RCMs provide high resolution climatic fields spatially and globally consistent with GCM scenarios.

The currently available results are not sufficient for comprehensive climate change impact assessments, but are used in this study to compare the skill of statistical versus dynamic downscaling methods.

Statistical downscaling is based on relationships between low resolution GCM outputs and associated higher resolution observations over the same historical period.

Wood and others [1] proposed a two-step statistical downscaling method to address bias correction and spatial disaggregation (BCSD).

The spatial disaggregation step translates the adjusted GCM data on climate model resolutions to a basin-relevant resolution (observational resolution) by using interpolated spatial factors.

BCSD is a very efficient statistical downscaling technique for climate change assessments.

The CA assumption is that the relationships between large-scale and downscaled fields derived based on historical reanalysis data will also be valid in future climates.

Joint Variable Spatial Downscaling (JVSD)
Of removing and replacing the variable long term trends before and after the bias correction step, JVSD uses a differencing process to create stationary time series

and joint frequency distributions (for temperature and precipitation) between GCM control and future runs.

In keeping with the previous discussion, the bias correction process consists of (1) creating a differenced series of future temperature and precipitation; (2) finding the joint frequency of the contemporaneous differenced data values; (3) considering that this joint frequency is the same in the future differenced series as it is in the control differenced series; and (4) mapping each joint frequency point of the GCM Control distribution to a corresponding point on the joint frequency distribution of the observed differenced series (OBS).

The bias corrected monthly temperature and precipitation series for each GCM cell (denoted TS7 and TS8) are obtained by inverting the differencing operation on the bias corrected series: The JVSD spatial downscaling component is based on matching the bias-corrected temperature and precipitation patterns with similar observed patterns (historical analogues) over the assessment region (e.g., the ACF river basin).

Downscaling Results and Comparisons

The figures show that JVSD results compare favorably with observed precipitation and temperature data in that they reproduce fairly well the seasonal spatial distributions and coherence. (In generating the JVSD results, the corresponding historical month being downscaled has, of course, been excluded from the historical analogue data set.) Furthermore, specifically for the CGCM A1B run shown, the results in columns 3 and 4 indicate: (1) Temperature exhibits increasing trends over the southeast and the ACF basin for all seasons; Temperature increases are more significant in the 2050–2099 time period.

Comparing JVSD with bias correction and DDS indicates that the former is significantly different from the latter for both temperature and precipitation at 0.05 and 0.01 significance levels. (Buford temperature is the only exception where the two frequency distributions are not statistically different at a 0.01 significance level, but the test statistic is marginal.) This finding combined with the favorable JVSD (BC) comparison with observed data (in previous sections) leads to the conclusion that dynamic downscaling without some form of bias correction may not be adequate for climate change assessments.

ACF Climate Change Assessments

The lower quartile (LQ) of the monthly precipitation distribution increases in January, February, and October, and decreases in March through August.

July and August register the largest such decrease, raising concerns for summer water availability. (2) Buford Temperature: Mean monthly temperature increases in all months of the year with the most pronounced increases taking place from January through May and October through December.

The monthly upper temperature quartile increases for all months, with March and September registering the largest change (of approximately 3 °C) for the A2 scenarios and the second half of the century.

The monthly upper mean precipitation quartile increases for all months with the largest increase occurring in February through May. The monthly lower mean precipitation quartile shows a decreasing trend from January through August, with the most marked decline noted in June, July, and August.

Conclusions

This article introduces a new statistical downscaling technique, named Joint Variable Spatial Downscaling—JVSD, for the generation of high resolution gridded datasets suitable for regional watershed modeling and assessments.

JVSD as well as all other existing statistical downscaling methods assume that the spatial pattern of finer scale precipitation and temperature within a large GCM grid and the temporal distribution of (daily) precipitation or temperature within a month will remain the same.

Application of the method to the Apalachicola-Chattahoochee-Flint (ACF) river basin (for all IPCC GCM scenarios) leads to the following conclusions: Mean monthly temperature exhibits increasing trends over the ACF basin for all seasons and all A1B and A2 scenarios.

In the southern ACF watersheds, mean precipitation generally exhibits a mild decline, except in late winter when it shows an increase.

For the northern ACF watersheds, mean precipitation increases are noted in winter (as in the south) but also early spring.

Acknowledgement

A machine generated summary based on the work of Zhang, Feng; Georgakakos, Aris P. (2011 in Climatic Change).

D4PDF: Large-Ensemble and High-Resolution Climate Simulations for Global Warming Risk Assessment

https://doi.org/10.1186/s40645-020-00367-7

Abstract-Summary

A large-ensemble climate simulation database, which is known as the database for policy decision-making for future climate changes (d4PDF), was designed for climate change risk assessments.

It contains the results of ensemble simulations conducted over a total of thousands years respectively for past and future climates using high-resolution global (60 km horizontal mesh) and regional (20 km mesh) atmospheric models.

Several sets of future climate simulations are available, in which global mean surface air temperatures are forced to be higher by 4, 2, and 1.5 K relative to preindustrial levels.

Nonwarming past climate simulations are incorporated in d4PDF along with the past climate simulations.

The atmospheric models satisfactorily simulate the past climate in terms of climatology, natural variations, and extreme events such as heavy precipitation and tropical cyclones.

Data users can obtain statistically significant changes in mean states or weather and climate extremes of interest between the past and future climates via a simple arithmetic computation without any statistical assumptions.

The database is helpful in understanding future changes in climate states and in attributing past climate events to global warming.

Introduction

Increasing the climate simulation period was highly desirable for the impact projection and assessment, and for adaptation to future extreme hazards.

They developed future scenarios at several warming levels in the twenty-first century by performing numerous members of climate simulations using high-resolution global and regional atmospheric models forced by observed and future sea surface temperatures projected by CMIP5 participating models.

To realize this, d4PDF is constituted by the high-resolution and large-ensemble climate simulations, aiming at re-examining previous results on climate changes including past and future extreme events and at drawing highly reliable conclusions in both climate and assessment studies.

Both UKCP18 and d4PDF enable probabilistic evaluations of occurrence of extreme events, although the former treats a wider range of uncertainties in future climates by using a coupled model rather than an atmospheric model, more CMIP5 model states, and physics parameter ensemble.

Review

Large-ensemble database d4PDF was designed to be able to estimate future changes in severe weather and climate events occurring rarely, such as heavy precipitation and tropical cyclones, as well as changes in mean climate states.

2 most early conducted a study with d4PDF-G. They attributed historical changes in daily temperature and precipitation extremes to global warming and pointed out that spatial distributions and anthropogenic impacts of record-breaking events are sensitive to climate models and boundary conditions used in the experiments.

Two studies commonly confirmed that the frequency of extreme rainfall events in Baiu will increase in the future simulations of d4PDF-R. In particular, the former study focused on changes in typical atmospheric synoptic patterns and reported that Baiu heavy rainfall events mainly occur in western Japan in the past climate; however, the areas extended eastward in future climates with increasing trends of accumulated precipitation amount per rainfall event.

Conclusions

The large-ensemble and high-resolution climate simulation database called d4PDF was developed for extensive use in impact/risk assessment and policy decision-making for adaptation measures.

Large-ensemble climate simulation studies have advanced our understanding of past and future climate changes and have provided statistical reliability for assessing the impacts of significant weather and climatic phenomena in the future.

Although d4PDF comprises climate simulations at several warming levels with uncertainties in the future projections of the CMIP5 participating models, its range of uncertainties is limited.

Those in the model simulations are shifted north in many models, because the resolution adopted by most CMIP5 climate models is too coarse to resolve the regional atmosphere and ocean climates in this region.

The current resolution of the JMA operational regional model is a 2 km mesh, which is expected to reach 1 km or less in the future.

Acknowledgement
A machine generated summary based on the work of Ishii, Masayoshi; Mori, Nobuhito (2020 in Progress in Earth and Planetary Science).

Effect of Empirical Correction of Sea-Surface Temperature Biases on the CRCM5-Simulated Climate and Projected Climate Changes Over North America

https://doi.org/10.1007/s00382-018-4596-2

Abstract-Summary
Dynamical downscaling (DD) consists in using archives of Coupled Global Climate Models (CGCM) simulations as the atmospheric and sea-surface boundary conditions (BC) to drive nested, Regional Climate Model (RCM) simulations.

Biases in the CGCM-generated driving BC, however, can have detrimental impacts on RCM performance.

It is well documented for the historical period that CGCM-simulated sea-surface temperatures (SST) suffer substantial biases, especially important near coastal regions.

The CGCM-simulated sea-surface temperatures (SST) are first empirically corrected by subtracting their systematic biases; the corrected SST are then used as ocean surface BC for an atmosphere-only GCM (AGCM) simulation; finally this AGCM simulation provides the atmospheric lateral BC to drive an RCM simulation.

This is what we refer to as the 3-step approach CGCM–AGCM–RCM of DD, which can be compared to the traditional 2-step approach CGCM–RCM consisting of driving an RCM simulation directly by CGCM-generated BC.

The results show that, in current climate, the seasonal-mean 2-m temperature fields simulated with the 3-step DD have generally smaller biases with respect to the observations than those simulated with the 2-step DD; in fact the performance of the 3-step DD simulations often approaches that of the reanalyses-driven simulation.

Differences are particularly important for temperature: over the bulk of the North American continent, the 3-step DD projects more warming in winter and less in summer.

Extended
Dynamical downscaling (DD) using fine-mesh limited-area nested regional climate models (RCMs) is one of the techniques used to generate high-resolution climate data in order to assess the anticipated climate changes at regional and local scales (Giorgi and Gutowski [3]; Rummukainen and others [4]; Rockel [5]).

Introduction
Over Europe, the SST correction had little impact on the simulated 2-m temperature and precipitation biases, but the improved atmospheric lateral BC (as a consequence of using an intermediate-resolution AGCM) had a large impact in reducing biases in the historical period for summer and autumn seasons; for the other seasons, however, results were different (Déqué and others [6]).

The correction of CGCM-generated SST, followed by an intermediate AGCM simulation using the corrected SST as lower BC, yields improved input for the RCM simulation while keeping physical consistency between the corrected SST and the driving corresponding atmospheric variables.

We investigate the effect of applying an empirical correction of CGCM-generated SST on RCM simulations for current and future periods under RCP8.5, using the fifth-generation Canadian Regional Climate Model (CRCM5) over the CORDEX North-America domain on a grid mesh of 0.22°.

Experiments Setup
The third and final step is to use the atmospheric fields from the AGCM simulation, together with the corrected SST, as lateral atmospheric BC and surface ocean BC, respectively, for driving an RCM simulation over the region of interest: in the present case, the North American CORDEX domain.

The corrected SST fields are then used as ocean surface BC for 2 AGCM simulations (referred to as Agcm_e1 and Agcm_e2, the subscript e being used as a reminder of the empirical correction applied to CGCM-simulated SST fields and the numbers as indicators of the corresponding CGCM outputs), from 1949–2100 under historical and RCP8.5 emission scenario.

The Agcm_e# simulations will provide the atmospheric lateral BC for the CRCM5 simulations over the North American CORDEX domain for the same period (1949–2100), using the corrected SST fields; these simulations will be referred to as Rcm/Agcm_e1 and Rcm/Agcm_e2, and they will be compared to those driven by the corresponding CGCM-driven CRCM5 simulations following the usual two-step dynamical downscaling, noted as Rcm/Cgcm1 and Rcm/Cgcm2.

Historical Climate Simulations
The skill of the 3-step DD (Rcm/Agcm_e1 and Rcm/Agcm_e2) and 2-step DD (Rcm/Cgcm1 and Rcm/Cgcm2) CRCM5 historical simulations are assessed with

respect to the CRU reference observational dataset, and compared to the ERA-driven hindcast simulation (Rcm/ReAn).

In JJA the Rcm/Agcm_e1 is overall better than the reference simulation (Rcm/ReAn), which is indicative of compensation between structural errors of CRCM5 and errors inherited from BC forcing.

In JJA both 3-step DD simulations Rcm/Agcm_e1 and Rcm/Agcm_e2 exhibit superior performance than that of the hindcast simulation Rcm/ReAn; this counter-intuitive result is assuredly indicative of compensation between structural errors of the RCM and of that of the driving BC.

The 4th row shows the difference between the Rcm/Cgcm# and Rcm/Agcm_e# historical simulations to illustrate the impact of the 3-step DD.

Climate-Change Projections

Rows 1 and 3 show the projected change from the Cgcm# as well as from the Rcm following the 2-step (Rcm/Cgcm#) and 3step (Rcm/Agcm_e#) DD technique.

Differences between the projected climate changes from the 2-step (Rcm/Cgcm#) and 3-step (Rcm/Agcm_e#) DD technique are shown in the right panel of rows 2 and 4.

The projected changes of the 3-step DD simulations are quite different from the corresponding 2-step ones, the difference being larger in the case of Rcm/Cgcm1 and Rcm/Agcme_e1 (right panel, second row).

Also that the 3-step DD simulations (Rcm/Agcm_e1 and Rcm/Agcm_e2) give different projections of seasonal mean precipitation while having the same AGCM model, but different SSTs (right panels, first and third rows); the projected precipitation differences between the 2-step and 3-step DD simulations are generally larger for the Cgcm1 suite of models.

Conclusions

For current climate, the 3-step DD simulations (Rcm/Agcm_e1 and Rcm/Agcm_e2) have in general smaller seasonal-mean temperature biases with respect to the CRU analysis of observations than the 2-step DD simulations; the biases in fact approach those of the reanalyses-driven simulation (RCM/ReAn).

Differences in projected climate change between the two methods of DD are, at least, as important as the differences between the driving and driven model of the 2-step DD simulations and often larger; this is the situation for the two studied variables in the two seasons and the two CGCM models and their suite of simulations.

In the case of precipitation, 3-step DD simulations are also rather similar in both seasons for the period of current climate, but different in the future, the difference in the future being more important in winter than in summer.

The differences in climate changes projected by the two DD approaches (with and without empirical correction of SST biases) are another indication of the uncertainties in downscaled climate simulations.

Acknowledgement
*A machine generated summary based on the work of Hernández-Díaz, Leticia;
Nikiéma, Oumarou; Laprise, René; Winger, Katja; Dandoy, Samuel (2019 in Climate
Dynamics).*

3-Step Dynamical Downscaling with Empirical Correction of Sea-Surface Conditions: Application to a CORDEX Africa Simulation

https://doi.org/10.1007/s00382-016-3201-9

Abstract-Summary
Dynamical downscaling of climate projections over a limited-area domain using a
Regional Climate Model (RCM) requires boundary conditions (BC) from a Coupled
Global Climate Model (CGCM) simulation.

It is in this context that an empirical method involving the bias correction of
the sea-surface conditions (SSCs; sea-surface temperature and sea-ice concentration) simulated by a CGCM has been developed: The 3-step dynamical downscaling
approach.

The SSCs from a CGCM simulation are empirically corrected and used as lower
BC over the ocean for an atmosphere-only global climate model (AGCM) simulation,
which in turn provides the atmospheric lateral BC to drive the RCM simulation.

We analyse the impact of this strategy on the simulation of the African climate,
with a special attention to the West African Monsoon (WAM) precipitation, using the
fifth-generation Canadian Regional Climate Model (CRCM5) over the CORDEX-
Africa domain.

The most remarkable effect of this approach is the positive impact on the simulation of all aspects of the WAM precipitation, mainly due to the correction of
SSCs.

That the climate-change projections under RCP4.5 scenario obtained with the 3-
step approach are substantially different from those obtained with usual downscaling
approach in which the RCM is directly driven by the CGCM output; in the WAM
region most of the differences in the projected climate changes came mainly from
the empirical correction of SST.

Extended
We analyse the results of the application of the "3-step DD" over the CORDEX Africa
domain using the fifth-generation Canadian Regional Climate Model (CRCM5).

We analyse the results of CRCM5 historical simulations comparing the skill of
the 3-step DD simulation (Rcm/Agcm_e) with that of the 2-step DD simulation
(Rcm/Cgcm), as well as with hindcast ERA-driven simulation (Rcm/ReAn) over the
CORDEX Africa domain.

Introduction

Bruyère and others [7] corrected the mean bias of the CGCM while retaining its synoptic and climate variability by first decomposing the CGCM-simulated data as well as the atmospheric and SST reanalyses into a mean seasonally-varying climatological component and a perturbation component, and then constructing the BC for driving the RCM simulation by replacing the CGCM climatological mean component by that of the reanalyses.

Their study shows that the simulation of precipitation is significantly improved for current climate when using the bias-corrected CGCM-SST as lower BC.

Sea-surface conditions simulated by a CGCM are empirically corrected and used as lower BC over the ocean for an intermediate-resolution AGCM simulation, which in turn will generate the atmospheric lateral BC to the RCM downscaling simulation, hence the name "3-step DD" (CGCM-AGCM-RCM), in contrast with the usual "2-step DD" (CGCM-RCM) used for example for CORDEX (Giorgi and others [8]; Jones and others [9]).

Methodology

The third and final step is to use the atmospheric fields from the AGCM simulation, together with the corrected sea-surface fields, as lateral atmospheric BC and surface ocean BC respectively, for driving an RCM simulation over the region of interest: in the present case, the CORDEX-Africa domain.

This Agcm_e simulation will provide the atmospheric lateral BC for the CRCM5 simulation over the CORDEX-Africa domain for the same period (1949–2100), using the corrected sea-surface fields; this simulation will be referred to as Rcm/Agcm_e, and it will be compared to that driven by the Cgcm following the usual two-step dynamical downscaling, noted as Rcm/Cgcm.

These include: a CRCM5 simulation driven by the atmospheric lateral BC from the CGCM but with the corrected sea-surface BC (noted Rcm/Cgcm_*e); two other AGCM-driven CRCM5 simulations: one with the uncorrected CGCM-simulated sea-surface fields (noted Rcm/Agcm_u) and another with the ERA-Interim analysed sea-surface fields (Rcm/Agcm_o); and finally a reanalysis-driven hindcast simulation (Rcm/ReAn).

Historical Climate Simulation Results

The most striking effect of the 3-step DD with the empirical correction of the SSTs can be seen in the WA-S region (panel b), where the bimodality of the annual cycle present in the observations and the reanalysis-driven simulation (Rcm/ReAn) that was lost in the Rcm/Cgcm simulation is recovered with the empirically corrected SST in Rcm/Agcm_e.

For the other regions, the 3-step DD (Rcm/Agcm_e) shows an improved representation of the annual cycle of precipitation, nearly as good as the hindcast simulation driven by the reanalysis (Rcm/ReAn) and even sometimes coincidentally better (e.g., CA-NH region, panel d).

Comparison with the sensitivity experiments simulations Rcm/Agcm_u and Rcm/Cgcm_*e (not shown) suggests that the improvement in the representation of the diurnal cycle of precipitation in the three equatorial regions (WA-S, CA-NH and CA-SH) is a result of the correction of the sea-surface conditions.

Climate-Change Projections

The comparison with the sensitivity test simulation Rcm/Agcm_u (not shown) revealed that the differences in the projected 2-m temperature changes between the two DD techniques came mainly from the empirical correction of the sea-surface correction in JFM, but in JAS the use of the intermediate-resolution AGCM contributed substantially also.

For JAS there is a substantial difference between the projected precipitation changes obtained with the two DD methods in the latitude band between 0 and 15°N, with the 3-step DD projected changes being smaller in magnitude and spatial extension than the 2-step DD projected changes.

The comparison with the sensitivity test simulation Rcm/Agcm_u (not shown) revealed again that the differences in the projected precipitation changes between the two DD techniques came mainly from the empirical correction of the sea-surface correction, not from the use of the intermediate-resolution AGCM.

Conclusions

The sea-surface conditions (SSC; sea-surface temperature, SST; and sea-ice concentration, SIC) from a simulation of a coarse-mesh coupled global climate model (CGCM), in this case the Earth System Model of the Max-Planck-Institut für Meteorologie (MPI-ESM-LR), were empirically corrected by subtracting the bias calculated over the historical period, and used as sea-surface boundary condition (BC) for a simulation of an atmosphere-only GCM (AGCM), in this case an intermediate-resolution global version of CRCM5.

Several sensitivity test experiments were also conducted in order to analyse the respective impact on the CRCM5 simulations of (1) the bias correction of SSC, (2) the change of driving atmospheric model (AGCM vs. CGCM), and (3) the combination of both modifications as in the 3-step DD.

The comparison with the sensitivity test simulations revealed that the differences in the projected climate changes between the two DD techniques came mainly from the empirical correction of the SSC, although there were regions and variables for which the use of the intermediate-resolution AGCM also contributed substantially.

Acknowledgement

A machine generated summary based on the work of Hernández-Díaz, Leticia; Laprise, René; Nikiéma, Oumarou; Winger, Katja (2016 in Climate Dynamics).

Dynamical Downscaling with the Fifth-Generation Canadian Regional Climate Model (CRCM5) Over the CORDEX Arctic Domain: Effect of Large-Scale Spectral Nudging and of Empirical Correction of Sea-Surface Temperature

https://doi.org/10.1007/s00382-017-3912-6

Abstract-Summary

As part of the CORDEX project, the fifth-generation Canadian Regional Climate Model (CRCM5) is used over the Arctic for climate simulations driven by reanalyses and by the MPI-ESM-MR coupled global climate model (CGCM) under the RCP8.5 scenario.

The analysis shows that SN is effective in reducing the spring MSLP bias, but otherwise it has little impact.

We have also conducted another experiment in which the CGCM-simulated sea-surface temperature (SST) is empirically corrected and used as lower boundary conditions over the ocean for an atmosphere-only global simulation (AGCM), which in turn provides the atmospheric lateral boundary conditions to drive the CRCM5 simulation.

This approach, so-called 3-step approach of dynamical downscaling (CGCM-AGCM-RCM), which had considerably improved the CRCM5 historical simulations over Africa, exhibits reduced impact over the Arctic domain.

The most notable positive effect over the Arctic is a reduction of the T2m bias over the North Pacific Ocean and the North Atlantic Ocean in all seasons.

Future projections using this method are compared with the results obtained with the traditional 2-step dynamical downscaling (CGCM-RCM) to assess the impact of correcting systematic biases of SST upon future-climate projections.

Introduction

An increase in precipitation, large reductions in sea ice and glacier volume, sea level rise and the thawing of permafrost are expected consequences in the Arctic of the projected global warming (IPCC [10]).

From the side of regional climate modelling, a comprehensive study of RCM hindcast simulations was done as a part of the Arctic Regional Climate Model Intercomparison Project (ARCMIP; Curry and Lynch [11]).

Dynamical downscaling of the CGCM projections over the Arctic by RCMs shows warming up to 6.5 K in the mean tropospheric temperature over Barents, Kara Seas and the Beaufort Sea during winter (Rinke and Dethloff [12]), the regions corresponding to the areas with the maximum projected sea ice loss.

The Arctic is also one of the recommended domains for the COordinated Regional climate Downscaling EXperiment (CORDEX), an international coordinated sets of experiments for hindcast, historical simulations and climate projections under RCP4.5 and RCP8.5.

Methodology

An empirical bias correction of sea-surface conditions (SSC) has been tested for CRCM5 simulations over the African CORDEX domain (HD16); it was shown that for the West African monsoon, the skill of the historical simulations, driven by an AGCM with the empirical correction of the CGCM-simulated sea-surface conditions, was substantially improved, approaching in fact that of hindcast simulations driven by reanalyses.

The atmospheric fields from this AGCM simulation is then used as lateral atmospheric BC, together with the corrected SST as surface ocean BC, for driving an RCM simulation over the region of interest: in the present case, the CORDEX Arctic domain.

Regarding the study of the effect of the empirical correction of SSTs, two CRCM5 simulations are performed, spanning the 1979–2100 time period under historical and RCP8.5 emission scenario, one is driven at the boundaries by the CGCM, and another by the AGCM with corrected SST.

Hindcast Climate Simulations Driven by Reanalyses

Hindcast simulations (noted CRCM5/ERA) are used to evaluate structural biases of CRCM5 over the CORDEX Arctic domain, upon assuming that LBC and SSC derived from reanalysis are quasi perfect, therefore the identified biases result from the structure of the model: its formulation, approximations and parameterizations.

The CRCM5 simulation driven by ERA-Interim using the technique of large-scale spectral nudging will be referred to as CRCM5(SN)/ERA.

In case of 2-m temperature (T2m), the different observationally-based gridded datasets exhibit substantial differences amongst themselves, and hence the apparent biases of CRCM5 simulation vary considerably depending on the reference data set used.

The apparent precipitation bias when compared to GPCP is larger in winter than in summer, possibly because of the difficulty for the model to adequately simulate winter clouds and precipitation, or for the satellite-based GPCP to distinguish clouds from snow cover.

Historical Climate Simulations Driven by GCM

We analyse the results of CRCM5 historical simulations driven by MPI-ESM-MR (CRCM5/CGCM) and compare it with the results of 3-step dynamical downscaling with the empirically corrected SST (CRCM5/AGCM_e).

The CRCM5/CGCM SST was interpolated from the MPI-ESM-MR and reflects the aforementioned biases over open oceans.

The CRCM/AGCM_e SST consists of the CGCM SST empirically corrected using ERA-Interim, and hence the bias should vanish in principle.

Areas with non-vanishing CRCM/AGCM_e SST bias reflect where the SST correction could not be applied due to conflicting values of SIC in CGCM and ERA-Interim.

The difference between CRCM5 simulations with and without SN as well as CRCM5 simulations driven by CGCM and AGCM_e is relatively modest.

Comparing CRCM5/CGCM and CRCM5/AGCM_e, the SST correction reduces some of the biases over the land and the remaining biases are comparable to those seen with CRCM5/ERA.

Climate Change Projections
The largest SIC decline is projected to occur during autumn, with more than 60% reduction over a vast area of the Arctic Ocean, particularly in East Siberian, Laptev and Kara seas, resulting in a nearly late summer ice-free Arctic Ocean by the end of twenty-first century.

Over the Beaufort sea, the CGCM projects a larger warming than CRCM5/AGCM_e and CRCM5/CGCM.

Some studies such as Koenigk and others [13] have suggested that precipitation and temperature changes in the future are somehow linearly related; the projected precipitation increase over Kara sea and Bering sea, where the largest warming is projected, tends to confirm that hypothesis.

There is a noteworthy difference between the projected precipitation changes by CRMC5/CGCM and CRCM5/AGCM_e in summer, with CRCM5/AGCM_e suggesting a widespread reduction (0.5 mm/day) for continental areas surrounding the Arctic Ocean.

Summary and Conclusions
The skill of CRCM5 hindcast simulations driven by ERA-Interim reanalysis, the effectiveness of the large-scale spectral nudging (SN), and future climate-change projections using 2- and 3-step dynamical downscaling over the CORDEX Arctic domain were investigated.

The precipitation bias of CRCM5 simulations when compared to the GPCP dataset is more pronounced in coastal areas.

Positive impact of SST correction are seen through a reduction of the T2m bias over North Pacific Ocean and North Atlantic Ocean in all seasons, as well as over the Bering Sea in summer.

Future projections of CGCM were compared with those performed with CRCM5 driven by CGCM (CRCM5/CGCM) and driven by the AGCM using bias-corrected SST (CRCM5/AGCM_e).

Both CRCM5 simulations project less warming over the Arctic Ocean for fall compared to the CGCM simulation.

For summer, CRCM/AGCM_e projects widespread decreasing precipitation unlike CGCM and CRCM5/CGCM, but the decrease is modest, about 0.5 mm/day.

Acknowledgement
A machine generated summary based on the work of Takhsha, Maryam; Nikiéma, Oumarou; Lucas-Picher, Philippe; Laprise, René; Hernández-Díaz, Leticia; Winger, Katja (2017 in Climate Dynamics).

Assessment of Climate Change Impact Over California Using Dynamical Downscaling with a Bias Correction Technique: Method Validation and Analyses of Summertime Results

https://doi.org/10.1007/s00382-020-05200-x

Abstract-Summary

This study explores climate-change influences on future air pollution-relevant meteorological variables (e.g., temperature, wind, humidity, boundary layer heights) and atmospheric phenomena (e.g., heat wave, marine air penetration, droughts) over California by the 2050s.

The Community Earth System Model simulation results from Coupled Model Intercomparison Project Phase 5 under an emission scenario that most closely aligns with California's climate change goals were bias-corrected with respect to North American Regional Reanalysis data to reduce biases in both the climatological mean and inter-annual variations.

Our downscaled results projected a future increase of approximately 1 K in summer mean surface temperature over California under this single future climate realization.

Water vapor mixing ratio is also projected to increase over California and off the coast.

There are discernable decreases in boundary layer heights over the mountain ranges surrounding the central valley of California, while increases in boundary layer heights are observed over other regions in California.

The occurrence of marine air penetration events over the northern California is also projected to increase in the future.

Extended

The results from this study create a potential future scenario in a high resolution.

Introduction

Given the seasonal variation of both PM and ozone, as well as the differences in the underlying meteorological drivers influencing these two types of pollution, we need to evaluate different meteorological parameters and phenomena in summer [e.g., temperature extremes and marine air penetration (MAP)] versus winter (precipitation characteristics and stagnation events) when we analyze climate change impacts on meteorological conditions relevant to air quality in California.

The primary goal of this study is to explore the future projection of air pollution-relevant meteorological variables (e.g., temperature, wind, precipitation and ventilation) and important atmospheric phenomena (e.g., heat wave events, MAP events and droughts) over California in greater detail using the aforementioned dynamical downscaling with bias correction (i.e., the mean bias correction plus the adjustment of inter-annual variations).

To evaluating the dynamical downscaling method with bias correction applied to the CESM data, this paper also presents the climate change impact on summertime meteorological conditions and atmospheric phenomena.

Models and Methodology

Following Xu and Yang [14], bias correction was applied to three-dimensional (3D) air temperature, zonal and meridional wind, water vapor mixing ratio, and geopotential height at each model grid and on each vertical level for each 6-hourly CESM output.

Figures S2 and S3 show monthly mean T2 at 00 UTC (4 pm PST) from the original CESM, bias-corrected CESM and NARR data for July averaged over the validation periods.

These bias correction coefficients are applied to adjust the biases of the CESM data for the future simulation period (2046–2055) in this study.

To WRF downscaling simulations driven by bias-corrected CESM data, WRF simulations were also conducted driven by NARR data and the original CESM data for the present 10-year period.

Evaluation of the Dynamical Downscaling Results with the Bias Correction Method

Driven by bias-corrected CESM data, D_BCCESM over (under) predicts RH2 for 5 (4) out of 9 sub-regions; and the magnitude of the difference is reduced more than 50% for SAC, BA, MC, SJV and SC.

After applying bias correction to the CESM data (i.e., D_BCCESM), the positive difference below 500 hPa was reduced by more than 50%, especially for the lowest 2 layers close to the surface where the differences were reduced from 1.72 K and 1.98 K to 0.28 K and 0.82 K, respectively.

The large negative biases above 500 hPa were reduced remarkably and the magnitudes of the differences were smaller than 0.4 K. The over-predictions of Qv from D_OCESM were reduced for almost all layers after using bias-corrected CESM data (i.e., D_BCCESM), especially for the lowest two layers.

Summertime Climate Change Projections Over California

Heat wave events are defined as a period of at least three consecutive days in which the regional-mean Tmax exceeds the 90th percentile threshold of summertime regional-mean Tmax for the present decade (Tc_max_90_reg) calculated based on WRF downscaled results.

The average durations of heat wave events are expected to increase by 1 day and 0.9 days in the future decade for MD and SJV, respectively, while a twofold increase is projected for the maximum duration of heat wave events for both regions.

Both the frequency and duration of heat wave events are expected to increase in the future decade for almost all sub-regions analyzed in this section, within which, MD, SJV and SC are projected to experience greater impacts than other sub-regions.

It was demonstrated in the previous section that California will likely experience more extreme hot days and heat wave events in future decades on top of the overall warming trend during summer months.

Concluding Remarks

The regional mean summertime surface temperature is projected to increase over 0.9 K for 7 out of 9 geographical sub-regions over California in the future decade, among which the increase over GBV and MD is over 1.1 K. The projected temperature increase is statistically significant at the 95% confidence level over almost the entire analysis domain.

Our WRF downscaling results project an over twofold increase in the number of heat wave events over MD and SC for the future simulation period.

This is the first comprehensive climate change study using dynamically downscaled bias-corrected CESM results from CMIP5 on to a 4 × 4 km^2 grid over California via WRF model, followed by detailed analysis on potential climate change impacts on meteorological variables and phenomena over different geographical sub-regions in California.

Acknowledgement

A machine generated summary based on the work of Zhao, Zhan; Di, Pingkuan; Chen, Shu-hua; Avise, Jeremy; Kaduwela, Ajith; DaMassa, John (2020 in Climate Dynamics).

Which Complexity of Regional Climate System Models is Essential for Downscaling Anthropogenic Climate Change in the Northwest European Shelf?

https://doi.org/10.1007/s00382-017-3761-3

Abstract-Summary

Climate change impact studies for the Northwest European Shelf (NWES) make use of various dynamical downscaling strategies in the experimental setup of regional ocean circulation models.

Projected change signals from coupled and uncoupled downscalings with different domain sizes and forcing global and regional models show substantial uncertainty.

We investigate influences of the downscaling strategy on projected changes in the physical and biogeochemical conditions of the NWES.

Our results indicate that uncertainties due to different downscaling strategies are similar to uncertainties due to the choice of the parent global model and the downscaling regional model.

Downscaled change signals reveal to depend stronger on the downscaling strategy than on the model skills in simulating present-day conditions.

Introduction

The conventional approach to address regional climate change impacts in the ocean is the dynamical downscaling of a global climate projection by a regional ocean model.

The importance to account for coupled air–sea interaction has been proposed in many studies on regional ocean modeling and dynamical downscaling (e.g. Schrum and others [15]; Rummukainen [16]; Tian and others [17]; Van Pham and others [18]; Gröger and others [19]; Wang and others [20]).

A high sensitivity in the response of physical and biogeochemical conditions in the ocean to changes in the atmospheric wind and thermal forcings is characteristic for shelf and marginal seas but cannot be accounted for by uncoupled models because they are driven with bulk formulae which do not allow the atmosphere to adjust to the downscaled ocean conditions (e.g. Ådlandsvik [21]; Holt and others [22]; Olbert and others [23]; Gröger and others [24]; Mathis and Pohlmann [25]; Tian and others [26]; Tinker and others [27]).

What is missing so far is a consistent test of the various downscaling strategies, ranging from a downscaling with a fully coupled regional climate system model over forcing a regional ocean model with data from a dynamical atmosphere downscaling to directly using forcing fields from the global model output.

Methods

With this experiment, we mimic a stand-alone regional ocean model that uses MPI-ESM output not only for the atmospheric forcing but also to prescribe oceanic conditions at the open lateral boundaries of its North Sea-Baltic Sea domain.

In the fifth experiment RF (REMO forcing), MPIOM is forced with REMO output from an atmosphere-only downscaling of ECHAM6, which has been driven by atmospheric lateral boundary data, sea surface temperature (SST), sea ice conditions and sea surface velocity from the global MPI-ESM simulation.

RFroc mimics a state-of-the-art uncoupled dynamical downscaling with a regional ocean model driven by downscaled high resolution atmospheric forcing and prescribed lateral boundary conditions in the ocean extracted from the parent global climate simulation.

To rule out remaining effects of model drift in the projected climate change signals, control simulations under constant preindustrial atmospheric CO_2 concentration (285 ppm) have been performed for the global and regional coupled setups.

Results

The change signal of the northeastern North Atlantic water masses is advected into the North Sea, causing a stronger salinity drop in MPI-ESM than in Cref in the northern and central North Sea as well as in the English Channel.

In EF, the decrease in North Sea primary production is generally stronger because of a weaker warming trend and a stronger nutrient decline in the northeastern North Atlantic, while the downscaled trend in RF is quite similar to the coupled reference simulation Cref.

When water temperature, salinity and nutrient concentrations are restored in EFroc and RFroc towards MPI-ESM anomalies, the less intense SPG nutrient decline in MPI-ESM weakens the downscaled negative trends in North Sea primary production and enhances the positive trends, compared to EF and RF, respectively.

Discussion

The experiments with three-dimensional restoring in the ocean EFroc and RFroc have further confirmed that the oceanic forcing at the open lateral boundaries of stand-alone regional ocean models is of secondary importance for downscaling SST change signals in the North Sea.

Experiments EFroc and RFroc have shown that downscalings by regional ocean models forced with salinity anomalies of a global simulation suffer from inconsistencies between the change signals of the global simulation and a fully coupled downscaling.

A larger domain size including the northeastern North Atlantic might reduce the influence of the global forcing data, as also indicated by Schrum and others [28] where different forcing global models yielded a similar spread in downscaled North Sea salinity changes as obtained in our study from different downscaling strategies.

Conclusions

When regional ocean models are used, however, the prescribed oceanic boundary conditions can introduce large errors to the downscaled change signals, even in coupled simulations.

We therefore recommend the coupling domain of the downscaling model system to include the northeastern North Atlantic region in order to allow for local ocean–atmosphere responses related to changes in the NAC pathway and transport rates.

The downscaling strategies compared here cover a wide range of complexity, decreasing gradually from a fully coupled ocean–atmosphere climate system model to a stand-alone ocean model forced with boundary conditions extracted from other general and regional circulation models of different grid resolutions.

The resulting change signals in the North Sea reveal to depend stronger on the individual downscaling strategy than on the fidelity of simulated present-day conditions.

Inter-model comparisons of climate projections regionalized for the NWES show large uncertainties in the downscaled change signals.

Acknowledgement
A machine generated summary based on the work of Mathis, Moritz; Elizalde, Alberto; Mikolajewicz, Uwe (2017 in Climate Dynamics).

Uncertainties in Predicting Impacts of Climate Change on Hydrology in Basin Scale: A Review

https://doi.org/10.1007/s12517-020-06071-6

Abstract-Summary
Modelling the hydrological impacts of climate change is generally done in various stages and has uncertainty associated with each of them.

These include scenario uncertainty in climate scenario selection, model uncertainty in climate simulation by global climatic models (GCMs), uncertainties while downscaling GCMs, biases in downscaled data, erroneous input to the hydrological model, and uncertainty in the structure and parameterisation of the hydrological model.

The present paper aims at reviewing the uncertainties involved at each stage of climate change impact assessment of hydrology.

Climate scenario uncertainties would be smaller than those associated with the choice of GCMs.

GCMs shall be downscaled by statistical or dynamical methods (regional climatic models (RCMs)) before using them for regional studies.

Taking into account the uncertainties associated with climate impact studies can help formulate effective adaptation strategies.

Introduction

Land use and climate change are two key factors which impact the quantity and quality of water available (Kaushal and others [29]).

Climate change as a result of global warming influences the availability and distribution of water.

Some studies have also analysed the effect of climate change on water quality (e.g. Azadi and others [30]).

Since climate change disrupts the hydrological cycle, the pattern of precipitation and temperature changes which in turn, affects the water resources of an area.

There are uncertainties involved at every stage of hydrological impact assessment of climate change at a local scale.

A crucial aspect of assessing the predictability of the hydrological impact of climate change is the quantification of the total uncertainty associated with the results (Maraun and others [31]).

The climate change effects on hydrology have become a priority area for water management strategies.

Uncertainties in Climate Impact Studies

Hydrological models require high-resolution meteorological forcing data for climate impact assessments (Barnett and others [32]; Asong and others [33]).

Teutschbein and Seibert [34] analysed the performance of six bias correction methods, namely linear scaling, variance scaling, local intensity scaling, power transformation, distribution transfer and the delta-change approach and found that distribution mapping performed the best in streamflow simulation of the five Swedish catchments considered in the study.

Kim and others [35] employed a method to reduce the distributional parametric uncertainty that comes along with observational and climate model data.

Two main approaches for selecting hydrological models for climate impact studies are (1) using multi-model ensembles without considering model performance and (2) selecting a single model by evaluating its performance.

There are a few studies which have analysed the uncertainties related to hydrological model choice in the climate impact studies (Jiang and others [36]; Maurer and others [37]; Bastola and others [38]; Najafi and others [39]; Surfleet and others [40]; Velázquez and others [41]; Vansteenkiste and others [42]; Karlsson and others [43]).

Discussion

Impact studies of climate change involve various uncertainties.

Uncertainties are considered barriers to climate change adaptation.

Addressing these uncertainties can significantly improve future predictions and thus improve climate adaptation options (Kusangaya and others [44]).

Uncertainties related to climate model projections are often the largest source of uncertainty in climate impact studies (Minville and others [45]).

Selection of the climate model is a critical step in climate impact studies (Lutz and others [46]).

In the prospect of longer planning, the uncertainties related to climate scenario selection are of prime importance.

For future climate, the basic assumption of stationarity in bias correction procedures makes the selection process questionable.

Model evaluation enhances the scientific reliability and acceptance of modelled results of climate change impact studies.

Climate models and hydrological models are expected to be better integrated in the future.

Conclusions

A range of possible climate scenarios is recommended for impact studies as future emissions would be mostly dependent on mitigation policies taken in the future, which are uncertain.

Use of single best or average case climate scenario may be misleading.

GCMs are often the largest contributors to uncertainty, followed by climate scenarios and impact models.

GCM simulations need downscaling for their use in regional or local impact studies.

Decent performance of hydrological models in the past increases the confidence of projected impacts and decreases the uncertainty related to hydrological models, but does not guarantee good performance in the future climatic condition.

Impact studies use either a multi-model ensemble or a single model after evaluating their performance.

Acknowledgement

A machine generated summary based on the work of Jose, Dinu Maria; Dwarakish, Gowdagere Siddaramaiah (2020 in Arabian Journal of Geosciences).

Evaluations of High-Resolution Dynamically Downscaled Ensembles Over the Contiguous United States

https://doi.org/10.1007/s00382-017-3645-6

Abstract-Summary
This study uses Weather Research and Forecast (WRF) model to evaluate the performance of six dynamical downscaled decadal historical simulations with 12-km resolution for a large domain (7200 km × 6180 km) that covers most of North America.

The initial and boundary conditions are from three global climate models (GCMs) and one reanalysis data.

The GCMs employed in this study are the Geophysical Fluid Dynamics Laboratory Earth System Model with Generalized Ocean Layer Dynamics component, Community Climate System Model, version 4, and the Hadley Centre Global Environment Model, version 2-Earth System.

We evaluate the model performance for seven surface variables and four upper atmospheric variables based on their climatology and extremes for seven subregions across the United States.

We find that the use of bias correction and/or nudging is beneficial in many situations, but employing these when running the RCM is not always an improvement when compared to the reference data.

The use of an ensemble mean and median leads to a better performance in measuring the climatology, while it is significantly biased for the extremes, showing much larger differences than individual GCM driven model simulations from the reference data.

Extended
We find that there is clear advantage by using the downscaled simulations over the raw GCM counterparts, especially in mountainous and convection dependent regions and for higher percentiles in the PDF distribution.

The use of bias correction does not cause large differences between the two CCSM4 driven WRF runs in relative errors, which have ranks that are mostly near the middle of pack.

The use of bias correction (WCB) reduces the cold bias over the Northwest, Southwest as well as Southeast regions by 0.5–2 °C, but increase the bias over the Midwest and Northeast regions by 1 °C in comparison with the run without bias correction (WCNB).

The value of using three separate GCMs to create an ensemble over North America will allow for an improved ability to capture the future uncertainties.

Introduction
This information is particularly needed in regions where global model's resolution cannot capture the orographic variations sufficiently enough to represent observed

changes accurately and for an improved ability at modeling regional extremes associated with rare climate events.

Uncertainties in the RCM simulations can come from a number of sources, including physics parameterization, model representation of internal variability, the choice of emission scenarios for projecting future changes in climate, and the differences in the global climate models used to drive the RCM (Giorgi and Bi [47]; Mearns and others [48]).

Halmstad and others [49] investigate the NARCCAP six simulations for the historical period over the Villamette River Basion, Oregon, and find that the weather research and forecasting (WRF) regional climate model performs better in extreme precipitation than its boundary conditions when driven by the Community Climate System Model, version 3, but performs worse than the boundary conditions when driven by the Canadian Climate Centre Coupled General Circulation Model version 3.

Model Description and Reference Data

Among the six WRF simulations, the NCEP-R2 driven run was conducted first, along with which we conducted sensitivity experiments considering different nudging strength, microphysics, convective parameterizations and spin-up time.

While the study by Wang and Kotamarthi [50] compare the impacts of bias correction using CCSM4 to drive the WRF model, they only focus on precipitation over different regions of North America.

This study compares not only the effect of bias correction, but also the effect of spectral nudging and different lateral boundary conditions on the model performance.

To explore the impacts of spectral nudging on model performance when bias correction is applied, we conducted two WRF runs driven by GFDL-ESG2G, with spectral nudging turned on in one of the simulations and turned off in the other simulation.

We also use PRISM monthly precipitation data set as the reference data to evaluate the model and understanding the uncertainty of model's performance to different reference data.

Results

The two CCSM4 driven WRF simulations (with and without bias correction) also underestimate the extreme maximum temperature for all seven climate regions, but show smaller cold bias than do mean and median.

In the two runs where WGN and WCB use both spectral nudging and bias correction, there are large differences in all seven regions, indicating that the GCM used to force the WRF makes a larger difference than the use of bias correction and nudging does for maximum temperature.

Similar to the maximum temperature, the tail in the PDF curve for GCM driven simulations are too close to the mean and underestimate the intensity of extreme cold temperature (with warm bias) for many of the regions.

The RMSE for the two models (WGN and WCBC) that use both bias correction and nudging is higher for the North Plains, South Plains, and Midwest regions, but they both show smaller bias in the same region for extreme precipitation than the other simulations.

Discussion and Summary

Few downscaling projects have compared CMIP5 GCM-driven dynamical downscaled model performance for variables other than surface temperature and precipitation, especially as an ensemble (Fowler and others [51]; Lee and others [52, 53]).

One of the challenges in a study like this is to compare the model output to best reference data set available, but in reality, the "ground truth" for variables, such as precipitation, often have sources of biases and error themselves (Cosgrove and others [54]).

If there is a known overall bias in the dynamically downscaled method for a specific region in all members of the ensemble, that can now be accounted for when making projections of future climate change.

Since most of the uncertainty in future climate comes from choices such as the climate model used and the emission scenario (Déqué and others [55]), our multi-climate model ensemble, while employing bias correction and spectral nudging, can prove valuable at analyzing the uncertainties in future climate extremes.

Acknowledgement

A machine generated summary based on the work of Zobel, Zachary; Wang, Jiali; Wuebbles, Donald J.; Kotamarthi, V. Rao (2017 in Climate Dynamics).

Improving Sea-Level Projections on the Northwestern European Shelf Using Dynamical Downscaling

https://doi.org/10.1007/s00382-019-05104-5

Abstract-Summary

The projections of ocean dynamic sea level presented in the IPCC AR5 were constructed with global climate models (GCMs) from the Coupled Model Intercomparison Project 5 (CMIP5).

We use the regional shelf seas model AMM7 to show that, depending on the driving CMIP5 GCM, dynamical downscaling can have a large impact on DSLC simulations in the NWES region.

Dynamical downscaling affects the simulated time of emergence of sea-level change (SLC) above sea-level variability, and can result in differences in the projected change of the amplitude of the seasonal cycle of sea level of over 0.3 mm/yr.

We find that the difference between GCM and downscaled results is of similar magnitude to the uncertainty of CMIP5 ensembles used for previous DSLC projections.

Our results support a role for dynamical downscaling in future regional sea-level projections to aid coastal decision makers.

Extended

We use the AMM7 (Coastal Ocean version 6) configuration of the primitive-equation modeling framework Nucleus for European Modelling of the Ocean (NEMO) V3.6 (Madec and NEMO Team [56]) to downscale long-term simulations of two CMIP5 GCMs.

Introduction

Projections of DSLC driven by climate change scenarios are commonly made with the output of coupled global climate models (GCMs, e.g. Slangen and others [57, 58]; Church et al. [59]; de Vries and others [60]; Kopp and others [61]; Palmer and others [62]).

Sea-level projections at a finer spatial resolution can be obtained by downscaling, which is a technique to obtain regional to local detail from larger scale information (Rummukainen [63]).

For the North Pacific, downscaled DSLC was computed with the regional ocean model ROMS (1/4° by 1/4°) for three different driving CMIP5 GCMs (Liu and others [64]).

We assess the importance of dynamical downscaling for the NWES region and quantify the uncertainties related to constructing regional DSLC projections with CMIP5 GCMs.

We do this by downscaling the simulations of two CMIP5 GCMs with a regional shelf seas model (the Atlantic Margin Model (AMM7), O'Dea and others [65]) and comparing the results with the original simulations for two different representative concentration pathways (RCPs, Meinshausen and others [66]).

Data and Methods

We use the AMM7 (Coastal Ocean version 6) configuration of the primitive-equation modeling framework Nucleus for European Modelling of the Ocean (NEMO) V3.6 (Madec and NEMO Team [56]) to downscale long-term simulations of two CMIP5 GCMs.

The lateral boundary conditions consist of monthly mean temperature and salinity, barotropic currents and SSH, which are derived from the GCMs and interpolated onto the AMM7 grid.

Discrepancies in mass transport across the boundaries of the NWES region between the GCMs and AMM7 can result from the interpolation of the lateral boundary conditions (e.g. barotropic currents) from the parent grid onto the AMM7 grid and from the different representations of bathymetry, atmosphere and river run-off.

To directly compare DSLC in the GCMs with DSLC in AMM7, we correct the DSLC output of AMM7 for the differences in regional mean DSLC resulting from the Boussinesq approximation and from discrepancies in mass transport due to artefacts of the downscaling setup.

The Impact of Dynamical Downscaling on Historical Simulations
In the Norwegian Sea, simulated MDT is too low in GCM-MPI, RCM-HAD and RCM-MPI, and does not agree spatially with the observations in GCM-HAD.

Similar to MDT, the biases of simulated SST with respect to the observations are larger for GCM-HAD than for GCM-MPI, and the improvement for GCM-HAD after downscaling is also larger.

Near the boundaries of the NWES region, biases of RCM-HAD and RCM-MPI with observations are larger than in the interior, and the downscaled simulations are closer to their driving GCMs due to the applied boundary conditions.

Comparing simulated interannual variability near TG stations to the observed TG data, GCM-HAD has a PCC of 0.7 and RMSE of 1.12 cm.

The comparison suggests that the impact of dynamical downscaling on simulations of interannual sea-level variability along the coast depends strongly on the driving GCM.

Dynamical downscaling generally improves the historical GCM simulations with respect to observations (i.e. reduces biases).

The Impact of Dynamical Downscaling on Projected DSLC
This difference is approximately 30% of the sterodynamic SLC (DSLC plus 'zostoga', Gregory and others [67]) simulated by GCM-HAD for the North Sea.

In the North Sea, GCM-MPI simulates slightly larger DSLC, but differences with RCM-MPI do not exceed 2.5 cm (7% of the sterodynamic SLC simulated by GCM-MPI).

The local steric change in GCM-HAD can be over 15 cm larger than in RCM-HAD in the northern North Sea.

SLC due to bottom pressure changes is up to 13 cm larger in GCM-HAD than in RCM-HAD in the North Sea.

Since sea-level variability and the timing of SLC differ between GCM and RCM, the effect of dynamical downscaling on ToE on the NWES can be large, even if differences in DSLC by the end of the twenty-first century are relatively small.

Projected Changes in the Seasonal Sea-Level Cycle
The large differences in trends between the GCM and downscaled simulations suggest that RCMs should be used for accurate projections of the change in seasonal amplitude in the NWES region.

Dynamical downscaling is important to better project the variability of the amplitude of the seasonal sea-level cycle in the NWES region.

Discussion and Conclusions
The objective of this study was to explore the use of dynamical downscaling with the regional model AMM7 to refine the CMIP5 GCM simulations of the ocean dynamic component of sea-level variability and long-term change for the NWES region.

Since the horizontal resolution of HadGEM2-ES is more typical for the CMIP5 ensemble than the horizontal resolution of MPI-ESM-LR, we expect the results of dynamical downscaling for HadGEM2-ES to be representative of other CMIP5 models as well.

Sea-level projections for the NWES constructed with an ensemble of GCMs could be improved by weighting or excluding models based on their bathymetry and land mask or skill at reproducing observations regionally (e.g. McSweeney and others [68]).

Our results show the importance of improving the representation of coastal regions in GCMs for regional sea-level projections for the NWES, and support a role for dynamical downscaling in improving projections for coastal regions.

Acknowledgement

A machine generated summary based on the work of Hermans, Tim H. J.; Tinker, Jonathan; Palmer, Matthew D.; Katsman, Caroline A.; Vermeersen, Bert L. A.; Slangen, Aimée B. A. (2020 in Climate Dynamics).

West African Monsoon Intraseasonal Activity and Its Daily Precipitation Indices in Regional Climate Models: Diagnostics and Challenges

https://doi.org/10.1007/s00382-016-3016-8

Abstract-Summary

This paper analyses an ensemble of simulations from six regional climate models (RCMs) taking part in the coordinated regional downscaling experiment, the ECMWF ERA-Interim reanalysis (ERAI) and three satellite-based and observationally-constrained daily precipitation datasets, to assess the performance of the RCMs with regard to the intraseasonal variability.

RCMs that fail to represent the seasonal-mean position and amplitude of the meridional gradient of PW show the largest discrepancies with respect to seasonal-mean observed precipitation.

The study investigates the skill of the models with respect to hydro-climatic indices related to the occurrence, intensity and frequency of precipitation events at the intraseasonal scale.

Although most of these indices are generally better reproduced with RCMs than reanalysis products, this study indicates that RCMs still need to be improved (especially with respect to their subgrid-scale parameterization schemes) to be able to reproduce the intraseasonal variance spectrum adequately.

Extended

The study investigates the frequency-time distribution of rainfall predicted by RCMs through spectrum analysis.

Introduction

Fewer studies have targeted the ability of RCMs to reproduce the monsoon intraseasonal events such as rainfall onset and retreat, AEW intermittency, or the precipitation diurnal cycle.

Although a set of arguments tends to confirm the real added value of RCMs, Flaounas and others [69] have shown a decisive impact of physical parameterizations on the monsoon onset and precipitation intraseasonal variability, implying that substantial work is still needed to calibrate or to adapt RCMs for Africa.

A large part of the WAM precipitation is attributable to intraseasonal activity and, therefore, the ability of RCMs to represent the intraseasonal variability is closely related with their performance regarding the day-to-day rainfall.

Focusing on the climate of the recent past (1989–2008), this paper aims to assess the ability of the RCMs (from available CORDEX runs) to represent the WAM intraseasonal variability spectrum in comparison with observations.

Data and Methodology

In order to give some measure of the uncertainty in satellite-based observations, two other sets of satellite-based and observationally-constrained daily precipitation were used: TRMM 3B42 version 6 (Tropical Rainfall Measuring Mission; Huffman and others [70]) providing 3-hourly precipitations on a 0.25°-mesh grid covering the tropical regions since 1998.

On the WAM (e.g. Vellinga and others [71]; Fitzpatrick and others [72]), two main features can be used for detecting the onset: (1) the large-scale onset that focuses on the shift (or jump) of the sustained rainy band from its Guinean position (~5°N) to a Sudanese position (~10°N), (2) the local onsets are related to more agricultural purposes and represent the effective start of sustained rainfall at a specific location.

The large-scale precipitation onset has been quite widely analyzed, in particular in CORDEX simulations (among others, Gbobaniyi and others [73]; Nikulin and others [74]; Diaconescu and others [75]) and most RCMs tend to improve the onset occurrence in comparison with ERAI and GCMs.

Climatological Background: Column-Integrated Humidity and Monsoon Precipitation

Over the Sahel, rainfall seems to occur only where PW exceeds 30 mm (south of 20°N).

It seems important that RCMs reproduce this baroclinic mean state over the Sahel, which can be inferred through PW meridional gradient.

Regarding the precipitation field, these RCMs simulate the precipitation patterns north of 17°N relatively well despite the fact that they overestimate PW.

This excessive gradient in CRCM5 appears to be related to its overestimated precipitation south of or near 10°N, and almost no rainfall north of 15°N. More generally, despite significant discrepancies with respect to ERAI PW, most of the RCMs tend to improve the spatial distribution of precipitation shown by ERAI over the whole Sahel area.

It also seems clear that, when RCMs have a good representation of PW (such as RCA4), they reproduce the seasonal-mean rainfall more accurately.

WAM Intraseasonal Scales in RCMs

The large Eastern African low-frequency activity shown in all RCMs simulations (variance field in 10–90-day) is consistent with the analysis of Alaka and Maloney [76] showing that the MJO dry and wet phase oscillations induce strong PW fluctuations over this region.

Over the coastal region, CRCM5 driven by ERAI has approximately the same variance as ERAI (~103% of ERAI variance): 42% of this intraseasonal variance is due to low-frequency activity while the rest (~58%) is due to synoptic scales.

The wavelets computed on RCM simulations, as opposed to ERAI, reveal a more intense synoptic-scale activity (below 10 days), more often after June, and so seem more consistent with observations.

The synoptic scales (i.e. 3–10 day and 1–3 day, by order of importance) explain more than 80% of the precipitation intraseasonal variability, consistently with the conclusions of Mathon and others [77] on the influence of MCSs on the seasonal rainfall amount.

WAM Onset/Retreat and Daily Precipitation Indices

This section aims to evaluate some specific aspects of the WAM intraseasonal variability, namely its onset/retreat and the resulting precipitation indices related to the occurrence, duration and intensity of daily events during the monsoon season.

There is a straightforward relationship between a model's ability to handle the intraseasonal spectrum and its performance regarding the daily precipitation statistics that account for the occurrence of events (e.g. onset/retreat and wet/dry days), their duration and their intensity (e.g. dry/wet spells and high quantiles).

R1mm is related to the occurrence of rainy events (defined as daily precipitation ≥ 1 mm day^{-1}), while other indices such as SDII, R20mm or Rd3 account for both occurrence and intensity (i.e. mean intensity per wet day, percentage of days with daily precipitation ≥ 20 mm/day, or maximum amount of rainfall over three consecutive days, respectively).

Conclusion and Future Work

The 2009 summer season, characterized by intense rainfall in September and an associated flooding event, is examined in the light of ARC2 wavelet analysis and illustrates two aspects: (1) the poor representation of ERAI regarding the shorter scale events, leading to a relatively flat precipitation time series, and (2) in RCMs, the relative excess of high-frequency activity, despite low-frequency activity being overestimated.

The 1–10 day scale phenomena such as AEWs and the related convective activity have more realistic impact on the shape of the spectrum. (–) Like ERAI, RCMs still overestimate the low-frequency (10–90-day) activity and this has the effect of artificially increasing the intraseasonal variability in general.

This appears to be related to the improvement of meteorological scale phenomena (e.g. MCSs). (–) ERAI outperforms all RCMs in terms of wet day occurrence and seasonal mean precipitation correlation.

Acknowledgement

A machine generated summary based on the work of Poan, E. D.; Gachon, P.; Dueymes, G.; Diaconescu, E.; Laprise, R.; Seidou Sanda, I. (2016 in Climate Dynamics).

Optimal Nudging Strategies in Regional Climate Modelling: Investigation in a Big-Brother Experiment Over the European and Mediterranean Regions

https://doi.org/10.1007/s00382-012-1615-6

Abstract-Summary

The objective of this work is to gain a general insight into the key mechanisms involved in the impact of nudging on the large scales and the small scales of a regional climate simulation.

The main focus is on the sensitivity to nudging time, but the BBE approach allows to go beyond a pure sensitivity study by providing a reference which model outputs try to approach, defining an optimal nudging time.

The BBE approach to optimal nudging is used with a realistic model, here the weather research and forecasting model over the European and Mediterranean regions.

The impact of other numerical parameters, specifically the domain size and update frequency of the large-scale driving fields, on the sensitivity of the optimal nudging time is investigated.

Regarding the determination of a possible optimal nudging time, the conclusion is not the same for indiscriminate nudging (IN) and spectral nudging and depends on the update frequency of the driving large-scale fields τ_a.

For IN, the optimal nudging time is around $\tau = 3$ h for almost all cases.

For spectral nudging, the best results are for the smallest value of τ used for the simulations ($\tau = 1$ h) for frequent update of the driving large-scale fields (3 and 6 h).

The optimal nudging time is 3 for 12 h interval between two consecutive driving large-scale fields due to time sampling errors.

In terms of resemblance to the reference fields, the differences between the simulations performed with IN and spectral nudging are small.

The optimal nudging time itself is not sensitive to domain size.

Extended

For spectral nudging, the optimal nudging time τ varies between 1 and 3 h. Indeed, for $\tau_a = 3$ and 6 h, the highest model skills are found for $\tau = 1$ h. One must note here that $\tau = 1$ h is the smallest value used for the simulations.

The optimal nudging time could thus be smaller than 1 h but this has not been investigated, in part due to numerical instabilities produced for very small values of τ.

Introduction

Regarding LAMs specifically, previous studies have investigated the specific sensitivity of the predictions to the update frequency of the boundary conditions, the size and resolution of the domain of simulation in order to prevent these models to mislead (Bhaskaran et al. [78]; Noguer et al. [79]; Seth and Giorgi [80]; Denis et al. [81, 82]; Castro et al. [83]).

Salameh et al. [84] consider the impact of the nudging time on the root-mean-square error of the modelled small and large scales.

Compared to previous studies examining the sensitivity of RCM results to domain size and the frequency of update of boundary conditions (e.g. Bhaskaran et al. [78]; Noguer et al. [79]; Seth and Giorgi [80]; Denis et al. [81, 82]; Castro et al. [83]), we address here specifically the interaction between these parameters and the nudging parameters.

Castro et al. [83] studied the sensitivity of RAMS model simulations to domain size and also to the activation of IN but a single large value of the nudging time (24 h) was considered.

Numerical Setup

We use the BBE (Denis et al. [81]) to investigate the impact of the indiscriminate and spectral nudging on WRF skills to produce significant small scales from low resolution driving data.

The nudging technique consists in relaxing the model state towards the driving large-scale fields by adding a non-physical term to the model equation.

The aim of this study is to achieve the best dynamical consistency between the driving large-scale field and the small-scale field simulated with a nudged LAM as a RCM.

In terms of the downscaling typology presented in Castro et al. [83] our work could be relevant to types 2, 3 or 4 depending on which type of fields are used to drive the regional model.

Contrary to Omrani et al. [85], we can not use a wide range of values for the update frequency of the driving large-scale fields τ_a because of the diurnal cycle.

Results Over the EURO–CORDEX Domain

As for the zonal surface wind component, when nudging is applied (IN and SN simulations), the bias is almost supressed over land, whatever the season.

In the IN simulations and for $\tau = 1$ h, the production of small scales is inhibited because the LB fields are too tightly constrained by nudging to the large-scale driving fields.

In the SN simulations, the nudging does not affect the small scales but for example for $\tau_a = 12$ h, the smallest scales present in the large scale driving fields are probably not well resolved in time (Omrani et al. [85]).

We also note that in summer, the SN simulations have systematically a higher standard deviation compared to BB, however the IN simulations display a smaller standard deviation for all updating and nudging times (τ_a, τ).

Comparison with the Results Over the HyMeX/MED–CORDEX Domain
The HyMeX/MED–CORDEX domain being smaller than the EURO–CORDEX domain, the impact of nudging should be damped (Miguez-Macho et al. [86]; Leduc and Laprise [87]; Omrani et al. [88]).

Comparisons of LB–Med simulations with the results obtained over the EURO–CORDEX domain (LB–Euro simulations) are made on the overlapping domain (i.e. the HyMeX/MED–CORDEX).

One can also note that the spatial pattern of the surface temperature bias is not the same between NN simulations over the HyMeX/MED–CORDEX and EURO–CORDEX domains.

The center of the LB–Euro simulation domain is located over Central Europe whereas it is located over the Mediterranean Sea for the LB–Med simulations.

Nudging inhibits the impact of the domain size on the simulated surface temperature whatever the season.

This confirms the two previous results: nudging reduces the sensitivity of the model to the domain size and the control by the boundaries is significantly more important over the LB–MED simulations even though the LB–Med simulations still need to be nudged.

Conclusions and Perspectives
We have analyzed the impact, as a function of the nudging time, of indiscriminate and spectral nudging on an ensemble of simulations performed with the WRF model.

The ensemble of numerical simulations was performed on two different but overlapping domains to investigate the impact of the size of the domain on the sensitivity to nudging time.

Domain size and update frequency of the large-scale driving fields are the key parameters identified by Omrani et al. [85, 88] controlling how the quality of nudged simulations depends on the nudging time.

Regarding the determination of a possible optimal nudging time, the conclusion is not the same for indiscriminate nudging and spectral nudging and depends on the update frequency of the driving large-scale fields τ_a.

This difference can be explained by the different time scales involved in the dynamics of the temperature, wind and precipitation fields and advocates for possible different optimal nudging time regarding the various variables.

Acknowledgement
A machine generated summary based on the work of Omrani, Hiba; Drobinski, Philippe; Dubos, Thomas (2012 in Climate Dynamics).

Downscaling Large-Scale Climate Variability Using a Regional Climate Model: The Case of ENSO Over Southern Africa

https://doi.org/10.1007/s00382-012-1400-6

Abstract-Summary

This study documents methodological issues arising when downscaling modes of large-scale atmospheric variability with a regional climate model, over a remote region that is yet under their influence.

Regional simulations are performed with WRF model, driven laterally by ERA40 reanalyses over the 1971–1998 period.

We document the sensitivity of simulated climate variability to the model physics, the constraint of relaxing the model solutions towards reanalyses, the size of the relaxation buffer zone towards the lateral forcings and the forcing fields through ERA-Interim driven simulations.

The incidence of SST prescription is also assessed through additional integrations using a simple ocean mixed-layer model.

The model deficiencies are found to result from biased atmospheric forcings and/or biased response to these forcings, whatever the physical package retained.

These results confirm the significant contribution of nearby ocean SST to the regional effects of ENSO, but also illustrate that regionalizing large-scale climate variability can be a demanding exercise.

Introduction

More particularly, attention is given on the regional effects of El Niño Southern Oscillation (ENSO), the leading mode of large-scale interannual variability in the tropics.

Focusing on the seasonal timescale, this study aims at (1) showing how accurately an RCM, namely the Weather Research and Forecasting (WRF) model, is capable to downscale ENSO-associated variability over SA when driven laterally by global reanalyses supposed to contain realistic ENSO signal due to data assimilation; (2) assessing to what extent these results are dependent of the model physics and experimental setup.

Rather than exploring in detail the physical mechanisms through which global-scale ENSO variability impacts SA rainfall (which has attracted a large number of publications in the last 25 years, but still remains partly matter of debate), focus is given here on a quantification of our regional model skill, the main factors likely to modulate its capabilities and deficiencies, and its sensitivity to the experimental protocol.

Data and Experimental Setup

ERA40 reanalyses are generated by an a posteriori integration of the IFS (Integrated Forecasting System) atmospheric GCM with 6 hourly assimilations of satellite data, buoys and radiosondes, at a T159 spectral truncation (giving a horizontal resolution of nearly 125 km for the reduced Gaussian grid) with 60 vertical levels.

Over these years, additional simulations forced by ERA-Interim reanalyses (Simmons and others [89]; Dee and others [90]) allow documenting sensitivity to lateral forcings.

ERA-Interim incorporates many improvements in the model physics and analysis methodology: a new 4D-var assimilation scheme, likely to improve the quality of the data over regions where amounts of assimilated data are low and inconstant, a T255 horizontal resolution (roughly 80 km), better formulation of background error constraint, additional cloud parameters and humidity analysis, more data quality control and variational bias correction.

EI, forced by ERA-Interim reanalyses over the same domain and season but for years 1979–1998.

These simulations aim at disentangling atmospheric and SST forcings responsible for rainfall anomalies and biases.

Simulating Southern African Rainfall (1971–1998)
Spatial correlations between WRF simulated and observed rainfall range between 0.38 and 0.86 (in 1995–1996 and 1985–1986, respectively) and reach the 95% significance bound 25 years out of 28.

They are even negative for 10 years out of 28, denoting once again a moderate skill in simulating seasonal South African rainfall interannual variability.

ERA40 performs better on average, suggesting a stronger skill for simulating South African rainfall interannual variability.

These results are consistent with the fact that the weak co-variability between observed and WRF simulated rainfall in SA mostly relates to a poor simulation of regional ENSO effects.

This shows that, if the model is deficient for simulating year-to-year rainfall fluctuations over South Africa as a whole, it is mostly due to the fraction of climate variability that is directly imputable to ENSO.

Case Studies: Two Strong and Contrasted El Niño Events
Although anomalies in the regional atmospheric dynamics differ from ERA40, especially in the upper layers, WRF has reasonable skill in reproducing the rainfall departures recorded in SA during the strongest El Niño event of the century.

This highlights that SST anomalies over adjacent oceanic basins significantly influence SA rainfall, confirming Nicholson [91] and Nicholson and Kim [92] succeeds at reproducing dry (wet) conditions over SA (the SWIO), despite near climatological atmospheric forcings, illustrating the specific influence (i.e. the contribution) of regional SST anomalies during this season.

ATM_CLIM exp. (isolating the specific influence of SST) simulate abnormally wet (dry) conditions over the SWIO (SA) and SST_CLIM exp. (documenting in contrast the role of the lateral forcings) enhance the dry (wet) oceanic (continental) biases compared to CTRL and OML exps.

Such anticyclonic (cyclonic) anomalies account for the contrasted dry (wet) anomalies simulated there by SST_CLIM and OML exp. (ATM_CLIM).

Conclusion and Discussion

This moderate skill (1) is barely modified when changing the model convective and planetary boundary layer parameterizations, suggesting a weak influence of the physics; (2) is not improved with a larger buffer zone used for the prescription of lateral boundary conditions, nor with yearlong integrations simulating the whole ENSO life cycle from the Spring Barrier to its peak in austral summer; (3) is unchanged when using a simple ocean mixed-layer model to avoid prescribing SST surface forcings; (4) is slightly improved when the model prognostic variables are relaxed towards reanalyses above the PBL, suggesting either that large-scale mass convergence is perfectible in these datasets or that the regional model deficiencies are not primarily related to upper-atmospheric dynamics.

This result (1) emphasizes the importance of SST anomalies over adjacent oceanic basins (Nicholson [91]; Nicholson and Kim [92]), although the atmospheric component of ENSO (Cook [93]) is equally important to produce rainfall anomalies of realistic amplitude; (2) identifies lateral boundary conditions as the origin of most of the regional model low performance, either due to perfectible forcing fields and/or to a biased response to realistic forcings.

Acknowledgement

A machine generated summary based on the work of Boulard, Damien; Pohl, Benjamin; Crétat, Julien; Vigaud, Nicolas; Pham-Xuan, Thanh (2012 in Climate Dynamics).

Impact of the Lateral Boundary Conditions Resolution on Dynamical Downscaling of Precipitation in Mediterranean Spain

https://doi.org/10.1007/s00382-007-0242-0

Abstract-Summary

Conclusions on the General Circulation Models (GCMs) horizontal and temporal optimum resolution for dynamical downscaling of rainfall in Mediterranean Spain are derived based on the statistical analysis of mesoscale simulations of past events.

Three sets of simulations are designed using input data with 1°, 2° and 3° horizontal resolutions (available at 6 h intervals), and three additional sets are designed using 1° horizontal resolution with less frequent boundary conditions updated every 12, 24 and 48 h. The quality of the daily rainfall forecasts is verified against raingauge observations using correlation and root mean square error analysis as well as Relative Operating Characteristic curves.

For the whole Mediterranean Spain, model skill is not appreciably improved when using enhanced spatial input data, suggesting that there is no clear benefit in using high resolution data from General Circulation Model for the regional downscaling of precipitation under the conditions tested.

The analysis is particularized for six major rain bearing flow regimes that affect the region, and differences in model performance are found among the flow types, with slightly better forecasts for Atlantic and cold front passage flows.

A remarkable spatial variability in forecast quality is found in the domain, with an overall tendency for higher Relative Operating Characteristic scores in the west and north of the region and over highlands, where the two previous flow regimes are quite influential.

The findings of this study could be of help for dynamical downscaling design applied to future precipitation scenarios in the region, as well as to better establish confidence intervals on its results.

Introduction

This idea was originally based on the concept of the one-way nesting, in which large-scale meteorological fields from general circulation model runs provide initial and time-dependent meteorological lateral boundaries conditions (LBCs) for high-resolution regional climate models (RCMs) simulations.

Issues of major importance in this technique are: numerical nesting strategy, spatial resolution difference between the driving data and the nested model, spin-up, update frequency of the LBCs, model physics, domain size, etc. Many of these issues have been addressed in subsequent studies, like Denis et al. [81] and Beck et al. [94], concluding that in spite of the sensitivity of the results to the RCM set-up, the one way nesting strategy has skill in downscaling large-scale information to the regional scales.

We evaluate the sensitivity of mesoscale numerical simulations of rainfall for Mediterranean Spain to large-scale model input data resolution, to help to answer the question whether GCM higher resolution would provide improved dynamically downscaled information in that region in the context of climate change research.

Data Base and Methodology

Six simulations are performed for each heavy rainfall day by nesting the HIRLAM mesoscale model within large scale analyses.

Large-scale meteorological analyses used to nest the HIRLAM model are constructed from the European Centre for Medium Range Weather Forecasts (ECMWF) ERA-15 spectral reanalysis of geopotential height, temperature, relative humidity and horizontal wind components at eleven standard pressure levels.

The model is run over a 54 h period, starting at 00 UTC on the day before the cataloged heavy rainfall day.

A fourth, fifth and sixth experiments, referred to as $1° + 12$, $1° + 24$ and $1° + 48$ h, are run by using $1°$ resolution input data but less frequent—12, 24 and 48 h apart, respectively—boundary updates for the large-scale meteorological fields.

In HIRLAM, the time varying boundary conditions during model integration are defined by linear interpolation between large scale data at consecutive boundary update times.

Results and Discussion

Atlantic flows, mostly associated with large-scale low pressure systems, favour rainfall over the western and northern zones but are hardly effective in the east and southeast; Mediterranean air flows, less common and associated with smaller-scale disturbances, encourage rainfall in these latter zones but not in sheltered areas like western and central Andalusia.

The higher ROC scores over the western and northern regions (even exceeding 0.9 in mountainous areas of western Andalusia) can be associated with the relatively high forecast capability for Atlantic flow situations.

In order to alleviate this problem, the circulation types have been further simplified by considering only two categories: northern disturbances, associated with a significant Atlantic or northerly component at low levels, composed by A, C and N situations (85 days); and southern disturbances, associated with a dominant easterly flow component over Mediterranean Spain, composed by SW, S and SE situations (80 days).

Conclusions

This study represents an attempt to examine the problem of dynamical downscaling of precipitation over Mediterranean Spain—a highly vulnerable region according to most of the climate change precipitation scenarios (Meteorological Office [95]; Watson and Zinyowera [96])—with respect to its sensitivity to the spatial and temporal resolution of GCM input fields.

Our conclusions have been outlined from various sets of 165 mesoscale numerical simulations of heavy rainfall events in Mediterranean Spain, initialized with real meteorological grid analyses at six different spatial and temporal resolutions, under the following assumptions: (1) heavy rainfall events are representative of the whole fraction of rainfall days with respect to the model sensitivity—or insensitivity—to input data resolution; (2) the six considered resolutions ($1°$, $2°$ and $3°$ in space, plus $1° + 12$, $1° + 24$ and $1° + 48$ h in time) are sufficient to describe the actual envelope of sensitivities of the forecast system; and most importantly, (3) the use of smoothed meteorological analyses is equivalent to coarse grid GCM outputs.

Acknowledgement

A machine generated summary based on the work of Amengual, A.; Romero, R.; Homar, V.; Ramis, C.; Alonso, S. (2007 in Climate Dynamics).

Limitations of Time-Slice Experiments for Predicting Regional Climate Change Over South Asia

https://doi.org/10.1007/s00382-004-0509-7

Abstract-Summary

While time-slice simulations with atmospheric general circulation models (GCMs) have been used for many years to regionalize climate projections and/or assess their uncertainties, there is still no consensus about the method used to prescribe sea surface temperature (SST) in such experiments.

The response of the Indian summer monsoon to increasing amounts of greenhouse gases and sulfate aerosols is compared between a reference climate scenario and three sets of time-slice experiments, consisting of parallel integrations for present-day and future climates.

Different monthly mean SST boundary conditions have been tested in the present-day integrations: raw climatological SST derived from the reference scenario, observed climatological SST, and observed SST with interannual variability.

For future climate, the SST forcing has been obtained by superimposing climatological monthly mean SST anomalies derived from the reference scenario onto the present-day SST boundary conditions.

None of these sets of time-slice experiments is able to capture accurately the response of the Indian summer monsoon simulated in the transient scenario.

The monsoon response is also shown to depend on the simulated control climate, and can therefore be sensitive to the use of observed rather than model-derived SSTs to drive the present-day atmospheric simulation, as well as to any approximation in the prescribed radiative forcing.

Extended

The relative agreement between the monsoon response in TS1/TS3 and SG1 is probably the fortuitous consequence of differences in their control climates, and there is no reason to believe that prescribing bias-corrected instead of raw model-derived SSTs in global time-slice experiments warrants a better consistency with the climate response of the coupled model.

Introduction

Coupled AOGCMs are still computationally expensive and transient climate change experiments cannot be run as often as necessary to explore all the possible combinations of radiative forcings and models.

The most straightforward technique is probably to derive monthly mean SST and sea-ice boundary conditions directly from a coupled model, using two "time slices" of a transient scenario corresponding to present-day and future climates respectively (May and Roeckner [97]).

Most time-slice experiments have been based on control simulations forced with observed monthly mean SSTs, while the perturbed simulations were driven by future SSTs constructed as the sum of the observed SSTs and of the SST anomalies derived from a coupled model.

Such an experiment design is generally justified by the fact that interannual variability is relevant to study some aspects of climate change (for example changes in extremes), but does not show robust responses in coupled climate change experiments (IPCC [98]).

Model and Experiments

Both scenarios show identical present-day climatologies and very similar climate anomalies when comparing the last 50 years (2050–2099, hereafter referred to as P3) with the first 50 years (1950–1999, hereafter referred to as P1) of the integrations (Ashrit and others [99]).

Various reasons can be put forward to explain this pattern, including the significant retreat of the Arctic sea ice, the relatively stronger warming over land (not shown) that could influence the surrounding oceans through atmospheric advection, as well as the weaker influence of deep convection than in the tropics where the surface warming is vertically redistributed.

All time-slice experiments show exactly the same surface ocean warming as in the reference scenario, even in the high latitudes.

In the 10-year integrations, the concentrations have been fixed at the annual level prescribed in SG1 in the middle of P1 (year 1975) and P3 (year 2075) for present-day and future climate, respectively, so that the radiative forcing is slightly different from that prescribed in SG1.

Comparison Between Forced and Coupled Experiments
They show a less homogeneous response during the monsoon season, since rainfall is decreased in July (when SG1 shows a maximum increase) and increased in August and September, which seems to be related to a strong increase in the late monsoon flow (as suggested by positive IMI anomalies, while SG1 shows negative ones).

Despite the use of observed rather than model SSTs in the control climate simulation, TS3 minus TS1 anomalies is more consistent with the CCM response, at least for precipitation.

The t-test is applied to test the statistical significance of the anomalies found in SG1, as well as of the difference between the climate response in the time-slice versus coupled experiments.

Differences between TS3 minus TS1 anomalies and SG1 anomalies are only marginally significant, though they indicate a weaker increase in monsoon rainfall over the west coast of the Indian peninsula.

Discussion
There are several possible reasons for the different monsoon response in the coupled scenario and in the various time-slice experiments: (1) the response can be sensitive to the control climate that is not fully identical among the various experiments; (2) the SST feedback is suppressed in the time-slice experiments, while it is known that both local and remote air–sea interactions in the Asia–Pacific sector play an important role in the maintenance and evolution of the Asian summer monsoon; (3) interannual variability in SST is removed in TP1/TP3 and TS1/TS3 while the Indian monsoon shows a large interannual variability that is partly driven by SST fluctuations.

The relative agreement between the monsoon response in TS1/TS3 and SG1 is probably the fortuitous consequence of differences in their control climates, and there is no reason to believe that prescribing bias-corrected instead of raw model-derived SSTs in global time-slice experiments warrants a better consistency with the climate response of the coupled model.

Conclusions

While it was shown by Ashrit and others [99] that the CCM performs a reasonable simulation of present-day climate and produces a global warming that is consistent with the instrumental records and with former IPCC projections, the present study is aimed at evaluating the ability of the atmospheric component of the CCM, namely the ARPEGE model, to capture the CCM response in equilibrium or time-slice climate change experiments.

Changing the model would probably lead to different results, but would not necessarily challenge the main conclusion, namely the sensitivity of the model response to the experiment design, at least in the regions where the SST feedback and variability have a strong influence on the control climate simulated by the coupled model.

Prescribed SSTs should not necessarily include interannual variability: while the lack of SST variability can contribute to alter the atmospheric response to global warming, using time-varying SST boundary conditions could be even worse due to a possible unrealistic atmospheric sensitivity to interannual SST fluctuations in such forced experiments.

Acknowledgement

A machine generated summary based on the work of Douville, Hervé (2005 in Climate Dynamics).

Dynamical Downscaling of CMIP5 Global Circulation Models Over CORDEX-Africa with COSMO-CLM: Evaluation Over the Present Climate and Analysis of the Added Value

https://doi.org/10.1007/s00382-014-2262-x

Abstract-Summary

The results of the application of the consortium for small-scale modeling (COSMO) regional climate model (COSMO-CLM, hereafter, CCLM) over Africa in the context of the coordinated regional climate downscaling experiment.

We compare the results of CCLM to those of the driving GCMs over the present climate, in order to investigate whether RCMs are effectively able to add value, at regional scale, to the performances of GCMs.

CCLM is generally able to better represent the annual cycle of precipitation, in particular over Southern Africa and the West Africa monsoon (WAM) area.

By performing a singular spectrum analysis it is found that CCLM is able to reproduce satisfactorily the annual and sub-annual principal components of the precipitation time series over the Guinea Gulf, whereas the GCMs are in general not able to simulate the bimodal distribution due to the passage of the WAM and show a unimodal precipitation annual cycle.

Extended

CCLM is generally not able to correct the bias, which, in some cases, e.g., the band around 20°N, is even worsened as a result of the combination of both GCM' and CCLM' cold biases over the region.

By performing a SSA over the regions along the Gulf of Guinea it was shown that CCLM is able to better represent the sub-annual principal components of the precipitation time series, in turn reproducing satisfactorily the bimodal distribution of the annual cycle, whereas GCMs are not able to simulate this feature and they show a unimodal distribution.

The analysis of the projections of future climate change will be presented in a forthcoming work.

Introduction

Despite GCMs have demonstrated the ability to generally replicate the precipitation trend over the second half of the twentieth century, they may present significant deficiencies in simulating the African climate, especially complex systems like the West Africa Monsoon (WAM), which is driven by the interaction of atmosphere, ocean, and land-surface, initiated by differential heating of the ocean and land surface (e.g. Steiner and others [100]), and also strongly related to mid-tropospheric circulation such as the African Easterly Jet (AEJ) (Cook [101]).

When RCMs are driven by GCMs, however, the downscaled climate may present even larger biases, as the ones inherited through the lateral boundary conditions are added to those introduced by the RCM by means, for instance, of model errors and parameterizations (e.g. Dosio and Paruolo [102]; Hong and Kanamitsu [103]).

It is important first to generally evaluate the ability of CCLM to reproduce the general characteristics of the African climate (e.g., seasonal distribution of temperature and precipitation, and WAM climatology) and, second, to investigate whether the downscaled simulations add value to the ones by the driving GCMs.

Model Description, Setup and Observational Data

The three-dimensional non-hydrostatic regional climate model COSMO-CLM (CCLM) is used, in the same configuration as the'evaluation runs' (i.e., forced by the ERA-Interim reanalysis) described in Panitz and others [104].

The historical control runs, forced by observed natural and anthropogenic atmospheric composition, cover the period from 1950 until 2005, whereas the projections (2006–2100) are forced by two Representative Concentration Pathways (RCP) (Moss and others [105]; Vuuren and others [106]), namely, RCP4.5 and RCP8.5.

In this work we evaluate the performance of CCLM by comparing the results to those of the driving GCMs over the present climate (1989–2005).

High-quality observational datasets for Africa are scarce, and significant discrepancies exist amongst different datasets mainly due, for instance, to the limited number of gauge stations, retrieval, merging, and interpolation techniques (Huffman and others [107]; Nikulin and others [74]; Sylla and others [108]).

A combination of available ground observations, satellite products, and reanalysis is used, as done in Panitz and others [104], where a critical review of the available observational dataset for Africa was presented.

Results
CCLM results are much closer to the observations, especially to GPCP, although in summer CWD is still slightly overestimated over the area between 0°N and 15°N. It is interesting to note that the downscaled CCLM simulations are similar to the ERA-Interim driven run, especially over land, and seem to be somehow independent of the influence of the driving GCM (i.e., lateral boundary conditions) compared to, for instance, the precipitation mean intensity.

GCMs' results are quantitatively similar to observed values, especially over land, although in JFM CNRM-CM5 and EC-Earth underestimate CDD over South Africa, whereas in JAS over North-East Africa CDD is overestimated by HadGEM2-ES and MPI-ESM-LR and underestimated by CNRM-CM5 and EC-Earth.

Summary and Concluding Remarks
It was important first to generally evaluate the ability of CCLM to reproduce the general characteristics of the African climate (e.g., seasonal distribution of temperature and precipitation, and WAM climatology) and, second, to investigate whether the downscaled simulations add value to those of the driving GCMs.

It is found that, in general, the geographical distribution of mean sea level pressure, surface temperature and seasonal precipitation is strongly affected by the boundary conditions (i.e. driving GCM): for instance, GCMs show a marked cold bias, especially in JFM, which CCLM is only partially able to correct, especially in areas such as Central and South Africa where the evaluation runs showed a slight warm bias.

Acknowledgement
A machine generated summary based on the work of Dosio, Alessandro; Panitz, Hans-Jürgen; Schubert-Frisius, Martina; Lüthi, Daniel (2014 in Climate Dynamics).

Intercomparison of Statistical and Dynamical Downscaling Models Under the EURO- and MED-CORDEX Initiative Framework: Present Climate Evaluations

https://doi.org/10.1007/s00382-015-2647-5

Abstract-Summary
Given the coarse spatial resolution of General Circulation Models, finer scale projections of variables affected by local-scale processes such as precipitation are often needed to drive impacts models, for example in hydrology or ecology among other fields.

Downscaling can be performed according to two approaches: dynamical and statistical models.

If several studies have made some intercomparisons of existing downscaling models, none of them included all those families and approaches in a manner that all the models are equally considered.

Six Statistical Downscaling Models (SDMs) and five Regional Climate Models (RCMs) are compared in terms of precipitation outputs.

The 11 models are evaluated according to four aspects of the precipitation: occurrence, intensity, as well as spatial and temporal properties.

The results indicate that marginal properties of rain occurrence and intensity are better modelled by stochastic and resampling-based SDMs, while spatial and temporal variability are better modelled by RCMs and resampling-based SDM.

The indicators suit specific purpose and therefore the model evaluation results depend on the end-users point of view and how they intend to use with model outputs.

Building on previous intercomparison exercises, this study provides a consistent intercomparison framework, including both SDMs and RCMs, which is designed to be flexible, i.e., other models and indicators can easily be added.

This framework provides a tool to select the downscaling model to be used according to the statistical properties of the local-scale climate data to drive properly specific impact models.

Extended

Based on that, it is expected to perform consistent future intercomparison studies between SDMs as well as RCMs.

Introduction

The alternative approach to RCMs is based on Statistical Downscaling Models (SDMs) that rely on determining statistical relationships between large- and local-scale variables and do not try to solve the physical equations modelling the dynamic of the atmosphere.

SDMs are based on a static relationship, i.e. the mathematical formulation of the relation between predictand (i.e., the local-scale variable to simulate) and predictors (i.e., the large-scale information or data used as inputs in the SDMs) is supposed to be valid for any time period: not only for the current climate on which the relationship is calibrated, but also, for example, for future climates.

To perform a consistent intercomparison, we want to compare models outputs from all types of models (i.e., from the four approaches of SDMs and from RCMs, see Schmidli et al. [109] and observational data with similar resolution over a common area.

Data and Experimental Setup

In order to design the experiment rigorously, it is essential to keep in mind assumptions under which the SDMs are performed (Hewitson and Crane [110]): (1) the relationship between local-scale data and large-scale predictors is fixed in time (even if the statistical properties of the downscaled simulations can evolve in time), (2) the predictors fully represent the climate signal, (3) the large-scale variables are well reproduced by climate models, including reanalysis.

SPARSE CCA was carried out between a set containing the first PC of each of the seven potential predictors and a second set of variables comprising the precipitation on the EURO-CORDEX area.

As precipitation is usually not well represented by the GCMs, this variable is rarely employed as a predictor in Statistical Downscaling Models.

The variable selection pre-processing resulted in 12 predictors (2 first PCs for each of the 5 variables selected through the SPARSE CCA and precipitation).

Statistical and Dynamical Downscaling Models
In the first one, GAM has been calibrated with all values (i.e., including 0's) and then rain intensity has been directly simulated and the rain occurrence is dealt by thresholding the outputs at 1 mm.

Quantile–quantile based methods have been widely used for downscaling (e.g., Vrac et al. [111] and references therein) or to correct bias in model outputs thanks to observations CDFs (e.g., Gudmundsson et al. [112], and references therein) and the correspondences between predictors and predictands quantiles.

Those methods can be directly calibrated on models outputs (e.g., GCM or RCM).

The CDF-t model consists in relating local-scale (i.e., here E-OBS precipitation) CDF to the large-scale (i.e., here ERAi reanalysis precipitation) CDF.

The CDF-t and quantile–quantile methods are similar in philosophy, except that CDF-t takes into account the change in the large-scale CDF from the calibration to the projection (or validation) time period, while quantile–quantile projects the simulated large-scale values onto the historical CDF to compute and match quantiles.

Intercomparison Results
For the winter season the results are more or less the same for all the models except for the EURO-CNRM model where the biases are smaller and distributed in terms of sign all over the domain.

All the models except GAM show skills for reproducing the wet spells of E-OBS, especially the EURO-IPSL and ANALOG models.

In winter, the results are similar except that the mean wet spells biases absolute values are smaller for all the models (see auxiliary materials).

Similar results are found for winter although with smaller biases for all the models except MED-IPSL (see auxiliary materials).

Similar patterns means that the models have a good ability to reproduce the spatial variability of the observations.

In winter (see auxiliary materials), the spatial variability of all the models is better caught than in summer, except for GAM-so and SWG-s again.

Conclusions and Discussion
11 models (six SDMs and five RCMs) have been selected and their outputs confronted according to criteria characterizing the four following aspects of the rain: occurrence, intensity, as well as spatial and temporal properties.

From the SDMs fitted over the historical period (e.g., 1979–2008) to the observations (E-OBS) and reanalyses (ERAi) (i.e., basically similar to which has been done in this study), new time series driven this time by GCMs as predictors will be generated and evaluated.

It is not generalized until the 2000s (e.g. Palmer and Shukla [113]; Pavan and Doblas-Reyes [114]; Lambert and Boer [115]; Gillett et al. [116]; Jacob et al. [117]; Ruti et al. [118]; Solman et al. [119]; Gallardo et al. [120]) and is consolidated as the standard in studies of climate performed with dynamical models.

One can object that the ANALOG model could have been run with the same set of predictors that have been used for the other SDMs (i.e., the 12 PCs).

Acknowledgement
A machine generated summary based on the work of Vaittinada Ayar, Pradeebane; Vrac, Mathieu; Bastin, Sophie; Carreau, Julie; Déqué, Michel; Gallardo, Clemente (2015 in Climate Dynamics).

Application of SDSM and LARS-WG for Simulating and Downscaling of Rainfall and Temperature

https://doi.org/10.1007/s00704-013-0951-8

Abstract-Summary
Impact studies on the water basin and region are difficult, since general circulation models (GCMs), which are widely used to simulate future climate scenarios, do not provide reliable hours of daily series rainfall and temperature for hydrological modeling.

There is a technique named as "downscaling techniques", which can derive reliable hour of daily series rainfall and temperature due to climate scenarios from the GCMs output.

The LARS-WG and SDSM models obviously are feasible methods to be used as tools in quantifying effects of climate change condition in a local scale.

Although both models do not provide identical results, the time series generated by both methods indicate a general increasing trend in the mean daily temperature values.

The trend of the daily rainfall is not similar to each other, with SDSM giving a relatively higher change of annual rainfall compared to LARS-WG.

Extended
There is a small agreement on the "best" method of selection of predictor variables.

The LARS-WG and SDSM models are popular statistical downscaling models to downscale the GCMs model.

Introduction

To estimate future climate change resulting from the continuous increase of greenhouse gas concentration in the atmosphere, general circulation models (GCMs) are used.

This paper focuses on two statistical downscaling methods, which are stochastic weather generators and regression models.

The resulting weather generator models are then used to simulate daily series of indefinite length representatives of the altered climate (Wilks and Wilby [121]).

Regression models involve in establishing a linear or nonlinear relationship between subgrid scale, (e.g., rainfall or temperature) parameters, and grid scale (e.g., GCMs) variables (Wilby and Wigley [122]).

The LARS-WG and SDSM models are popular statistical downscaling models to downscale the GCMs model.

Both SDSM and LARS-WG models can be adopted with reasonable confidence as downscaling tools to undertake climate change impact assessment studies for the future (Dibike and Coulibaly [123]; Hashmi and others [124]).

Materials and Methods

The framework of this study consists of four steps, which are: (1) download and screen the GCM data for the province under different emission scenarios, (2) downscale the GCM data using the statistical downscaling models (SDSM and LARS-WG), (3) validate the statistical downscaling models with the observed data, and (4) project the rainfall, maximum (T_{max}), and minimum (T_{min}) temperatures corresponding to the climate change scenarios for the next 90 years as 2011–2040 (2025), 2041–2070 (2055), and 2071–2100 (2085).

Since p_i of very low daily rainfall (<1 mm) is high, and low rainfall gives little effect on the outcome of a process-based impact model, the study used the values recommended by Semenov and Stratonovitch [125], with $v_1 = 0.5$ mm and $v_2 = 1$ mm to approximate rainfall within the interval (0, 1), with p_i written as: An extremely long dry and wet series are considered in the model, with two values close to 1 are used in SEDs for wet and dry series, $p_{n-1} = 0.99$ and $p_{n-2} = 0.98$.

Results and Discussions

The downscaled mean daily and monthly temperature by SDSM and LARS-WG are slightly smaller in difference as compared to the observed temperature, which is about 0–0.07 °C.

Both of RMSE and R values indicate that SDSM perform better in simulating the monthly rainfall compared to the LARS-WG model.

The monthly T_{max} and T_{min} simulated by SDSM (including SDSM A2 and B2) and LARS-WG show a better simulation with the range of RMSE and R values of between 0.50 and 1.15 and 0.41 and 0.92 °C, respectively.

For stations in the central and southern regions, the changes in annual rainfall corresponding to the LARS-WG A2 and B2 models with respect to baseline are not similar with the changes of annual rainfall projected by SDSM A2 and SDSM B2.

An increase in T_{max} is shown in this station by 0.02 to 2.94 °C, simulated by the SDSM and LARS-WG models under the climate scenarios.

Discussion

The study found that the SDSM model was robust in downscaling performance of the time series of daily and monthly rainfall and temperature variables as compared to the LARS-WG model.

In terms of the characteristic of daily mean rainfall, the LARS-WG model is able to downscale very well, although it is slightly overestimated, the mean rainfall time series, in which similar results had been shown by Dibike and Coulibaly [123] and Fowler and others [51].

In the results, the HadCM3 A2 and B2 scenarios, which donated as SDSM A2 and SDSM B2, respectively, are able to generate the synthetic monthly rainfall and temperature data series very well, based on the RMSE and R indicators.

Based on the differences of the observed and generated rainfall and temperature data from the graphical comparisons, the SDSM A2 and B2 are consistent in showing a good agreement between observed and simulated data.

Conclusion

The SDSM model approximates the observed climate data series corresponding to the present climate series reasonably well as compared to the LARS-WG models.

The SDSM outputs show a relatively higher change in annual rainfall compared to the LARS-WG outputs.

LARS-WG produced a relatively higher change in T_{max} and T_{min} as compared to SDSM, with the higher value of 3.60 °C.

Acknowledgement

A machine generated summary based on the work of Hassan, Zulkarnain; Shamsudin, Supiah; Harun, Sobri (2013 in Theoretical and Applied Climatology).

Multisite Multivariate Modeling of Daily Precipitation and Temperature in the Canadian Prairie Provinces Using Generalized Linear Models

https://doi.org/10.1007/s00382-016-3004-z

Abstract-Summary

Based on the Generalized Linear Model (GLM) framework, a multisite stochastic modelling approach is developed using daily observations of precipitation and minimum and maximum temperatures from 120 sites located across the Canadian Prairie Provinces: Alberta, Saskatchewan and Manitoba.

Large scale atmospheric covariates from the National Center for Environmental Prediction Reanalysis-I, teleconnection indices, geographical site attributes, and observed precipitation and temperature records are used to calibrate these models for the 1971–2000 period.

Results of the study indicate that the developed models are able to capture spatiotemporal characteristics of observed precipitation and temperature fields, such as inter-site and inter-variable correlation structure, and systematic regional variations present in observed sequences.

A number of simulated weather statistics ranging from seasonal means to characteristics of temperature and precipitation extremes and some of the commonly used climate indices are also found to be in close agreement with those derived from observed data.

This GLM-based modelling approach will be developed further for multisite statistical downscaling of Global Climate Model outputs to explore climate variability and change in this region of Canada.

Extended

Based on the residual plots, distributional features of both Tmin and Tmax are relatively better described by the GLMs compared to the precipitation field when the entire domain is considered as one region.

Based on the analyses performed for other sites (not shown due to space constraints), the GLMs were able to reproduce the systematic regional variations and spatial structures of both mean and extreme weather states at the majority of the 120 sites.

Introduction

The dynamic downscaling techniques use Regional Climate Models (RCMs) to predict finer-scale climate variables when these models are driven by GCM outputs at their boundaries (Giorgi [126]).

For many water resources design and management related projects, particularly in large river basins, it is important to model simultaneous sequences of multiple variables (e.g., precipitation and temperature) over large heterogeneous areas, while maintaining physically plausible spatial, temporal and inter-variable relationships.

By Maraun and others [31], multisite generation offers many significant challenges primarily due to the need to model the joint distribution of, for example, precipitation simultaneously at all sites and inter-variable and inter-site dependence structures.

This study seeks to investigate the suitability of GLMs for multisite multivariate modelling of precipitation and temperature fields in the Canadian Prairie Provinces, with the aim of using these models for downscaling GCM outputs for climate change impact analysis.

Study Area and Data

Apart from the moderating effects due to regional changes in topography, atmospheric circulation also controls precipitation patterns (Borchert [127]).

Standardized daily values of large scale atmospheric covariates are derived for the 1961–2005 period from the National Center for Environmental Prediction and the National Center for Atmospheric Research (NCEP/NCAR) Reanalysis-I (Kalnay and others [128]) over a spatial domain encompassing latitudes 40°N to 70°N and longitudes 130°W to 70°W. In total, 21 large scale covariates (wind speed at 10-m, 500- and 850-hPa; U-component and V-component at 10-m, 500- and 850-hPa, vertical velocity, geo-potential height, specific humidity, and relative humidity at 850- and 500-hPa; total cloud cover, mean sea level pressure, precipitable water and 2-m air temperature) are explored.

The above mentioned observed temperature and precipitation datasets, large scale atmospheric covariates and indices of PDO, PNA and AO were used in Asong and others [129] to partition the study area into five homogeneous precipitation regions on which most of the analyses presented herein are based.

Methodology

For modelling precipitation, Tmax and Tmin, the statistical significance of the covariates is assessed simultaneously using likelihood ratio tests, adjusted for inter-site dependence following the approach described in Chandler and Bate [130], when extending a model by adding more covariate terms in the GLM framework.

GLMs are fitted separately to precipitation and temperature fields (i.e. Tmin and T_{max}) considering the entire study area as a single region and using observations from the 1971–2000 period.

The software provides a wide range of residual-based diagnostics to check that the fitted models are able to reproduce the systematic structure in the observations, as well as the distributional assumptions (e.g., precipitation intensities follow gamma distributions) and the assumed inter-site correlation structure (see Yang and others [131] for further details).

Seasonal maxima (minima) of daily Tmax (Tmin) are derived from observed data as well as from simulated sequences for the calibration and the two validation periods.

Results and Discussion

Additional results of the residual analysis by month and year for the same amounts and occurrence models (figures are not shown) suggest that a single model for the entire region is not adequate for describing daily precipitation sequences because the pattern of residuals do not satisfy the underlying distributional assumptions.

Based on the residual plots, distributional features of both Tmin and Tmax are relatively better described by the GLMs compared to the precipitation field when the entire domain is considered as one region.

These plots indicate a good correspondence between observed and simulated monthly precipitation totals for all regions for the calibration period.

For temperature related indices (i.e. tncw10 and txhw90), it can be stated that the GLMs are able to simulate well observed median values for both winter and summer, given that the observed values were within the inter-quartile range of the simulated distribution for most of the cases.

Summary and Conclusions

The suitability of the fitted GLMs for characterizing precipitation and temperature fields in terms of (1) simulating their mean values at the daily, monthly and seasonal scales, (2) characteristics of extreme values, (3) intervariable relationships and (4) selected climate indices are investigated using independent observations from pre- and post-calibration periods.

Following this approach, residual plots for each region show significant improvement in the performance of the fitted GLMs. (2) For both calibration and validation periods, there is generally good agreement between the simulated and observed values of various precipitation and temperature characteristics for each month of the year.

The simulated values of precipitation characteristics are more variable than those of temperature fields. (3) The fitted GLMs are able to capture spatial and inter-variable dependence structure.

Acknowledgement

A machine generated summary based on the work of Asong, Zilefac E.; Khaliq, M. N.; Wheater, H. S. (2016 in Climate Dynamics).

Evaluation of the Impacts of Climate Change on Streamflow Through Hydrological Simulation and Under Downscaling Scenarios: Case Study in a Watershed in Southeastern Brazil

https://doi.org/10.1007/s10661-020-08671-x

Abstract-Summary

This work aimed to analyse future climate scenarios for the Ribeirão do Lobo River Basin, located in the state of São Paulo, Brazil.

In all, five future scenarios were generated, with scenarios A, B, C and D projected based on the 5th report of the IPCC and scenario E based on the trend of climate data in the region.

Among the scenarios generated, scenario D, which considers an increase of 4.8 °C in air temperature and a reduction of 10% in rainfall, is responsible for the worst water condition in the basin and can reduce up to 72.41% of the average flow and up to 55.50%, 54.18% and 38.17% of the low flow parameters Q90%, Q95% and Q7, 10, respectively, until the end of the twenty-first century.

The E scenario also becomes a matter of concern, since it was responsible for greater increases in temperature and greater reductions in rainfall and, consequently, more drastic monthly reductions in streamflow, which may negatively impact water resources and affect the various uses of water in the Ribeirão do Lobo River Basin.

Introduction

Researchers have used hydrological models integrated with climate models in order to analyse future scenarios and assess the possible impacts of climate change on water resources in different environments and spatial scales.

Alvarenga and others [132] evaluated the future hydrological impacts due to climate changes in the Lavrinha hydrographic basin, located in the southeastern region of Brazil, and concluded that the streamflow is very sensitive to the increase in temperature and decrease in rainfall, and the simulation carried out by the authors projected a reduction in monthly streamflow between 20 and 77% over the twenty-first century.

Based on these assumptions, this study aimed to assess the impact of climate change on the streamflow of a hydrographic basin by means of hydrological simulations using the SMAP (Soil Moisture Accounting Procedure) model associated with future climate scenarios created by the PGECLIMA_R software.

Materials and Methods

The calibration took place in order to obtain the best values for the following model parameters: Tuin = initial moisture content (%) Ebin = initial basic flow (m³/s) ai = initial abstraction (mm) aap = field capacity (%) kkt = basic runoff recession constant (days) sat = soil saturation capacity (mm) k2t = surface runoff recession constant (days) Crec = ground water recharge parameter (%) The Tuin value was established at 20%, and for the determination of the Ebin value, the first value of the observed data hydrograph was considered, according to the National Electric System Operator (ONS) [133].

For the simulation of climate data, the methodology used considers a base period, for which the monthly parameterization of the statistical distributions of the climatological series is carried out.

For the initialization of the rest of the simulated subperiods, the low flow parameter Q95% of the previous periods was calculated in order to represent the minimum flow value and the initial state (Ebin).

Result and Discussions

Despite observing discrepant results for different studies in the southeastern region of Brazil, it is noteworthy that with the exception of Junior and Mauad [134], all authors detected average annual reductions of approximately 50% in the average flow of the studied hydrographic basins, a result that may be due to the increase in temperature and consequently to ETo, thus suggesting possible negative impacts on water resources for the southeastern region of Brazil, in the subperiod from 2060 to 2069.

It is noted that for both low flow parameters (Q95% and Q90%), in the subperiods from 2030 to 2039 and 2060 to 2069, scenario E was responsible for the worst condition, indicating that the average annual increase in temperature and, consequently, in evapotranspiration, even with the average annual increase in PREC, is responsible for the reduction in low flow parameters, pointing again the sensitivity of the modelling to changes in the air temperature.

Conclusions

The use of the PGECLIMA_R stochastic generator was extremely relevant for the simulation of future scenarios, contributing to the evaluation of the possible impacts of climate changes at a local level.

It is expected that the results generated from the creation of future climate scenarios and the integration of the models can effectively collaborate in decision-making, related to the planning and management, especially in long term, of water resources in the Ribeirão do Lobo River Basin, in addition to providing input for studies on the impact of these results presented on the water quality of the basin.

The usefulness of this work is highlighted as a case study, presenting and proving the applicability of the tools/models used in order to predict the impact of climate change on water availability, thus being able to be applied in other hydrographic basins of interest of researchers or managers linked to the management of water resources.

Acknowledgement

A machine generated summary based on the work of Neves, Gabriela Leite; Barbosa, Mariana Abibi Guimarães Araujo; Anjinho, Phelipe da Silva; Guimarães, Tainá Thomassim; das Virgens Filho, Jorim Sousa; Mauad, Frederico Fábio (2020 in Environmental Monitoring and Assessment).

Assessment of a Stochastic Downscaling Methodology in Generating an Ensemble of Hourly Future Climate Time Series

https://doi.org/10.1007/s00382-012-1627-2

Abstract-Summary

This study extends a stochastic downscaling methodology to generation of an ensemble of hourly time series of meteorological variables that express possible future climate conditions at a point-scale.

Marginal distributions of factors of change are computed for several climate statistics using a Bayesian methodology that can weight GCM realizations based on the model relative performance with respect to a historical climate and a degree of disagreement in projecting future conditions.

With either approach, the derived factors of change are applied to the climate statistics inferred from historical observations to re-evaluate parameters of the weather generator.

The re-parameterized generator yields hourly time series of meteorological variables that can be considered to be representative of future climate conditions.

The time series are generated in an ensemble mode to fully reflect the uncertainty of GCM projections, climate stochasticity, as well as uncertainties of the downscaling procedure.

The inferences of the methodology for the period of 2000–2009 are tested against observations to assess reliability of the stochastic downscaling procedure in reproducing statistics of meteorological variables at different time scales.

Introduction
Most of the techniques that have been presented in downscaling General Circulation Model (GCM) realizations have targeted regional spatial scales at the daily or even monthly time resolutions (Müller-Wohlfeil et al. [135]; Hay et al. [136]; Wilby et al. [137]; Barnett et al. [138]; Wood et al. [1]; Schmidli et al. [139]; Merritt et al. [140]; Leander and Buishand [141]; Burton et al. [142]).

The need to account for the uncertainty in climate change predictions, as obtained from a multi-model ensemble, can be also regarded as a fundamental task in downscaling studies.

Several studies expressed certain confidence that a GCM ensemble can provide more reliable projections of climate change or, at least, that the uncertainty is reasonably well captured by the variation among different models (Räisänen [143]; Knutti [144]).

The argument has been that while model weighting is promising in principle, the lack of correlation or information on the relation among error characteristics and climate change variables hampers the possibility of obtaining robust weights (Weigel et al. [145]; Giorgi and Coppola [146]; simple averaging might be preferred to avoid another level of uncertainty (Knutti et al. [147]; Christensen et al. [148]).

Methodology
The stochastic downscaling methodology allows one to derive the distributions of factors of change that are calculated as ratios or "delta" differences of climate statistics (Anandhi et al. [149]) for historical and future periods.

In Fatichi et al. 150, AWE-GEN was used to generate the time series of predicted mean/median future climate, using a single set of weather generator parameters corresponding to the means/medians of the PDFs of factors of change.

Transferring the complete uncertainty contained in the PDFs of factors of change into generated meteorological time series can be regarded as the ultimate step in a downscaling methodology, allowing one to account for a heterogeneous nature of climate predictions produced by different models.

The stochastic downscaling technique derives 170 PDFs of factors of change from an ensemble of climate models (Fatichi et al. [150]).

Data
Hourly air temperature, wind speed, relative humidity, and atmospheric pressure for the period of 1962 through 2010 were obtained for the Firenze Peretola station from the National Climatic Data Center (NCDC) (Peterson and Vose [151]).

The shortwave radiation and cloudiness parameters of the weather generator were estimated from the data for another station, Firenze Università (about 2.1 km distant from Firenze Ximeniano), available for the period of 2000 through 2009 (Fatichi et al. [152]).

Since not all of the models had outputs for the "validation" period of 2000–2009, the factors of change for this period were estimated using the methodology of interpolating transient factors of change presented by Burton et al. [142].

To Burton et al. [142], a single set of factors of change was used for the period of 2000 through 2009, which was assumed to be stationary.

Results

That the relative uncertainty of the simulated future climates does not change significantly for different periods, i.e., the range of the BWA ensemble is approximately the same for different periods.

The simulated and observed daily cycles of temperature for the period of 2000–2009 show that the overall direction and magnitude of the change are fairly well captured by the stochastic downscaling.

The observed and simulated delta changes of standard deviations of the daily cycle between the control scenario and the 2000–2009 period are very similar on average, +0.11 and +0.16 °C respectively, with data for most of the hours well within the BWA uncertainty bounds.

As for air temperature, the future relative humidity simulated with the SA approach is always very close to the changes predicted with the ensemble median (with the apparent exception for July for the period of 2081–2100).

Discussion and Conclusions

The novelty of this study is represented by a transfer of the uncertainty of climate change predictions inferred from an ensemble of climate models to an ensemble of hourly time series representing future climate conditions.

While the uncertainty derived with the presented methodology of Bayesian weighting (the BWA approach) or simple averaging (the SA approach) of multiple GCMs does not reflect all possible sources of uncertainty (for instance, it considers a single emission scenario and cannot incorporate some of climate model structural uncertainties), it represents important information for evaluations of climate change predictions (Knutti [144]; Knutti et al. [147]).

The changes in the precipitation regime are difficult to evaluate, given the uncertainty of stochastic realizations and climate model predictions.

In combination with the demonstrated validation of the stochastic methodology, this allows us to conclude that the presented results can be regarded as robust estimates of climate change for the location of Firenze, given the present knowledge of climate systems (climate model realizations) and data available for downscaling.

Acknowledgement

A machine generated summary based on the work of Fatichi, S.; Ivanov, V. Y.; Caporali, E. (2012 in Climate Dynamics).

Multi-site Multivariate Downscaling of Global Climate Model Outputs: An Integrated Framework Combining Quantile Mapping, Stochastic Weather Generator and Empirical Copula Approaches

https://doi.org/10.1007/s00382-018-4480-0

Abstract-Summary

The site-specific or variable-specific downscaling methods only produce climate change scenarios for a specific site or specific variable, which is inadequate to drive distributed hydrological models to investigate the spatio-temporal variability of climate change impacts at the catchment scale.

This study proposes an integrated framework combining quantile mapping (QM), stochastic weather generator (WG) and Empirical Copula (EC) approaches for multi-site multivariate downscaling of global climate model outputs from monthly, grid-scale to daily, station-specific scale.

In this hybrid scheme, the QM method is used to spatially downscale the monthly large-scale climate model outputs; then a stochastic WG is used to temporally down-scale the monthly data to daily data by adjusting the WG parameters according to the predicted changes from large-scale climate models; at last, the observed inter-site and inter-variable dependencies, the temporal persistence, as well as the inter-annual variability are restored using the EC approach.

The results show that the proposed downscaling approach is able to reconstruct the marginally distributional statistics, inter-site and inter-variable dependencies, and temporal persistence in the downscaled data for the validation period.

The proposed methodologies are useful for downscaling ensembles of large-scale climate model simulations and projections for distributed hydrological impact studies.

Extended

This study proposes an integrated multi-site multivariate downscaling framework to downscale outputs of large-scale climate models from monthly, grid-scale to daily, station-specific scales at multiple locations across a catchment.

In this hybrid scheme, the quantile mapping was used to spatially downscale the monthly precipitation from GCM dataset from GCM grid-scale to station-specific scale; then the spatially downscaled monthly GCM precipitation was temporally downscaled into daily data by adjusting the parameters of MulGETS.

The results show a good performance of the proposed downscaling framework in reconstructing the marginally distributional statistics of the climate variables, the inter-site and inter-variable dependencies, and the temporal persistences.

The proposed methodologies are generic, conceptually simple, and computation-ally inexpensive, hence can be applied to downscale ensembles of large-scale climate model projections for distributed hydrological impact studies for different regions.

Introduction
This is inadequate to investigate the spatial variability of hydrological impact under climate change, since a distributed input of downscaled climate simulations at multiple locations—with reliable representation of inter-site dependence—is required for hydrological modeling, especially for large catchments with diverse topographies.

The new framework integrates quantile mapping, stochastic weather generator, and the EC approaches into a hybrid scheme, which aims to reconstruct the observed marginally distributional statistics, inter-site and inter-variable dependencies, temporal persistence as well as the inter-annual variability in the distributed downscaled climate simulations at the catchment scale.

In this integrated framework, the quantile mapping method is used to spatially downscale the monthly outputs of large-scale climate models from a grid-scale to a site-specific scale; the site-specific downscaled monthly outputs are then temporally downscaled into daily data by adjusting the WG parameters; At the last step, the EC approach is used to reconstruct the inter-variable and inter-site dependencies, temporal persistence, as well as the inter-annual variability in the downscaled climate simulations.

Study Area and Data
Multi-site multivariate climate change scenarios have a greater significance for a larger basin than a smaller one for hydrological impact studies, hence a large basin (Daqing river basin) is used as an example to illuminate the application of the proposed downscaling approach.

The quality-controlled observed daily precipitation, maximum and minimum temperature series covering the period during 1957–2016 at eleven stations across Daqing river basin are obtained from China Meteorological Data Service Center (CMDC) (https://data.cma.cn/).

Large-scale model simulation data from two ESMs, i.e., the CCSM4 and MRI-CGCM3, are used to present and test the proposed methodology.

It is necessary to emphasize here that, for large-scale climate models, only monthly data are required by the proposed downscaling method.

A minimum of four large-scale climate model grid box is normally used for point-scale climate change impact study to avoid the risk of non-representative data (Li et al. [153]; Zhang et al. [154]).

Methodology
The first step spatially downscales the monthly large-scale climate model outputs to the site-specific scale based on quantile mapping approach; In the second step, the spatially downscaled monthly data are temporally downscaled into daily data using a single-site weather generator; The last step rebuilds the inter-site, inter-variable, and temporal dependencies based on the Empirical Copula method.

The observed data of the calibration period are directed used as the "reference" to build the EC template and subsequently used to arrange the spatially and temporally downscaled GCM data of the validation period, such that the observed temporal ordering structure as well as the inter-variable and inter-site dependencies (in particular, the Spearman rank correlation) are reconstructed during the validation period.

To explore this assumption, we compare the values of three forms of dependencies, i.e., inter-site correlation, Lag-1 serial autocorrelation, and inter-variable cross correlation across the stations and months between the calibration period and the validation period based on the observed data.

Results

We examine the capability of the post-processed spatially and temporally downscaled GCM data (GCM-STD-EC) in reproducing the observed inter-annual variability (represented by inter-annual standard deviation at both annual and monthly scales) for all the meteorological variables over the validation period.

For inter-annual variability, the GCM-STD-EC data are generally able to represent the standard deviation for maximum and minimum temperatures across most of the stations at both annual and monthly timescales, though underestimations are noticed for some stations.

The spatially and temporally downscaled GCM (GCM-STD) temperature data without adopting the EC approach fail to reproduce the inter-annual variability at neither annual nor monthly timescales (results not shown due to space limit).

For precipitation, the CCSM4-STD-EC data is able to reproduce the observed inter-annual variance at monthly timescale, whereas the MRI-CGCM3-STD-EC data tend to overestimate it for some months and stations.

Discussion

The multi-site, multivariate downscaling of large-scale climate model outputs is one of the most challenging tasks, considering the various targets/constraints it needs to achieve/satisfy, such as a good representation of the inter-site and inter-variable dependencies, a reliable reflection of the short-term temporal dependence, as well as a suitable consideration of the inter-annual variability.

For post-processing, the EC approach is implemented to reproduce the observed inter-variable and inter-site dependencies, the temporal persistence, as well as the inter-annual variability in the spatially and temporally downscaled climate model simulations.

To tackle this issue, Vrac [155] proposed a "rank resampling" approach for multivariate bias correction where the stationarity assumption of EC is relaxed by letting the climate model drive the temporal properties and their changes in time, whereas the dependence structures are assumed invariant.

Summary and Conclusion

This study proposes an integrated multi-site multivariate downscaling framework to downscale outputs of large-scale climate models from monthly, grid-scale to daily, station-specific scales at multiple locations across a catchment.

This framework integrates quantile mapping, stochastic weather generator, and the Empirical Copula approach into a hybrid scheme, in which the quantile mapping method is used to spatially downscale the monthly large-scale climate model outputs; then the downscaled monthly data is temporally downscaled into daily data by adjusting the parameters of the weather generators; at last, the observed inter-site and inter-variable dependencies, the temporal persistence, as well as the inter-annual variability is restored by using the Empirical Copula approach.

Acknowledgement
A machine generated summary based on the work of Li, Xin; Babovic, Vladan (2018 in Climate Dynamics).

Book Reading List

Climate Change Modeling Methodology
By Rasch, P. J. (Ed) *(2012).*
    The Earth's average temperature has risen by 1.4°F over the past century, and computer models project that it will rise much more over the next hundred years, with significant impacts on weather, climate, and human society. Many climate scientists attribute these increases to the build up of greenhouse gases produced by the burning of fossil fuels and to the anthropogenic production of short-lived climate pollutants. Climate Change Modeling Methodologies: Selected Entries from the Encyclopaedia of Sustainability Science and Technology provides readers with an introduction to the tools and analysis techniques used by climate change scientists to interpret the role of these forcing agents on climate.
    Please see https://www.springer.com/gp/book/9781461457664 for original source.

Climate Modelling
By A. Lloyd, E. (Ed), Winsberg, E. (Ed) *(2018).*
    This edited collection of works by leading climate scientists and philosophers introduces readers to issues in the foundations, evaluation, confirmation, and application of climate models. It engages with important topics directly affecting public policy, including the role of doubt, the use of satellite data, and the robustness of models.
    Please see https://www.springer.com/gp/book/9783319650579 for original source.

Demystifying Climate Models
By Gettelman, A., Rood, R. B. *(2016).*
    This book demystifies the models we use to simulate present and future climates, allowing readers to better understand how to use climate model results. In order to predict the future trajectory of the Earth's climate, climate-system simulation models are necessary. When and how do we trust climate model predictions? The book offers a framework for answering this question. It provides readers with a basic primer on

climate and climate change, and offers non-technical explanations for how climate models are constructed, why they are uncertain, and what level of confidence we should place in them.

Please see https://www.springer.com/gp/book/9783662489574 for original source.

Development and Evaluation of High Resolution Climate System Models
By Yu, R., Zhou, T., Wu, T., Xue, W., Zhou, G. *(2016)*.

This book is based on the project "Development and Validation of High Resolution Climate System Models" with the support of the National Key Basic Research Project under grant No. 2010CB951900. It demonstrates the major advances in the development of new, dynamical Atmospheric General Circulation Model (AGCM) and Ocean General Circulation Model (OGCM) cores that are suitable for high resolution modeling, the improvement of model physics, and the design of a flexible, multi-model ensemble coupling framework.

Please see https://www.springer.com/gp/book/9789811000317 for original source.

Introduction to Climate Modelling
By Stocker, T. *(2011)*.

A three-tier approach is presented: (i) fundamental dynamical concepts of climate processes, (ii) their mathematical formulation based on balance equations, and (iii) the necessary numerical techniques to solve these equations. This book showcases the global energy balance of the climate system and feedback processes that determine the climate sensitivity, initial-boundary value problems, energy transport in the climate system, large-scale ocean circulation and abrupt climate change.

Please see https://www.springer.com/gp/book/9783642007729 for original source.

Models for Tropical Climate Dynamics
By Khouider, B. *(2019)*.

This book is a survey of the research work done by the author over the last 15 years, in collaboration with various eminent mathematicians and climate scientists on the subject of tropical convection and convectively coupled waves. In the areas of climate modelling and climate change science, tropical dynamics and tropical rainfall are among the biggest uncertainties of future projections. This not only puts at risk billions of human beings who populate the tropical continents but it is also of central importance for climate predictions on the global scale. This book aims to introduce the non-expert readers in mathematics and theoretical physics to this fascinating topic in order to attract interest into this difficult and exciting research area.

Please see https://www.springer.com/gp/book/9783030177744 for original source.

Stochastic Climate Models
By Imkeller, P. (Ed), Storch, J. v. (Ed) *(2001)*.

The proceedings of the summer 1999 Chorin workshop on stochastic climate models captures well the spirit of enthusiasm of the workshop participants engaged in research in this exciting field. It is amazing that nearly 25 years after the formal theory of natural climate variability generated by quasi-white-noise weather forcing was developed, and almost 35 years after J. M. Mitchell first suggested this mechanism as the origin of sea-surf ace-temperature fluctuations and climate variability, there have arisen so many fresh perspectives and new applications of the theory.

Please see https://www.springer.com/gp/book/9783034895040 for original source.

Stochastic Climate Theory
By Dobrovolski, S. G. *(2000)*.

The author describes the stochastic (probabilistic) approach to the study of changes in the climate system. Climatic data and theoretical considerations suggest that a large part of climatic variation/variability has a random nature and can be analyzed using the theory of stochastic processes. This work summarizes the results of processing existing records of climatic parameters as well as appropriate theories: from the theory of random processes (based on the results of Kolmogorov and Yaglom) and Hasselmann's "stochastic climate model theory" to recently obtained results.

Please see https://www.springer.com/gp/book/9783540663102 for original source.

References

1. Wood AW, Leung LR, Sridhar V, Lettenmaier DP (2004) Hydrologic implications of dynamical and statistical approaches to downscaling climate model outputs. Clim Change 62:189–216. https://doi.org/10.1023/B:CLIM.0000013685.99609.9e
2. Shiogama H, Imada Y, Mori M, Mizuta R, Stone D, Yoshida K, Arakawa O, Ikeda M, Takahashi C, Arai M et al (2016) Attributing historical changes in probabilities of record-breaking daily temperature and precipitation extreme events. SOLA 12:225–231
3. Giorgi F, Gutowski WJ (2015) Regional dynamical downscaling and the CORDEX initiative. Annu Rev Environ Resour 40:467–490. https://doi.org/10.1146/annurev-environ-102 014-021217
4. Rummukainen M, Rockel B, Barring L, Christensen JH, Reckermann M (2015) Twenty-first-century challenges in regional climate modeling. Bull Am Meteorol Soc 96:135–138
5. Rockel B (2015) The regional downscaling approach: a brief history and recent advances. Curr Clim Change Rep. https://doi.org/10.1007/s40641-014-0001-3
6. Déqué M, Alias A, Dubois C, Somot S (2014) Some sources of bias in the Eurocordex historical runs. 3rd International lund regional-scale climate modelling workshop, Lund, Sweden. https://www.baltex-research.eu/RCM2014/index.html
7. Bruyère CL, Done JM, Holland GJ, Fredrick S (2014) Bias corrections of global models for regional climate simulations of high-impact weather. Clim Dyn 43:1847–1856. https://doi.org/10.1007/s00382-013-2011-6
8. Giorgi F, Jones C, Asrar G (2009) Addressing climate information needs at the regional level: The CORDEX framework. World Meteorol Organ Bull 58:175–183. https://wcrp.ipsl.jussieu.fr/RCD_Projects/CORDEX/CORDEX_giorgi_WMO.pdf

9. Jones C, Giorgi F, Asrar G (2011) The coordinated regional downscaling experiment: CORDEX. an international downscaling link to CMIP5. CLIVAR Exch 56:34–40
10. IPCC (2013) Climate change 2013: the physical science basis. Contribution of working group I to the fifth assessment report of the intergovernmental panel on climate change. In: Stocker TF, Qin D, Plattner GK, Tignor M, Allen SK, Boschung J, Nauels A, Xia Y, Bex V, Midgley PM (eds) Cambridge University Press, Cambridge, p 1535
11. Curry JA, Lynch AH (2002) Comparing Arctic Regional climate model. EOS 83:87. https://doi.org/10.1029/2002EO000051
12. Rinke A, Dethloff K (2008) Simulated circum-Arctic climate changes by the end of the 21st century. Glob Planet Change 62:173–186
13. Koenigk T, Berg P, Doscher R (2015) Arctic climate change in an ensemble of regional CORDEX simulations. Polar Res. https://doi.org/10.3402/polar.v34.24603(ISSN1751-8369)
14. Xu Z, Yang Z (2012) An improved dynamical downscaling method with gcm bias corrections and its validation with 30 years of climate simulations. J Clim. https://doi.org/10.1175/JCLI-D-12-00005.1
15. Schrum C, Hübner U, Jacob D, Podzun R (2003) A coupled atmosphere/ice/ocean model for the North Sea and the Baltic Sea. Clim Dyn 21:131–151
16. Rummukainen M (2010) State-of-the-art with regional climate models. Wiley interdisciplinary reviews. Clim Change 1(1):82–96
17. Tian T, Boberg F, Christensen O, Christensen J, She J, Vihma T (2013) Resolved complex coastlines and land-sea contrasts in a high-resolution regional climate model: a comparative study using prescribed and modelled SSTs. Tellus A 65:19
18. Van Pham T, Brauch J, Dieterich C, Frueh B, Ahrens B (2014) New coupled atmosphere-ocean-ice system COSMO-CLM/NEMO: assessing air temperature sensitivity over the North and Baltic Seas. Oceanologia 56(2):167–189
19. Gröger M, Dieterich C, Meier M, Schimanke S (2015) Thermal air-sea coupling in hindcast simulations for the North Sea and Baltic Sea on the NW European shelf. Tellus A 67:22
20. Wang S, Dieterich C, Döscher R, Höglund A, Hordoir R, Meier HEM, Samuelsson P, Schimanke S (2015) Development and evaluation of a new regional coupled atmosphere-ocean model in the North Sea and Baltic Sea. Tellus A 67:20
21. Ådlandsvik B (2008) Marine downscaling of a future climate scenario for the North Sea. Tellus A 60(3):451–458
22. Holt J, Wakelin S, Lowe J, Tinker J (2010) The potential impacts of climate change on the hydrography of the northwest European continental shelf. Prog Oceanogr 86:361–379
23. Olbert AI, Dabrowski T, Nash S, Hartnett M (2012) Regional modelling of the 21st century climate changes in the Irish Sea. Cont Shelf Res 41:48–60
24. Gröger M, Maier-Reimer E, Mikolajewicz U, Moll A, Sein D (2013) NW European shelf under climate warming: implications for open ocean—shelf exchange, primary production, and carbon absorption. Biogeosciences 10:3767–3792
25. Mathis M, Pohlmann T (2014) Projection of physical conditions in the North Sea for the 21st century. Clim Res 61:1–17
26. Tian T, Su J, Boberg F, Yang S, Schmitt T (2016) Estimating uncertainty caused by ocean heat transport to the North Sea: experiments downscaling EC-EARTH. Clim Dyn 46(1):99–110
27. Tinker J, Lowe J, Pardaens A, Holt J, Barciela R (2016) Uncertainty in climate projections for the 21st century northwest European shelf seas. Prog Oceanogr 148:56–73
28. Schrum C, Lowe J, Meier HEM, Grabemann I, Holt J, Mathis M, Pohlmann T, Skogen MD, Sterl A, Wakelin S (2016) Projected change—North Sea. In: Quante M, Colijn F (eds) North Sea region climate change assessment. Springer, Berlin, Heidelberg, pp 175–217
29. Kaushal SS, Gold AJ, Mayer PM (2017) Land use, climate, and water resources-global stages of interaction. Water (Switzerland) 9:815. https://doi.org/10.3390/w9100815
30. Azadi F, Ashofteh P, Loáiciga HA (2019) Reservoir water-quality projections under climate-change conditions. Water Resour Manag 33:401–421. https://doi.org/10.1007/s11269-018-2109-z

31. Maraun D, Wetterhall F, Ireson AM et al (2010) Precipitation downscaling under climate change: recent developments to bridge the gap between dynamical models and the end user. Rev Geophys 48:1–34
32. Barnett TP, Adam JC, Lettenmaier DP (2005) Potential impacts of a warming climate on water availability in snow-dominated regions. Nature 438:303–309. https://doi.org/10.1038/nature04141
33. Asong ZE, Ezzat Elshamy M, Princz D et al (2020) High-resolution meteorological forcing data for hydrological modelling and climate change impact analysis in the Mackenzie River Basin. Earth Syst Sci Data 12:629–645. https://doi.org/10.5194/essd-12-629-2020
34. Teutschbein C, Seibert J (2012) Bias correction of regional climate model simulations for hydrological climate-change impact studies: review and evaluation of different methods. J Hydrol 456:12–29. https://doi.org/10.1016/j.jhydrol.2012.05.052
35. Kim KB, Kwon H, Han D (2015) Bias correction methods for regional climate model simulations considering the distributional parametric uncertainty underlying the observations. J Hydrol 530:568–579. https://doi.org/10.1016/j.jhydrol.2015.10.015
36. Jiang T, Chen YD, Xu C, Chen X, Chen X, Singh VP (2007) Comparison of hydrological impacts of climate change simulated by six hydrological models in the. J Hydrol 336:316–333. https://doi.org/10.1016/j.jhydrol.2007.01.010
37. Maurer EP, Brekke LD, Pruitt T (2010) Contrasting lumped and distributed hydrology models for estimating climate change impacts on California watersheds. J Am Water Resour Assoc 46:1024–1035. https://doi.org/10.1111/j.1752-1688.2010.00473.x
38. Bastola S, Murphy C, Sweeney J (2011) The role of hydrological modelling uncertainties in climate change impact assessments of Irish river catchments. Adv Water Resour 34:562–576. https://doi.org/10.1016/j.advwatres.2011.01.008
39. Najafi MR, Moradkhani H, Jung IW (2011) Assessing the uncertainties of hydrologic model selection in climate change impact studies. Hydrol Process 25:2814–2826. https://doi.org/10.1002/hyp.8043
40. Surfleet CG, Tullos D, Chang H, Jung I (2012) Selection of hydrologic modeling approaches for climate change assessment: a comparison of model scale and structures. J Hydrol 465:233–248. https://doi.org/10.1016/j.jhydrol.2012.07.012
41. Velázquez JA, Schmid J, Ricard S, Muerth MJ, Gauvin St-Denis B, Minville M, Chaumont D, Caya D, Ludwig R, Turcotte R (2013) An ensemble approach to assess hydrological models' contribution to uncertainties in the analysis of climate change impact on water resources. Hydrol Earth Syst Sci 17:565–578. https://doi.org/10.5194/hess-17-565-2013
42. Vansteenkiste T, Tavakoli M, Ntegeka V, de Smedt F, Batelaan O, Pereira F, Willems P (2014) Intercomparison of hydrological model structures and calibration approaches in climate scenario impact projections. J Hydrol 519:743–755. https://doi.org/10.1016/j.jhydrol.2014.07.062
43. Karlsson IB, Sonnenborg TO, Refsgaard JC et al (2016) Combined effects of climate models, hydrological model structures and land use scenarios on hydrological impacts of climate change. J Hydrol 535:301–317
44. Kusangaya S, Warburton ML, van Garderen EA, Jewitt GPW (2014) Impacts of climate change on water resources in southern Africa: a review. Phys Chem Earth Parts A/B/C 67:47–54
45. Minville M, Brissette F, Leconte R (2008) Uncertainty of the impact of climate change on the hydrology of a nordic watershed. J Hydrol 358:70–83. https://doi.org/10.1016/j.jhydrol.2008.05.033
46. Lutz AF, ter Maat HW, Biemans H, Shrestha AB, Wester P, Immerzeel WW (2016) Selecting representative climate models for climate change impact studies: an advanced envelope-based selection approach. Int J Climatol 36:3988–4005. https://doi.org/10.1002/joc.4608
47. Giorgi F, Bi X (2000) A study of internal variability of a regional climate model. J Geophys Res Atmos 105:29503–29521. https://doi.org/10.1029/2000JD900269
48. Mearns LO et al (2012) The North American regional climate change assessment program: overview of phase I results. Bull Am Meteorol Soc 93(9):1337–1362. https://doi.org/10.1175/BAMS-D-11-00223.1

49. Halmstad A, Najafi MR, Moradkhani H (2013) Analysis of precipitation extremes with the assessment of regional climate models over the Willamette River Basin, USA. Hydrol Process 27:2579–2590. https://doi.org/10.1002/hyp.9376

50. Wang J, Kotamarthi VR (2015) High—resolution dynamically downscaled projections of precipitation in the mid and late 21st century over North America. Earth's Future 3:268–288. https://doi.org/10.1002/2015EF000304

51. Fowler HJ, Blenkinsop S, Tebaldi C (2007) Linking climate change modelling to impacts studies: recent advances in downscaling techniques for hydrological modelling. Int J Climatol 27:1547–1578. https://doi.org/10.1002/joc.1556

52. Lee JW, Hong SY, Chang EC, Suh MS, Kang HS (2014) Assessment of future climate change over East Asia due to the RCP scenarios downscaled by GRIMs-RMP. Clim Dyn 42:733–747. https://doi.org/10.1007/s00382-013-1841-6

53. Lee JW, Ham S, Hong SY, Yoshimura K, Joh M (2014b) Future changes in surface runoff over Korea projected by a regional climate model under A1B scenario. Adv Meteorol

54. BA Cosgrove et al (2003) Real-time and retrospective forcing in the North American Land Data Assimilation System (NLDAS) project J Geophys Res Atmos 108. https://doi.org/10.1029/2002JD003118

55. Déqué M et al (2007) An intercomparison of regional climate simulations for Europe: assessing uncertainties in model projections. Clim Change 81:53–70

56. Madec G, NEMO Team (2016) NEMO ocean engine. https://doi.org/10.5281/zenodo.1472492.

57. Slangen ABA, Katsman CA, van de Wal RSW, Vermeersen LLA, Riva REM (2012) Towards regional projections of twenty-first century sea-level change based on IPCC SRES scenarios. Clim Dyn 38(5–6):1191–1209. https://doi.org/10.1007/s00382-011-1057-6

58. Slangen ABA, Carson M, Katsman CA, van de Wal RSW, Koehl A, Vermeersen LLA, Stammer D (2014) Projecting twenty-first century regional sea-level changes. Climat Change 124:317–332. https://doi.org/10.1007/s10584-014-1080-9

59. Church JA, Clark PU, Cazenave A, Gregory JM, Jevrejeva S, Levermann A, Merrifield MA, Milne GA, Nerem RS, Nunn PD, Payne AJ, Pfeffer WT, Stammer D, Unnikrishnan AS (2013) Sea level change. In: Stocker TF, Qin D, Plattner GK, Tignor M, Allen SK, Boschung J, Nauels A, Xia Y, Bex V, Midgley PM (eds) climate change 2013: the physical science basis. contribution of working group I to the fifth assessment report of the intergovernmental panel on climate change. Cambridge University Press, Cambridge, United Kingdom and New York, USA

60. de Vries H, Katsman C, Drijfhout S (2014) Constructing scenarios of regional sea level change using global temperature pathways. Environm Res Lett. https://doi.org/10.1088/1748-9326/9/11/115007

61. Kopp RE, Horton RM, Little CM, Mitrovica JX, Oppenheimer M, Rasmussen DJ, Strauss BH, Tebaldi C (2014) Probabilistic 21st and 22nd century sea-level projections at a global network of tide-gauge sites. Earth's Future 2(8):383–406. https://doi.org/10.1002/2014EF000239

62. Palmer M, Howard T, Tinker J, Lowe J, Bricheno L, Calvert D, Gregory J, Harris G, Krijnen J, Pickering M, Roberts C, Wolf J (2018) UKCP18 Marine report

63. Rummukainen M (2010) State-of-the-art with regional climate models. WIREs Climate Change 1(1):82–96. https://doi.org/10.1002/wcc.8

64. Liu ZJ, Minobe S, Sasaki YN, Terada M (2016) Dynamical downscaling of future sea level change in the western North Pacific using ROMS. J Oceanogr 72(6):905–922. https://doi.org/10.1007/s10872-016-0390-0

65. O'Dea E, Furner R, Wakelin S, Siddorn J, While J, Sykes P, King R, Holt J, Hewitt H (2017) The CO5 configuration of the 7km Atlantic margin model: large-scale biases and sensitivity to forcing, physics options and vertical resolution. Geosci Model Dev 10(8):2947–2969. https://doi.org/10.5194/gmd-10-2947-2017

66. Meinshausen M, Smith SJ, Calvin K, Daniel JS, Kainuma MLT, Lamarque J-F, Matsumoto K, Montzka SA, Raper SCB, Riahi K, Thomson A, Velders GJM, van Vuuren DPP (2011) The RCP greenhouse gas concentrations and their extensions from 1765 to 2300. Clim Change 109:210–241. https://doi.org/10.1007/s10584-011-0156-z

67. Gregory JM, Griffies SM, Hughes CW, Lowe JA, Church JA, Fukimori I, Gomez N, Kopp RE, Landerer F, Le Cozannet G, Ponte RM, Stammer D, Tamisiea ME, van de Wal RSW (2019) Concepts and terminology for sea level: mean variability and change, both local and global. Surv Geophys 9:9–10. https://doi.org/10.1007/s10712-019-09525-z

68. McSweeney CF, Jones RG, Lee RW, Rowell DP (2015) Selecting CMIP5 GCMs for downscaling over multiple regions. Clim Dyn 44:3237–3260. https://doi.org/10.1007/s00382-014-2418-8

69. Flaounas E, Bastin S, Janicot S (2011) Regional climate modelling of the 2006 West African monsoon: sensitivity to convection and planetary boundary layer parameterisation using wrf. Clim Dyn 36(5–6):1083–1105. https://doi.org/10.1007/s00382-010-0785-3

70. Huffman GJ, Adler RF, Bolvin DT, Gu G, Nelkin EJ, Bowman KP, Stocker EF, Wolff DB (2007) The TRMM multi-satellite precipitation analysis: quasi-global, multi-year, combined-sensor precipitation estimates at fine scale. J Hydrometeorol 8:33–55

71. Vellinga M, Arribas A, Graham R (2013) Seasonal forecasts for regional onset of the West African monsoon. Clim Dyn 40:3047–3070. https://doi.org/10.1007/s00382-012-1520z

72. Fitzpatrick RGJ, Bain C, Knippertz P, Marsham JH, Parker DJ (2015) The West African monsoon onset: a concise comparison of definitions. J Clim. https://doi.org/10.1175/JCLI-D-15-0265.1

73. Gbobaniyi E, Sarr A, Sylla MB, Diallo I, Lennard C, Dosio A, Dhiédiou A, Kamga A, Klutse NAB, Hewitson B, Nikulin G, Lamptey B (2013) Climatology, annual cycle and interannual variability of precipitation and temperature in CORDEX simulations over West Africa. Int J Climatol 34:2241–2257. https://doi.org/10.1002/joc.3834

74. Nikulin G, Jones C, Giorgi F, Asrar G, Buchner M, Cerezo-Mota R, Christensen OB, Déqué M, Fernandez J, Hansler A, van Meijgaard E, Samuelsson P, Sylla MB, Sushama L (2012) Precipitation climatology in an ensemble of CORDEX-Africa regional climate simulations. J Clim 25(18):6057–6078

75. Diaconescu EP, Gachon P, Scinocca J, Laprise R (2015) Evaluation of daily precipitation statistics and monsoon onset/retreat over western Sahel in multiple data sets. Clim Dyn 45:1325–1354. https://doi.org/10.1007/s00382-014-2383-2

76. Alaka GD, Maloney E (2012) The influence of the MJO on upstream precursors to African easterly waves. J Clim 25:3219–3236

77. Mathon V, Laurent H, Lebel T (2002) Mesoscale convective system rainfall in the Sahel. J Appl Meteorol 41:1081–1092

78. Bhaskaran B, Jones RG, Murphy JM, Noguer M (1996) Simulations of the Indian summer monsoon using a nested regional climate model: domain size experiments. Clim Dyn 12:573–587

79. Noguer M, Jones R, Murphy J (1998) Sources of systematic errors in the climatology of a regional climate model over Europe. Clim Dyn 14:691–712

80. Seth A, Giorgi F (1998) The effects of domain choice on summer precipitation simulation and sensitivity in a regional climate model. J Clim 11:2698–2712

81. Denis B, Laprise R, Caya D, Côté J (2002) Downscaling ability of one-way-nested regional climate models: The Big-Brother experiment. Clim Dyn 18:627–646

82. Denis B, Laprise R, Caya D (2003) Sensitivity of a regional climate model to the resolution of the lateral boundary conditions. Clim Dyn 20:107–126

83. Castro C, Pielke R Sr, Leoncini G (2005) Dynamical downscaling: assessment of value retained and added using the regional atmospheric modeling system (RAMS). J Geophys Res 110:D05108. https://doi.org/10.1029/2004JD004721

84. Salameh T, Drobinski P, Dubos T (2010) The effect of indiscriminate nudging time on large and small scales in regional climate modelling: application to the mediterranean basin. Q J R Meteorol Soc 136:170–182

85. Omrani H, Drobinski P, Dubos T (2012) Spectral nudging in regional climate modeling: how strongly should we nudge? Q J R Meteorol Soc 138:1808–1813

86. Miguez-Macho G, Stenchikov GL, Robock A (2004) Spectral nudging to eliminate the effects of domain position and geometry in regional climate model simulations. J Geophys Res 109:D13104–D13104

87. Leduc M, Laprise R (2009) Regional climate model sensitivity to domain size. Clim Dyn 32:833–854
88. Omrani H, Drobinski P, Dubos T (2012) Investigation of indiscriminate nudging and predictability in a nested quasi-geostrophic model. Q J Roy Meteorol Soc 138:158–169
89. Simmons A, Uppala S, Dee D, Kobayashi S (2007) ERA-interim: new ECMWF reanalysis products from 1989 onwards. ECMWF Newslett 110:25–35
90. Dee DP, Uppala SM, Simmons AJ, Berrisford P, Poli P, Kobayashi S, Andrae U, Balmaseda MA, Balsamo G, Bauer P, Bechtold P, Beljaars ACM, van de Berg L, Bidlot J, Bormann N, Delsol C, Dragani R, Fuentes M, Geer AJ, Haimberger L, Healy SB, Hersbach H, Holm EV, Isaksen L, Kallberg P, Kohler M, Matricardi M, McNally AP, Monge-Sanz BM, Morcrette J-J, Park B-K, Peubey C, de Rosnay P, Tavolato C, Thepaut J-N, Vitart F (2011) The ERA-Interim reanalysis: configuration and performance of the data assimilation system. Q J R Meteorol Soc 137:553–597. https://doi.org/10.1002/qj.828
91. Nicholson SE (1997) An analysis of the ENSO signal in the tropical Atlantic and western Indian oceans. Int J Climatol 17:345–375
92. Nicholson SE, Kim J (1997) The relationship of the El Niño-Southern Oscillation to African rainfall. Int J Climatol 17:117–135
93. Cook KH (2001) A Southern Hemisphere wave response to ENSO with implications for southern Africa precipitation. J Atmos Sci 15:2146–2162
94. Beck A, Ahrens B, Stadlbacher K (2004) Impact of nesting strategies in dynamical downscaling of reanalysis data. Geophys Res Lett 31:L19101
95. Meteorological Office (UK) (2001) Some recent results from the Hadley Centre. The Hadley Centre, Met Office, Bracknell, UK
96. Watson RT, Zinyowera MC (eds) (2001) The regional impacts of climate change—an assessment of vulnerability. IPCC Special Report IPCC, Geneva, Switzerland
97. May W, Roeckner E (2001) A time-slice experiment with the ECHAM4 AGCM at high resolution: the impact of horizontal resolution on annual mean climate change. Clim Dyn 17:407–420
98. IPCC (2001) Climate change 2001. In: Houghton JT et al (eds) The scientific basis. Cambridge University Press, Cambridge
99. Ashrit RG, Douville H, Rupa Kumar K (2003) Response of the Indian monsoon and ENSO-monsoon teleconnection to enhanced greenhouse effect in the CNRM coupled model. J Meterol Soc Japan 81:779–803
100. Steiner AL, Pal JS, Bell JL, Diffenbaugh NS, Boone A, Sloan LC, Giorgi F (2009) Land surface coupling in regional climate simulations of the West African monsoon. Clim Dyn 33(6):869–892. https://doi.org/10.1007/s00382-009-0543-6
101. Cook K (1999) Generation of the African easterly jet and its role in determining West African precipitation. J Clim 12:1165–1184
102. Dosio A, Paruolo P (2011) Bias correction of the ENSEMBLES high-resolution climate change projections for use by impact models: evaluation on the present climate. J Geophys Res 116(D16):1–22. https://doi.org/10.1029/2011JD015934
103. Hong SY, Kanamitsu M (2014) Dynamical downscaling: fundamental issues from an NWP point of view and recommendations. Asia Pac J Atmos Sci 50:83–104. https://doi.org/10.1007/s13143-014-0029-2
104. Panitz HJ, Dosio A, Büchner M, Lüthi D, Keuler K (2014) COSMO-CLM (CCLM) climate simulations over CORDEX-Africa domain: analysis of the ERA-Interim driven simulations at 0.44 and 0.22 resolution. Clim Dyn. https://doi.org/10.1007/s00382-013-1834-5
105. Moss RH et al (2010) The next generation of scenarios for climate change research and assessment. Nature 463:747–756
106. Van Vuuren DP, Edmonds J, Kainuma M et al (2011) The representative concentration pathways: an overview. Clim Change 109:5–31. https://doi.org/10.1007/s10584-011-0148-z
107. Huffman GJ, Adler RF, Bolvin DT, Gu G (2009) Improving the global precipitation record: GPCP version 2.1. Geophys Res Lett 36(17):L17808. https://doi.org/10.1029/2009GL040000

108. Sylla MB, Giorgi F, Coppola E, Mariotti L (2012) Uncertainties in daily rainfall over Africa: assessment of gridded observation products and evaluation of a regional climate model simulation. Int J Climatol. https://doi.org/10.1002/joc.3551

109. Schmidli J, Goodess CM, Frei C, Haylock MR, Hundecha Y, Ribalaygua J, Schmith T (2007) Statistical and dynamical downscaling of precipitation: an evaluation and comparison of scenarios for the European Alps. J Geophys Res Atmos 112(D4). https://doi.org/10.1029/200 5JD007026

110. Hewitson BC, Crane RG (1996) Climate downscaling: techniques and application. Clim Res 07:97–110

111. Vrac M, Drobinski P, Merlo A, Herrmann M, Lavaysse C, Li L, Somot S (2012) Dynamical and statistical downscaling of the french mediterranean climate: uncertainty assessment. Nat Hazards Earth Syst Sci 12(9):2769–2784. https://doi.org/10.5194/nhess-12-2769-2012 . https://www.nat-hazards-earth-syst-sci.net/12/2769/2012/

112. Gudmundsson L, Bremnes JB, Haugen JE, Engen-Skaugen T (2012) Technical note: downscaling RCM precipitation to the station scale using statistical transformations—A comparison of methods. Hydrol Earth Syst Sci 16(9):3383–3390. https://doi.org/10.5194/hess-16-3383-2012. https://www.hydrol-earth-syst-sci.net/16/3383/2012/

113. Palmer TN, Shukla J (2000) Editorial. Q J R Meteorol Soc 126(567):1989–1990. https://doi.org/10.1002/qj.49712656701

114. Pavan V, Doblas-Reyes FJ (2000) Multi-model seasonal hindcasts over the Euro-Atlantic: skill scores and dynamic features. Clim Dyn 16(8):611–625. https://doi.org/10.1007/s00382 0000063

115. Lambert SJ, Boer GJ (2001) Cmip1 evaluation and intercomparison of coupled climate models. Clim Dyn 17(2–3):83–106. https://doi.org/10.1007/PL00013736

116. Gillett NP, Zwiers FW, Weaver AJ, Stott PA (2003) Detection of human influence on sea-level pressure. Nature 422(6929):292–294. https://doi.org/10.1038/nature01487

117. Jacob D, Bärring L, Christensen O, Christensen J, de Castro M, Déqué M, Giorgi F, Hagemann S, Hirschi M, Jones R, Kjellström E, Lenderink G, Rockel B, Sánchez E, Schär C, Seneviratne S, Somot S, van Ulden A, van den Hurk B (2007) An inter-comparison of regional climate models for europe: model performance in present-day climate. Clim Change 81(1):31–52. https://doi.org/10.1007/s10584-006-9213-4

118. Ruti PM, Williams JE, Hourdin F, Guichard F, Boone A, Van Velthoven P, Favot F, Musat I, Rummukainen M, Domínguez M, Gaertner MA, Lafore JP, Losada T, Rodriguez de Fonseca MB, Polcher J, Giorgi F, Xue Y, Bouarar I, Law K, Josse B, Barret B, Yang X, Mari C, Traore AK (2011) The west african climate system: a review of the amma model inter-comparison initiatives. Atmos Sci Lett 12(1):116–122. https://doi.org/10.1002/asl.305

119. Solman S, Sanchez E, Samuelsson P, da Rocha R, Li L, Marengo J, Pessacg N, Remedio A, Chou S, Berbery H, Le Treut H, de Castro M, Jacob D (2013) Evaluation of an ensemble of regional climate model simulations over South America driven by the era-interim reanalysis: model performance and uncertainties. Clim Dyn 41(5–6):1139–1157. https://doi.org/10.1007/s00382-013-1667-2

120. Gallardo C, Gil V, Hagel E, Tejeda C, de Castro M (2013) Assessment of climate change in Europe from an ensemble of regional climate models by the use of Köppen-Trewartha classification. Int J Climatol 33(9):2157–2166. https://doi.org/10.1002/joc.3580

121. Wilks DS, Wilby RL (1999) The weather generation game: a review of stochastic weather models. Prog Phys Geog 23(3):329–357

122. Wilby RL, Wigley TML (1997) Downscaling general circulation model output: a review of methods and limitations. Prog Phys Geogr 21:530–548. https://doi.org/10.1177/030913339 702100403

123. Dibike YB, Coulibaly P (2005) Hydrologic impact of climate change in the Saguenay watershed: comparison of downscaling methods and hydrologic models. J Hydrol 307(1–4):145–163

124. Hashmi MZ, Shamseldin AY, Melville B (2009) Statistical downscaling of precipitation: state-of-the-art and application of Bayesian multi-model approach for uncertainty assessment. Hydrol Earth Syst Sci 6:6535–6579

125. Semenov MA, Stratonovitch P (2010) Use of multimodel ensembles from global climate models for assessment of climate change impacts. Clim Res 41:1–14
126. Giorgi F (2006) Regional climate modeling: status and perspectives. J Phys IV 139:101–118
127. Borchert JA (1950) The climate of the central North American grassland. Ann Assoc Am Geogr 40:1–39
128. Kalnay E et al (1996) The NCEP/NCAR 40-year reanalysis project. Bull Am Meteorol Soc 77(3):437–471
129. Asong ZE, Khaliq MN, Wheater HS (2015) Regionalization of precipitation characteristics in the Canadian Prairie Provinces using large-scale atmospheric covariates and geophysical attributes. Stoch Env Res Risk Assess 29(3):875–892
130. Chandler RE, Bate SM (2007) Inference for clustered data using the independence log-likelihood. Biometrika 94(1):167–183
131. Yang C, Chandler RE, Isham VS, Wheater HS (2005) Spatial-temporal rainfall simulation using generalized linear models. Water Resour Res 41:1–13
132. Alvarenga LA, de Mello CR, Colombo A, Cuartas LA, Chou SC (2016) Hydrological responses to climate changes in a headwater watershed. Ciência E Agrotecnologia 40(6):647–657. https://doi.org/10.1590/1413-70542016406027716
133. ONS. (2008). Metodologia para a Previsão de Vazões uma Semana à Frente na Bacia do Alto/Médio rio Grande. Nota Técnica no.139/2008. https://www2.aneel.gov.br/aplicacoes/consulta_publica/documentos/NT%20139-2008%20R1.pdf. Accessed 30 Nov 2017.
134. Junior P, Mauad F (2015) Simulação dos Impactos das Mudanças Climáticas na Vazão da Bacia do Ribeirão do Feijão—SP. Revista Brasileira de Recursos Hídricos 20(3):741–751. https://doi.org/10.21168/rbrh.v20n3.p741-751
135. Müller-Wohlfeil DI, Bürger G, Lahmer W (2000) Response of a river catchment to climate change: application of expanded downscaling to northern Germany. Clim Change 47:61–89
136. Hay LE, Clark MP, Wilby RL, Gutowski WJ Jr, Leavesley GH, Pan Z, Arritt RW, Takle ES (2002) Use of regional climate model output for hydrologic simulations. J Hydrometeorol 3:571–590
137. Wilby RL, Dawson CW, Barrow EM (2002) SDSM—a decision support tool for the assessment of regional climate change impacts. Environ Model Softw 17(2):145–157
138. Barnett T, Malone R, Pennell W, Stammer D, Semtner B, Washington W (2004) The effects of climate change on water resources in the west: introduction and overview. Clim Change 62:1–11
139. Schmidli J, Frei C, Vidale PL (2006) Downscaling from GCM precipitation: a benchmark for dynamical and statistical downscaling methods. Int J Climatol 26:679–689
140. Merritt WS, Alila Y, Barton M, Taylor B, Cohen S, Neilsen D (2006) Hydrologic response to scenarios of climate change in sub watersheds of the Okanagan basin, British Columbia. J Hydrol 326:79–108. https://doi.org/10.1016/j.jhydrol.2005.10.025
141. Leander R, Buishand TA (2007) Resampling of regional climate model output for the simulation of extreme river flows. J Hydrol 332:487–496. https://doi.org/10.1016/j.jhydrol.2006.08.006
142. Burton A, Fowler HJ, Blenkinsop S, Kilsby CG (2010) Downscaling transient climate change using a Neyman-Scott Rectangular Pulses stochastic rainfall model. J Hydrol 381:18–32. https://doi.org/10.1016/j.jhydrol.2009.10.031
143. Räisänen J (2007) How reliable are climate models? Tellus A 59:2–29. https://doi.org/10.1111/j.1600-0870.2006.00211.x
144. Knutti R (2008) Should we believe model predictions of future climate change? Philos Trans R Soc A 366:4647–4664. https://doi.org/10.1098/rsta.2008.0169
145. Weigel AP, Knutti R, Liniger MA, Appenzeller C (2010) Risks of model weighting in multimodel climate projections. J Climate 23:4175–4190. https://doi.org/10.1175/2010JCLI3594.1
146. Giorgi F, Coppola E (2010) Does the model regional bias affect the projected regional climate change? An analysis of global model projections. Clim Change 100:787–795. https://doi.org/10.1007/s10584-010-9864-z

147. Knutti R, Furrer R, Tebaldi C, Cermak J, Meehl GA (2010) Challenges in combining projections from multiple climate models. J Climate 23:2739–2758
148. Christensen JH, Kjellström E, Giorgi F, Lenderink G, Rummukainen M (2010) Assigning relative weights to regional climate models: exploring the concept. Clim Res 44:179–194. https://doi.org/10.3354/cr00916
149. Anandhi A, Frei A, Pierson DC, Schneiderman EM, Zion MS, Lounsbury D, Matonse AH (2011) Examination of change factor methodologies for climate change impact assessment. Water Resour Res 47(W03501). https://doi.org/10.1029/2010WR009104
150. Fatichi S, Ivanov VY, Caporali E (2011) Simulation of future climate scenarios with a weather generator. Adv Water Resour 34(4):448–467
151. Peterson TC, Vose RS (1997) An overview of the global historical climatology network temperature database. Bull Am Meteorol Soc 78:2837–2849
152. Fatichi S, Ivanov VY, Caporali E (2010) Simulating hydro-meteorological variables across a range of temporal scales with a weather generator. In: International workshop advances in statistical hydrology, Taormina, Italy
153. Li Z, Liu WZ, Zhang XC, Zheng FL (2011) Assessing the site-specific impacts of climate change on hydrology, soil erosion and crop yields in the Loess Plateau of China. Clim Change 105(1–2):223–242
154. Zhang XC, Liu WZ, Li Z, Chen J (2011) Trend and uncertainty analysis of simulated climate change impacts with multiple GCMS and emission scenarios. Agric For Meteorol 151(10):1297–1304
155. Vrac M (2018) Multivariate bias adjustment of high-dimensional climate simulations: the rank resampling for distributions and dependences (R^2D^2) bias correction. Hydrol Earth Syst Sci 22(6):3175

Chapter 3
Response and Alternative Theories in Climate Change

Introduction by Guido Visconti

The response theory applied to study climate change has its origins in a couple of papers published by Cecil Leith in the 1970s. It is based on the Fluctuations Dissipation Theorem (FDT) whose classical application is to the Brownian motion. In this case, the random motion of a particle, in the fluid, is forced by the thermal movement of the molecules. During the motion the energy of the particle is dissipated by the viscosity of the fluid and converted to heat that contribute to maintain the temperature of the fluid. Following the words of Leith, "In this analogy the detailed motions of gas molecules correspond to the weather and the statistical properties of a gas such as temperature and pressure correspond to the climate" with some additional problems. For practical purposes climate is defined as the average of weather in some location and this must include the mean but also the standard deviation from the mean. Beside the mean must be carried out over some time interval and this is a crucial point. As a matter of fact, climate changes with time so that the interval must be long enough but not too long that would eliminate long term variations. In statistical mechanics such a problem is circumvented by recurring to the concept of ensemble mean. The ensemble is constituted by a large number (infinity?) of identical systems. This concept is not very practical to define climate because it would require studying a large number of Earths, subject to the same conditions including forcing. There is an alternative definition of climate as the distribution of the different variables (temperature, precipitation, etc.) with the weather being a sample from that distribution.

The problem was solved by Leith assuming that the system recovers from a small natural anomaly in the same way that it does from one induced by external forcing and this leads to expressing the sensitivity of the model (that not necessarily coincides with that of climate) as a matrix response function. As observed in a lucid paper by Thomas Bell some years later this matrix is of fundamental importance and in theory can be calculated but not "easily accessible to observation". Consequently, the only

G. Visconti (ed.), *Climate, Planetary and Evolutionary Sciences*, https://doi.org/10.1007/978-3-030-74713-8_3

way to apply FDT theorem to climate was to do everything in the model, that is, calculate the sensitivity of the model to different perturbations.

We have to consider however that the climate system, just because it is under perturbation, is not in a steady state and it is known that fluctuations in a perturbed system have different spectral character with respect to an unperturbed one. For these reasons FDT is not strictly applicable to such a system and the Linear Response Theory (LRT) was developed to solve this problem. In practice LRT boils down to calculate the Green function for the system and then it is applied to predict the signal from an assigned perturbation. The Green function can be calculated either from a single General Circulation Model (GCM) or from an ensemble of such models like those of CMIP (Coupled Model Intercomparison Project) and it is calculated as the time derivative of the mean response to an assigned perturbation. Usually reference is to an increase of 1% per year of CO_2 or to an abrupt doubling of the same greenhouse gas. Then the same function can be used to evaluate the climate signal for any perturbation.

It is to note the fundamentally LRT gives average value of the Global Mean Surface Temperature (GMST) so it is fine for evaluating the signal from an assigned scenario of greenhouse gas emission but cannot reproduce the details of a GCM run that predicts the behavior of several climate variables and their geographical distribution. Beside it assumes the GCM runs as reference neglecting all the philosophical and technical problems that plague GCMs. As a matter of fact, the results as far as temperature is concerned are quite reasonable even examining the latitudinal distribution. However, the results for precipitation are quite poor making clear that the method has not the necessary physical details and mechanisms that produce precipitation. However, the method remains very much suitable for all the exercises of the IPCC (Intergovernmental Panel on Climate Change) and actually LRT has been used also to develop a so called Stochastic Space State Model that combines emission scenario with the climate forcing introducing some stochastic forcing in both carbon dioxide and temperature. This produces not just the average value of temperature changes but the temperature distribution for the different scenarios.

This application of LRT is quite interesting but at the least now cannot predict the geographical distribution of the climate changes and so does not constitute an input for example to evaluate regional climate change. Sometimes you got the impression that theoretical (mathematical) work on climate change is taking revenge on several years of honest and neglected work of the GCM practitioners. There are some people that insist on putting climate studies in some sophisticated mathematical framework. This follows an old habit of the physicists that believe in the old reductionist practice.

Machine-Generated Summaries

Keywords: *Ensemble, parameter, error, space, projection, theory, ppe, member, approach, small, future, parameter space, surface, run, scale.*

Finding Plausible and Diverse Variants of a Climate Model.
Part 1: Establishing the Relationship Between Errors
at Weather and Climate Time Scales

https://doi.org/10.1007/s00382-019-04625-3

Abstract-Summary

In this first part, the extent to which climate biases develop at weather forecast timescales is assessed with two PPEs, which are based on 5-day forecasts and 10-year simulations with a relatively coarse resolution (N96) atmosphere-only model.

The study confirms more robustly than in previous studies that investigating the errors on weather timescales provides an affordable way to identify and filter out model variants that perform poorly at short timescales and are likely to perform poorly at longer timescales too.

The use of PPEs also provides additional information for model development, by identifying parameters and processes responsible for model errors at the two different timescales, and systematic errors that cannot be removed by any combination of parameter values.

Extended

In this first part, we build on ideas from previous studies to use model performance at short timescales (here 5 days) to filter the parameter space (Rodwell and Palmer [1]), and to use 5-day forecast errors to infer something about model errors at longer timescales (e.g. Ma and others [2]).

In part II, we show how this result can be exploited, in an application which is to select a number of model variants capable of providing plausible simulations of historical climate and diverse projections of future climate change.

Introduction

This paper is the first of two aimed at designing a "small" perturbed parameter ensemble (PPE) of plausible simulations based on a relatively expensive global climate model that can be used in producing climate projections for adaptation planning.

The second paper (Karmalkar and others [3]; hereafter Part II) focuses on a methodology that uses this relationship in model performance across weather and climate timescales to identify a small PPE of plausible simulations by screening out parameter combinations.

Some national projections like those from the Netherlands (van den Hurk and others [4]) or Australia (CSIRO and Bureau of Meteorology [5]) use the CMIP5 multimodel ensemble to represent modelling uncertainty, but there are advantages to also providing a PPE derived from a single climate model.

We want to achieve this with coupled ocean–atmosphere simulations but for this two part study we limit ourselves to atmosphere-only models to test the basic concept of screening out poorly performing parameter combinations whilst maintaining a diversity of credible process behaviour and future climate response.

Experimental Design and Elicitation

Design of a PPE first requires selection of a model configuration to perturb, then eliciting prior probability distributions for the parameters to perturb as chosen by the parameterisation experts, and finally deciding how to sample the parameters.

SHELF also allows for the elicitation to be completed after the meeting as long as the experts understand what is required of them and have experience with one example of a parameter already.

To elicit the plausible range for each parameter, the most important aspect is to explain to the experts that the simulations will be evaluated against a wide range of observational metrics so that (1) PPE members can be ruled out as implausible and (2) the final uncertainty quantification can be based on constrained parameter ranges.

Many parameter ranges were based on the experts' own analyses of very high resolution process models such as Large Eddy Simulations or Cloud Resolving Models.

Data and Methods

For TAMIP, in this study, we focus on evaluating the mean forecast error across the 16 start dates for day 2 and day 5 of the simulations.

Using the average errors across all 16 initial conditions makes the results more robust, though it limits us to relating the TAMIP MSEs to annual mean MSEs from the longer term ATMOS simulations.

The high correlations for surface air temperature and precipitation suggest that the error growth of each variable is largely due to the same parameters at days 2 and 5.

The day 5 errors for outgoing shortwave radiation in the clear sky are slightly less than the day 2 errors, but this comes from the Arctic and Antarctic where sea ice is fixed over the 5-day forecast to its initial value, and so should be considered a feature of the design of the TAMIP experiment.

Results

The similarity between the error patterns at the two timescales also applies for most variables and single model variants, with only a few variables (in particular 250 hPa eastward wind and specific humidity) showing correlations close to zero for an appreciable fraction of the ensemble.

The fraction might be reduced with smoother TAMIP patterns as described above, but until this is tested, the distributions for these variables show that the link between the two timescales is not robust across all parameter space, and so it cannot be assumed that the forecast errors are indicative of the climate biases.

The negative uncentred correlation in this region reflects the development of process errors on timescales longer than 5 days that lead to the change the sign of the ensemble mean bias noted above.

For model tuning where the search is only across a more focussed parameter space, there is still information in the patterns of 5-day forecast errors about the mean climate biases.

Discussion

The results show that this size of PPE, which is several times the number of parameters, offers new insights into the relationship between errors at different timescales and the underlying processes, and is potentially valuable for model tuning and prioritising which model errors need to be reduced by model development.

The impact of our results is discussed below in terms of the emergent relationships between errors on the two timescales and how the influences of the parameters affects these, followed by the implications for experimental design and then model development.

Strong emergent relationships exist between the model errors at 5-day and 5-year timescales, they can be exploited to inform the efficient design of a PPE suitable for predictions across multiple timescales.

It is very likely that good performance at the 5-day timescale would also be an important indicator of credibility of climate model projections of climate variability and extremes, although our simulations were not long enough to support investigation of such links in the present paper.

Acknowledgement

A machine generated summary based on the work of Sexton, D. M. H.; Karmalkar, A. V.; Murphy, J. M.; Williams, K. D.; Boutle, I. A.; Morcrette, C. J.; Stirling, A. J.; Vosper, S. B. (2019 in Climate Dynamics).

Finding Plausible and Diverse Variants of a Climate Model. Part II: Development and Validation of Methodology

https://doi.org/10.1007/s00382-019-04617-3

Abstract-Summary

Exploratory work towards developing a strategy to select variants of a state-of-the-art but expensive climate model suitable for climate projection studies.

The strategy combines information from a set of relatively cheap, idealized perturbed parameter ensemble (PPE) and CMIP5 multi-model ensemble (MME) experiments, and uses two criteria as the basis to select model variants for a PPE suitable for future projections: (a) acceptable model performance at two different timescales, and (b) maintaining diversity in model response to climate change.

This relationship is used to filter out parts of parameter space that do not give credible simulations of present day climate, while minimizing the impact on ranges in forcings and feedbacks that drive model responses to climate change.

We use statistical emulation to explore the parameter space thoroughly, and demonstrate that about 90% can be filtered out without affecting diversity in global-scale climate change responses.

This leads to the identification of plausible parts of parameter space from which model variants can be selected for projection studies.

Comparisons with the CMIP5 MME demonstrate that our approach can produce a set of plausible model variants that span a relatively wide range in model response to climate change.

Extended

This work in progress will be documented in future papers, underpinned by the proof of concept developments described here.

Introduction

Multi-model ensembles (MMEs) comprising climate model simulations carried out by various institutions all over the world (e.g. CMIP3, CMIP5 archive) have been used widely to provide a range in climate change projections (Meehl et al. [6]; Taylor et al. [7]).

As models become more sophisticated, for example, due to increases in their horizontal and vertical resolutions to improve representation of various aspects of climate variability and extremes (Scaife et al. [8]), creating a large ensemble for probabilistic projections becomes increasingly expensive.

The PPE is based on the atmospheric component of the Hadley Centre Global Environmental Model version 3 (HadGEM3-A; Hewitt et al. [9]) and is described in detail in Part I. This paper is heuristic in the sense that we describe a set of atmosphere-only PPE simulations and evaluation techniques capable of informing the subsequent definition of a climate projection system, but without progressing to the final step of evaluating our identified atmosphere model variants in coupled (AOGCM) simulations with a dynamic ocean component.

Principles of Methodology

The selection is based on the following criteria: (1) Assessment of model performance: The plausible variants must have satisfactory performance at Numerical Weather Prediction (NWP; 5 days) and climate (5 or more years) timescales. (2) Diversity in model response to climate change: The selected variants should explore the range of forcings and feedbacks from the entire plausible sub-region of parameter space identified in (1), as far as possible.

The seamless assessment approach, based on the idea that one can diagnose and characterize model errors by assessing performance at different timescales ranging from weather to climate, is very useful in this regard.

The spread in model responses to increasing GHGs is mainly determined by uncertainties in radiative forcings and climate feedback processes (Bony et al. [10]; Webb et al. [11]).

The success of a PPE in either matching or augmenting MME ranges in relevant aspects of climate response depends on the underlying model (Yokohata et al. [12]) and the experimental design.

Experimental Details

We used the Latin hypercube sampling technique (McKay et al. [13]), which ensures that the prior probability of each parameter is sampled evenly over the 21-dimensional parameter space, to create a 250-member ensemble that allows perturbing all 21 independent parameters simultaneously.

This was determined from the model development exercise for GA4.0 (Walters et al. [14]), and brought the total ensemble size to 251.

These AMIP-style experiments (Gates et al. [15]), which are forced at the ocean–atmosphere interface using observed estimates of SSTs and sea ice, are suitable for studying the impact of poorly constrained atmospheric and land surface parameters on uncertainties in the performance and response of the model.

In order to diagnose forcings and feedbacks, the second phase of the Cloud Feedback Model Intercomparison Project (CFMIP) proposed a set of idealized experiments (CFMIP-2 Experimental Design; Bony et al. [16]).

PPE Results

While a majority of variables show positive relationships over land and oceans, surface variables such as surface air temperature (tas) and downwelling longwave radiation at the surface (rlds) have much stronger correlations over land because cross-ensemble variations over the oceans are heavily constrained by the use of prescribed SSTs in both experiments.

We must include: (1) variables such as temperature and precipitation that are commonly used and important for understanding impacts of future climate change. (2) Variables that show strong relationships between TAMIP and ATMOS errors, which will allow us to find model variants that perform well at both time scales. (3) Variables that have a ratio of parametric uncertainty to structural uncertainty greater than 1, where the latter denotes the component of error that cannot be reduced by changing parameter values (Rougier [17]).

One of the most important variables, surface air temperature (tas) was not chosen, in spite of its TAMIP-ATMOS errors being correlated, because it shows relatively small spread across the ensemble (see Part I).

Filtering Out Parts of Parameter Space

Although our numbers of completed ensemble members (194 for TAMIP and 80 for ATMOS) are relatively small samples of a 21-dimensional parameter space, the Latin Hypercube design fills the space efficiently enough to allow us to build an emulator for the six assessment metrics at each of the two timescales to predict selected model output variables at untried combinations of parameter values.

We determine MSE-based tolerances to rule out parts of parameter space based on performance benchmarks relative to CMIP5, emulator predictions for model crashes and on maintaining ranges of the diversity metrics.

To quantify uncertainty associated with internal variability, we calculated the variance in MSE in a 16-member ATMOS ensemble of the standard version of the model produced by varying only the initial conditions. (4) Emulator error In addition to best-estimate predictions for the ATMOS and TAMIP MSE of untried parameter combinations, our emulators provide estimates of the associated uncertainties.

Plausible Model Variants: Selection and Validation

Once the parameter space is reduced to contain only acceptable and diverse variants of the atmosphere model, the challenge is to pick a small subset of variants—called 'plausible', suitable for (notional) use to provide climate change projections using the AOGCM configuration of the model.

The algorithm did have difficulty in picking model variants at the extremes of diversity metrics, due to the presence of relatively few acceptable variants to pick from.

The second criterion may appear counter-intuitive, since a more obvious aim might be to pick the best performing 50 model variants (subject to spanning a range of forcing and feedbacks), rather than variants that sample a range of performance across each of the individual assessment metrics.

Tests showed that without the second criterion, the 50 chosen variants would not include enough of the better performing models across each of the 12 assessment metrics.

Emergent Properties

AMIP-style experiments, where SSTs are not allowed to adjust to changes in the atmosphere, can potentially result in wide variations in the net TOA radiation flux, and provide a suitable design to expose the full consequences of atmospheric modelling errors on this metric.

Also a risk that excessively restricting the range of acceptable values, through comparison with a set of highly tuned multi-model results, could artificially restrict the range of outcomes consistent with uncertainties in the large set of processes that contribute to global energy balance (e.g., Collins et al. [18]).

Our discovery of a range of TOA net fluxes outside the CMIP5 range occurs because the overall ranges in net TOA fluxes and albedo for the 'acceptable' model variants are not reduced significantly compared to the emulated range across the full model parameter space.

Conclusions

The methodology—that includes (1) an assessment of model performance at weather and climate timescales for a variety of metrics, (2) the use of benchmarking information from structurally different models [specifically the CMIP5 multi-model ensemble (MME)] and (3) the maintenance of diversity in forcings and feedbacks— allows us to reduce significantly the prior parameter space specified by modelling experts to a sub-region suitable for the selection of ensemble projection system members.

The seamless assessment approach, in particular, shows that the large parameter space can be efficiently explored by running the climate model in weather forecast mode using 5 day "Transpose AMIP" (TAMIP) experiments, in conjunction with the statistical technique of emulation.

In simulations of present day climate, the PPE explores the ranges of skill spanned by the CMIP5 MME for most key climate variables, often with a few model variants better than the best performer from CMIP5 and a few variants worse than the worst CMIP5 performer.

Discussion

The presence of strong relationships between weather and climate errors for many variables will enable us to use inexpensive NWP hindcasts as an efficient way of pre-screening the parameter space to exclude parts giving rise to physically unrealistic model behavior, before investing in longer climate simulations either in atmospheric or coupled mode.

While an initial assessment of seasonal mean errors showed strong relationships with their annual mean counterparts, this does not necessarily imply that seasonal errors could not play a useful role in refining future assessments of model performance.

We also build emulators separately for TAMIP and ATMOS runs, but given that there is a strong relationship between model performance across weather and climate timescales, it may be more optimal to consider building emulators in future work that linked the two timescales.

Acknowledgement

A machine generated summary based on the work of Karmalkar, Ambarish V.; Sexton, David M. H.; Murphy, James M.; Booth, Ben B. B.; Rostron, John W.; McNeall, Doug J. (2019 in Climate Dynamics).

Multivariate Probabilistic Projections Using Imperfect Climate Models Part I: Outline of Methodology

https://doi.org/10.1007/s00382-011-1208-9

Abstract-Summary

This method combines information from a perturbed physics ensemble, a set of international climate models, and observations.

This is important if different sets of impacts scientists are to use these probabilistic projections to make coherent forecasts for the impacts of climate change, by inputting several uncertain climate variables into their impacts models.

Unlike a single metric, multiple metrics reduce the risk of rewarding a model variant which scores well due to a fortuitous compensation of errors rather than because it is providing a realistic simulation of the observed quantity.

The method also has a quantity, called discrepancy, which represents the degree of imperfection in the climate model i.e. it measures the extent to which missing processes, choices of parameterisation schemes and approximations in the climate model affect our ability to use outputs from climate models to make inferences about the real system.

Discrepancy also provides a transparent way of incorporating improvements in subsequent generations of climate models into probabilistic assessments.

The set of international climate models is used to derive some numbers for the discrepancy term for the perturbed physics ensemble, and associated caveats with doing this are discussed.

Introduction

Perturbed physics ensembles (PPEs) provide an alternative strategy for exploring uncertainty in climate modelling (Murphy et al. [19]; Stainforth et al. [20]; Webb et al. [21]; Yokohata et al. [12]), by generating ensembles where each member differs from the standard version of a climate model by having a different set of values for the model parameters.

Murphy et al. [19] used an interpolation technique so that the model variants sampled by the PPE were used to predict the climate sensitivity and the relative skill in simulating some observable aspects of the climate system (in that case, fields of multiannual means of multiple climate variables) for untried points in parameter space.

The method used by Murphy et al. [19] demonstrates several features: use of probability to represent uncertainty; emulation, a technique in which a statistical model is trained on a PPE and then used to predict the output of untried model variants; using observations to constrain the probabilistic projection to higher quality parts of parameter space.

Data

These were the model parameters, observations, model output that corresponds to the observations and prediction variables, the true climate (of which the observations and the model output are uncertain estimates), and the discrepancy which is a link between the model output and the true climate.

Webb et al. [21] used the data from the first stage to choose 128 members that the method of Murphy et al. [19] predicted to be relatively credible model variants that spanned a wide range of parameter space and climate sensitivity.

As the spread of the PPE at larger spatial scales is generally much greater than internal variability, the leading eigenvectors are mainly driven by the changes to model parameters and are therefore representative of the major changes in the physics of the climate model across the PPE.

This is estimated from a 600-year long control integration of HadSM3, the model variant with standard values for the parameters (Barnett et al. [22]).

Outline of the Calculations

The first stage of any Bayesian analysis is the specification of the uncertain objects which make up the joint probability distribution e.g. the model data, the observations.

The second term in the integrand is called the likelihood function of x given some observed values and is equal to the probability of obtaining the observed values given those values of x. The third term is the prior distribution of the values for the model parameters.

The general problem of predicting several model outputs constrained by several observations requires an emulator for each element in o, and for each prediction variable in y_f.

The close relationship between the emulated and modelled values is not guaranteed as some PPEs sample parameter space in a way that is very different to a prior distribution of where the best input is.

Specification of Discrepancy Distribution

In searching for points in the HadSM3 parameter space which best match the physics of a multimodel ensemble member, it is important to base the search for analogues on a wide range of climate variables, in order to reduce the risk that a fortuitous match could be found through a compensation of errors.

For each multimodel ensemble, we used four different points in parameter space from the initial 100,000, chosen randomly from the leading good fits of the initial sample, and estimate four different best analogues.

Variations within the set of four analogues for each multimodel ensemble member are small compared to variations between members, though there are examples (e.g. when attempting to find analogues for the UIUC model), which confirm the importance of sampling initial conditions in the optimisation of the best analogues.

Results and Discussion

For $N_h = 6$, 93% of the sampled values had lower probability density than the actual observed values, indicating that the joint prior distribution, which combines climate model data, emulators, parametric uncertainty, the discrepancy, and the choices we make like the number of dimensions used to represent the observed quantities, compares adequately with the actual observed values and we do not expect strong sensitivity in the results or "surprises" as O'Hagan and Forster [23] call them.

By removing the historical discrepancy, fewer sampled points receive a relatively high weight so that the effective sample size becomes smaller by a factor of 4; this leads to a less smooth PDF which underestimates the range of climate sensitivity in comparison with the full posterior distribution, and would lead to an increased risk of poor decisions based on this PDF.

Conclusions

We have simply allowed the multimodel data to determine the relationship between historical and future discrepancy, with the unavoidable caveat that structural errors common to all current climate models are not included.

We believe that including a defensible estimate of discrepancy leads to a more realistic quantification of prediction uncertainties, and allows us to obtain an improved estimate of the spread of possible future climate outcomes consistent with current modelling technology and understanding of climate feedback processes, because we have combined information from a PPE and a multimodel ensemble.

The final advantage is that the framework, especially the emulator, allows us to assess the robustness of our results to a number of key methodological choices, including the prior distribution of model parameters, discrepancy, the set of multi-model ensemble members, and the choice of the observational metrics used to constrain the prediction.

Acknowledgement

A machine generated summary based on the work of Sexton, David M. H.; Murphy, James M.; Collins, Mat; Webb, Mark J. (2011 in Climate Dynamics).

Multivariate Probabilistic Projections Using Imperfect Climate Models. Part II: Robustness of Methodological Choices and Consequences for Climate Sensitivity

https://doi.org/10.1007/s00382-011-1209-8

Abstract-Summary

A method for providing probabilistic climate projections, which applies a Bayesian framework to information from a perturbed physics ensemble, a multimodel ensemble and observations, was demonstrated in an accompanying paper.

This information allows us to account for the combined effects of more sources of uncertainty than in any previous study of the equilibrium response to doubled CO_2 concentrations, namely parametric and structural modelling uncertainty, internal variability, and observational uncertainty.

Such probabilistic projections are dependent on the climate models and observations used but also contain an element of expert judgement.

Two expert choices in the methodology involve the amount of information used to (1) specify the effects of structural modelling uncertainty and (2) represent the observational metrics that constrain the probabilistic climate projections.

We are therefore confident that, despite sampling sources of uncertainty more comprehensively than previously, the improved multivariate treatment of observational metrics has narrowed the probability distribution of climate sensitivity consistent with evidence currently available.

The main caveat is that the handling of structural uncertainty does not account for systematic errors common to the current set of climate models and finding methods to assess the impact of this provides a major challenge.

Introduction

The method in Part 1 uses a Bayesian framework based on Rougier [17] where a joint probability distribution is constructed to contain probabilistic information about the uncertain objects in the climate projection problem: model parameters; observations; the true climate, consisting of the future that we want to predict and the past which we can compare with the observations; and model output, corresponding to our choice of observed climate variables and also variables we want to predict.

By assuming that this set of structural differences are exchangeable with the structural differences of our climate model with the real system i.e. they are effectively sampled from the same distribution, we can pool these prediction errors over the multimodel ensemble and use them to inform the mean and covariance of the discrepancy term.

Effect of Dimensionality Used to Represent Historical Climate

ESS is a measure of how effectively the observational information restricts the prior parameter space to regions of parameter space that are consistent with the observations used to constrain the PDF; so if all weight was assigned to one sample, ESS would be 1 (though this would be a strong indication that the posterior PDF would not be robust if the full sample was repeated); if all samples were assigned equal weight, ESS would be equal to the sample size, indicating no constraint at all from the observations.

For the observational constraint, an increase in N_h makes it harder to randomly select a point in parameter space that is a reasonable match to the observed values of all N_h historical eigenvectors, making the weights less evenly distributed, and so ESS decreases.

This indicates that interactions between estimated model errors (obtained by projecting emulated and observed values of relevant climate variables onto our N_h eigenvectors) and the off-diagonal terms in the discrepancy covariance matrix (representing relationships between the structural component of model errors in different variables) can play a significant role in determining variations in the weights across parameter space, and hence the ESS.

Sensitivity Tests

We check the sensitivity of our probabilistic climate projections to a number of subjective choices that affect the expert prior probability on the model parameters, the discrepancy term, and number of eigenvectors of historical climate variables used to estimate the relative likelihood of points in parameter space.

Based on these sensitivity tests, the median of the posterior PDF of climate sensitivity is between 3.2 and 3.3 K, with the 5th percentile between 2.2 and 2.4 K, and the 95th percentile is between 4.1 and 4.5 K. We can make a direct comparison between our results and those of our previous study Murphy et al. [19].

Considering first the prior PDFs, Murphy et al. [19] used a uniform sampling of parameter space, and found a 95th percentile of the prior PDF of climate sensitivity of 5.3 K. This compares to 5.0 K for the prior PDF in the present study, when an equivalent uniform sampling is tried as one of our sensitivity tests.

Conclusions

To make an expert assessment about the likely range of equilibrium climate sensitivity, Meehl et al. [24] used PDFs of climate sensitivities from two main categories of study (see their Box 10.2 for details).

This is because the AR4 assessment included evidence based on observational constraints offered by past climate change (the first category identified above), which is not considered in our study.

Extensions of our approach to include constraints based on historical climate change are feasible, and offer the prospect of a transparent, quantitative and testable synthesis of much of the evidence from both major categories assessed in AR4.

In our method, the prior knowledge can be based on expert elicitation (e.g. the prior distribution for the model parameters), or on our judgement (e.g. that the other climate models sample a distribution of climate processes not explored by the variants of HadSM3, and so provides a meaningful way to inform the discrepancy term).

Acknowledgement

A machine generated summary based on the work of Sexton, David M. H.; Murphy, James M. (2011 in Climate Dynamics).

Climate Model Errors, Feedbacks and Forcings: A Comparison of Perturbed Physics and Multi-Model Ensembles

https://doi.org/10.1007/s00382-010-0808-0

Abstract-Summary

Ensembles of climate model simulations are required for input into probabilistic assessments of the risk of future climate change in which uncertainties are quantified.

Model-error characteristics derived from time-averaged two-dimensional fields of observed climate variables indicate that the perturbed physics approach is capable of sampling a relatively wide range of different mean climate states, consistent with simple estimates of observational uncertainty and comparable to the range of mean states sampled by the multi-model ensemble.

The perturbed physics approach is also capable of sampling a relatively wide range of climate forcings and climate feedbacks under enhanced levels of greenhouse gases, again comparable with the multi-model ensemble.

Extended

The perturbed physics ensembles described here, together with others documented elsewhere, are combined with a statistical emulator of the model parameter space (see e.g., Rougier and others 25 for an example) and a "time-scaling" technique (Harris and others [26] which maps equilibrium to transient responses taking into account any errors that may arise because of a mismatch between the patterns of transient and equilibrium.

The perturbed physics approach can sample a wide range of different model "errors" in two-dimensional time-averaged climate fields for a number of different variables that for many variables are comparable with uncertainties in the observations and comparable with the errors in the members of the multi-model archive.

The perturbed physics approach can sample a wide range of global-mean feedbacks under climate change.

Introduction

The main motivation for this paper is to document the design and characteristics of a number of perturbed physics ensembles that have been produced as part of an extensive programme of research at the Met Office Hadley Centre to produce regional climate projections (e.g. Murphy and others [27, 28]) and to contrast aspects of those perturbed physics ensembles with corresponding multi-model ensembles.

We might naively assume that the multi-model ensemble contains members with a wide range of different error characteristics, whereas the perturbed-physics approach produces members with very similar baseline climates and thus very similar errors.

We know that the perturbed physics approach is capable of producing model variants with a wide range of different feedbacks strengths under climate change (e.g. Webb and others [21]; Sanderson and others [29]).

Question 4 is highly relevant when we use ensembles of climate model projections to generate predictions of climate change expressed in terms of PDFs which provide a measure the uncertainty (or credibility) in that prediction.

Climate Model Ensembles and Variables

For more complex versions of the model (e.g. using a dynamical ocean component rather than a mixed-layer, q-flux or slab component) fewer ensemble members are possible because of the extra resources required to spin-up model versions and run scenario experiments.

In the design of the ensemble, an attempt was made to minimise the average of the root mean squared error of a number of time-averaged model fields while sampling a wide range of surface and atmospheric feedbacks under climate change.

The model versions are therefore suitable for quantifying uncertainty and examining feedbacks, etc. This ensemble uses the fully coupled version of HadCM3 but with perturbations only to parameters in the atmosphere component (an updated version of the ensemble described in Collins and others [18]).

For historical reasons, the sea-ice scheme in HadCM3 is contained in the atmosphere component of the model and parameters in the scheme are perturbed in line with the equivalent S-PPE-M ensemble.

Model "Errors"

In the slab-ocean multi model ensemble, S-MME, we see a similar range of land SAT biases as in the case of the perturbed physics ensembles, but a somewhat wider range of RMS errors.

Both SST bias and RMS errors are of a similar magnitude in slab-ocean perturbed physics and multi model ensembles and are in many cases smaller than those errors seen in the non-flux-adjusted CMIP3 coupled models (AO-MME).

Global mean biases in precipitation in the slab-model ensembles follow a similar pattern to those in global land surface air temperature and SST in the different ensembles, except that the S-MME has a relatively wider range of biases than any of the other slab-ocean perturbed physics ensembles.

For the surface sensible heat flux, the range of both biases and RMS errors is generally smaller in the perturbed physics ensembles in comparison with the multi-model ensembles.

Feedbacks and Forcings

In the case of the AO-PPE-O ensemble, with identical HadCM3 atmosphere components but perturbations to parameters in the ocean model, there is a similarly small spread.

Despite the fact that the volcanic forcing time series of stratospheric optical depth is precisely the same in each member of the perturbed physics ensemble, the spread in total negative volcanic radiative forcing is comparable with the spread in the multi-model case in which different input forcing data are used.

LW forcing in 1995–2004 is centred around 2.4 W m^{-2} in both the multi-model and perturbed-physics ensembles, with a range of 1.5–3.1 W m^{-2} in the AO-MME case and a smaller range of 2.1–2.7 W m^{-2} in the AO-PPE-A case (in both cases the range is greater than would be expected from natural variability).

Relating Model Errors to Feedbacks

Having examined model errors and climate change feedbacks in the multi-model and perturbed physics ensembles, we now examine the relationships between them.

To improve models we need to know how to target research to do this, i.e., by quantifying the relationship between error and climate feedback, we may learn which improvements to different aspects of the model simulations will lead to the most progress in reducing uncertainty in predictions.

The only variable in which there is a reasonably high correlation between errors and feedbacks in both perturbed physics and multi model ensembles are the biases in the global mean cloud amount (coefficients around 0.6–0.7, see also Yokohata and others [30]).

For the perturbed physics ensembles there are weak to moderately strong correlations for a number of variables suggesting that the combination of those (and other) variables into a single metric would be a way of constraining the climate feedback parameter.

Discussion and Conclusions

The perturbed physics approach can sample a wide range of different model "errors" in two-dimensional time-averaged climate fields for a number of different variables that for many variables are comparable with uncertainties in the observations and comparable with the errors in the members of the multi-model archive.

It is possible to produce quite different baseline climates with the perturbed physics approach such that the ensemble-mean appears as the "best" model in comparison with any individual ensemble member.

For regional measures, and for variables not examined here such as variability or extremes, there may be differences between perturbed physics and multi model ensembles which do not fit with these general conclusions.

In our companion work on producing probabilistic climate change projections, we combine perturbed physics and multi-model ensemble information together with observations and estimates of uncertainty in observations to produce projects based on as much information about the climate system as is possible (Murphy and others [27, 28]).

Acknowledgement

A machine generated summary based on the work of Collins, Matthew; Booth, Ben B. B.; Bhaskaran, B.; Harris, Glen R.; Murphy, James M.; Sexton, David M. H.; Webb, Mark J. (2010 in Climate Dynamics).

Predicting Climate Change Using Response Theory: Global Averages and Spatial Patterns

https://doi.org/10.1007/s10955-016-1506-z

Abstract-Summary

The provision of accurate methods for predicting the climate response to anthropogenic and natural forcings is a key contemporary scientific challenge.

Response theory allows one to practically compute the time-dependent measure supported on the pullback attractor of the climate system, whose dynamics is non-autonomous as a result of time-dependent forcings.

We assess strengths and limitations of the response theory in predicting the changes in the globally averaged values of surface temperature and of the yearly total precipitation, as well as in their spatial patterns.

We also show how it is possible to define accurately concepts like the inertia of the climate system or to predict when climate change is detectable given a scenario of forcing.

Extended

Response theory allows to practically compute such a time-dependent measure starting from the invariant measure of a suitably chosen reference autonomous dynamics.

Introduction

One needs to consider that the study of climate faces, on top of all the difficulties that are intrinsic to any nonequilibrium system, the following additional aspects that make it especially hard to advance its understanding: the presence of well-defined subdomains—the atmosphere, the ocean, etc. —featuring extremely different physical and chemical properties, dominating dynamical processes, and characteristic time-scales; the complex processes coupling such subdomains; the presence of a continuously varying set of forcings resulting from, e.g., the fluctuations in the incoming solar radiation and the processes—natural and anthropogenic—altering the atmospheric composition; the lack of scale separation between different processes, which requires a profound revision of the standard methods for model reduction/projection to the slow manifold, and calls for the unavoidable need of complex parametrization of subgrid scale processes in numerical models; the impossibility to have detailed and homogeneous observations of the climatic fields with extremely high-resolution in time and in space, and the need to integrate direct and indirect measurements when trying to reconstruct the past climate state beyond the industrial era; the fact that we can observe only one realization of the process.

Pullback Attractor and Climate Response
After a sufficiently long time, related to the slowest time scale of the system, at each instant the statistical properties of the ensemble of simulations do not depend anymore on the choice of the initial conditions.

A prominent example of this procedure is given by how simulations of past and historical climate conditions are performed in the modeling exercises such as those demanded by the IPCC [31, 32], where time-dependent climate forcings due to changes in greenhouse gases, volcanic eruptions, changes in the solar irradiance, and other astronomical effects are taken into account for defining the radiative forcing to the system.

In order to construct the time dependent measure following directly the definition of the pullback attractor, we need to construct a different ensemble of simulations for each choice of $F(x, t)$.

In other terms, from the knowledge of the time dependent measure of one specific pullback attractor, we can derive the time dependent measures of a family of pullback attractors.

A Climate Model of Intermediate Complexity: The Planet Simulator—PLASIM
A detailed study of the impact of changing oceanic heat transports on the dynamics and thermodynamics of the atmosphere can be found in [33].

We remark that previous analyses have shown that using a spatial resolution approximately equivalent to T21 allows for obtaining an accurate representation of the major large scale features of the climate system.

While the lack of a dynamical ocean hinders the possibility of having a good representation of the climate variability on multidecadal or longer timescales, the climate simulated by PLASIM is definitely Earth-like, featuring qualitatively correct large scale features and turbulent atmospheric dynamics.

We are confident of the thermodynamic consistency of our model, which is crucial for evaluating correctly the climate response to radiative forcing resulting from changes in the opacity of the atmosphere.

Results
One expects that coarse grained (in space) quantities will have a better signal-to-noise ratio and will allow for performing higher precision climate projection using response theory.

We will begin by looking into globally averaged quantities, and then address the problem of predicting the spatial patterns of climate change.

We dedicate some additional care in studying the climate response in terms of changes in the globally averaged surface temperature.

We would like to be able to assess when not only the projected change in the ensemble average is distinguishable from the statistics of the control run, but, rather, when an actual individual simulation is incompatible with the statistics of the unperturbed climate, because we live in one of such realizations, and not on any averaged quantity.

The methods of response theory allow us to treat seamlessly also the problem of predicting the climate response for (spatially) local observables.

A Critical Summary of the Results

The performance of response theory in predicting the change in the globally averaged surface temperature and precipitation is rather good at all time horizons, with the predicted response falling within the ensemble variability of the direct simulations for all time horizons except for a minor discrepancy in the time window 40–60 y. Additionally, our results confirm the presence of a strong linear link in the form of modified Clausius–Clapeyron relation between changes in such quantities, as already discussed in the literature.

Response theory provides an excellent tool also for predicting the change in the zonal mean of the surface temperature, except for an underestimation of the warming in the very high latitude regions in the time horizons of 40–60 y. This is, in fact, the reason for the small bias found already when looking at the prediction of the globally averaged surface temperature.

Challenges and Future Perspectives

The ab-initio construction of the linear response operator has proved elusive because of the difficulties associated with dealing effectively with both the unstable and stable directions in the tangent space.

What is extremely interesting about BVs is that (1) their growth factors are strongly dependent on the region of the phase space where the system is; and (2) the choice of the reference norm of the perturbation and of the time interval between two successive renormalization procedures (breeding period) effects strongly the properties of the dominant instabilities specifically active on the chosen time scales.

Of meteo-climatic relevance it has been shown that a relatively low number of BVs is extremely effective for reconstructing the properties of the unstable space, and that BVs contain useful information on spatially localized features, so that it may be worth trying to construct an approximation to the Ruelle response operator using the BVs.

Using such results in a reduced state space might provide a novel and effective method for approaching the problem of climate response.

Acknowledgement

A machine generated summary based on the work of Lucarini, Valerio; Ragone, Francesco; Lunkeit, Frank (2016 in Journal of Statistical Physics).

Beyond Forcing Scenarios: Predicting Climate Change Through Response Operators in a Coupled General Circulation Model

https://doi.org/10.1038/s41598-020-65297-2

Abstract-Summary

Global Climate Models are key tools for predicting the future response of the climate system to a variety of natural and anthropogenic forcings.

We show how to use statistical mechanics to construct operators able to flexibly predict climate change.

We perform our study using a fully coupled model—MPI-ESM v.1.2—and for the first time we prove the effectiveness of response theory in predicting future climate response to CO_2 increase on a vast range of temporal scales, from inter-annual to centennial, and for very diverse climatic variables.

The change in the Atlantic Meridional Overturning Circulation (AMOC) and of the Antarctic Circumpolar Current (ACC) is accurately predicted.

We are able to predict accurately the temperature change in the North Atlantic.

Introduction

Global climate models (GCMs) are currently the most advanced tools for studying future climate change; their future projections are key ingredients of the reports of the Intergovernmental Panel on Climate Change (IPCC) and are key for climate negotiations [34].

For IPCC-class GCMs, future climate projections are usually constructed by defining a few climate forcing scenarios, given by changes in the composition of the atmosphere and in the land use, each corresponding to a different intensity and time modulation of the equivalent anthropogenic forcing.

No rigorous prescription exists for translating the climate change projections if one wants to consider different time modulations of a given forcing, e.g. a faster or slower CO_2 increase.

Response Theory and Climate Change

The FDT has recently been key to inspiring the theory of emergent constraints, which are tools for reducing the uncertainties on climate change by looking at empirical relations between climate response and variability of some given observables [35, 36].

Response theory is a generalisation of the FDT that allows one to to predict how the statistical properties of general—near or far from equilibrium, deterministic or stochastic—systems change as a result of forcings.

Encouragingly, response theory has recently been shown to have a great potential for predicting climate change in multi-model ensembles of CMIP5 atmosphere–ocean coupled GCMs outputs [37].

The response of a a slow (oceanic) climatic observable of interest has been investigated so far in relation to the change in the dynamical properties of some other climatic observable, by constructing a linear regression between the predictand and predictor using the properties of the natural variability of the system [38–41].

Predicting Climate Change Using the Ruelle Response Theory

We then focus on two key aspects of the large-scale ocean circulation, namely the Atlantic Meridional Overturning Circulation (AMOC) [42, 43] and the Antarctic Circumpolar Current (ACC) [44], and show that we can achieve excellent skill in predicting the the slow modes of the climate response.

In current conditions, the ocean is well-known to absorb a large fraction of the Earth's energy imbalance due to global warming and to store it through its large thermal inertia, up to time scales defined by the deep ocean circulation [45].

Results

Of the presence of slow oceanic time scales, the Green function significantly departs from a simple exponential relaxation behavior, which is sometimes adopted to describe the relaxation of the climate system to forcings [46, 47].

On short time scales, we have a reduction of AMOC, as a result of the negative value of the Green function.

On longer time scales (>100 y), a negative feedback acts as a restoring mechanism, associated with a positive sign in the Green function.

On decadal scales, we have a loss in the correlation between wind stress and ACC, corresponding to the Green function turning negative after about 30 y. Beyond these time scales, we have time-wise coherent response of the AMOC and ACC, underlying the response of the global ocean circulation.

This has profound implications for setting the time scales of the ACC and AMOC response.

Discussion and Conclusions

In all considered cases response theory successfully predicts the time-dependent change.

Ruelle's response theory provides a relatively simple yet robust and powerful set of diagnostic and prognostic tools to study the response of climatic observables to external forcings.

The availability of a large number of ensemble members allows for constructing more accurate Green functions and for studying effectively the response of a broader class of climatic observables.

A promising application is the definition of functional relations between the response of different observables of a system to forcings, in the spirit of some recent investigations (see, e.g. Zappa and others [48]).

Being based on a perturbative approach, response theory (linear and nonlinear) has, by definition, only a limited range of applicability (e.g. one cannot use it to treat arbitrarily strong forcings).

Appendix A: Methods

The susceptibility gives a spectroscopic description of the properties of the response of the observable, and its analysis can give interesting information on the most relevant time scales and related processes that determine the response of the observable.

From the Green function, one could in principle compute the susceptibility and perform a spectral analysis of the properties of the response.

This translates into the fact that, despite the Green function and the susceptibility being strictly connected, to obtain a satisfactory estimate for the latter require a statistics orders of magnitude larger than for the former, and possibly different and dedicated numerical estimation approaches [49–51].

An analysis of the susceptibility in experiments similar to what done in this work was attempted in Ragone and others [52], but using ten times more ensemble members.

While the analysis of the detailed frequency response of a climate model remains a very interesting and promising topic, it has to likely wait until experiments with at least several hundreds ensemble members will be available.

Acknowledgement

A machine generated summary based on the work of Lembo, Valerio; Lucarini, Valerio; Ragone, Francesco (2020 in Scientific Reports).

Improving Prediction Skill of Imperfect Turbulent Models Through Statistical Response and Information Theory

https://doi.org/10.1007/s00332-015-9274-5

Abstract-Summary

Statistical uncertainty quantification (UQ) to the response to the change in forcing or uncertain initial data in such complex turbulent systems requires the use of imperfect models due to the lack of both physical understanding and the overwhelming computational demands of Monte Carlo simulation with a large-dimensional phase space.

The systematic development of reduced low-order imperfect statistical models for UQ in turbulent dynamical systems is a grand challenge.

The forty mode Lorenz 96 (L-96) model which mimics forced baroclinic turbulence is utilized as a test bed for the calibration and predicting phases for the hierarchy of computationally cheap imperfect closure models both in the full phase space and in a reduced three-dimensional subspace containing the most energetic modes.

For reduced-order model for UQ in the three-dimensional subspace for L-96, the systematic low-order imperfect closure models coupled with the training strategy provide the highest predictive skill over other existing methods for general forced response yet have simple design principles based on a statistical global energy equation.

The systematic imperfect closure models and the calibration strategies for UQ for the L-96 model serve as a new template for similar strategies for UQ with model error in vastly more complex realistic turbulent dynamical systems.

Introduction

A conceptual framework intermediate between detailed dynamical physical modeling and purely statistical analysis based on empirical information theory has been proposed (Majda and Gershgorin [53, 54]; Gershgorin and Majda [55]) to address imperfect model fidelity and sensitivity problems.

In Majda and Gershgorin [56], a direct link by utilizing fluctuation–dissipation theorem (FDT) for complex systems together with the framework of empirical information theory for improving imperfect models is developed.

We investigate and develop systematic strategies for improving the imperfect model prediction skill for complex turbulent dynamical systems by employing ideas in both the information-theoretic framework and linear response theory mentioned above.

Following the direct link between the linear response and empirical information theory demonstrated in Majda and Gershgorin [56] for models with equilibrium fidelity, it is shown that they can be seamlessly combined into a precise systematic framework to improve imperfect model sensitivity through measuring the information error of the linear response operator in the training phase with unperturbed statistics.

Theories for Improving Imperfect Model Prediction Skill

The information theory offers a least biased measure for quantifying the error between the imperfect model prediction and the truth; and the linear response theory gives an important tool relating the model responses to stationary state statistics of the dynamical system.

With the help of these theories (Majda and Gershgorin [53, 56]), one systematical process to tune model parameters in a training phase to possibly achieve the optimal model with sensitivity to all kinds of perturbations is discussed.

It is reasonable to claim that an imperfect model with precise prediction of this linear response operator should possess uniformly good sensitivity to different kinds of perturbations.

Considering all these good features of the linear response operator, information barrier due to model sensitivity to perturbations can be overcome by minimizing the information error in the imperfect model kicked response distribution relative to the true response from observation data (Majda and Gershgorin [56]).

L-96 System as a Test Bed and Its Statistical Dynamics

To quantify these uncertainties, we are interested in resolving the statistical features of this dynamical system, especially the first two-order moments.

The equations will never be closed under this process by calculating the dynamical equations for each order moments.

Even though we derive the moment equations above from the L-96 system under homogeneous assumption for the sake of analysis and will focus on them in the following discussions, the moment equations are actually quite representative and are easy to be extended to general nonlinear systems with conservative quadratic forms.

We can focus on the simplified moment equations and investigate the statistical properties inside this system.

Pointwise statistics by only considering the variance at each grid point, and ignoring the correlations between different grids may not be sufficient for accurate model predictions.

This will end up with large information barrier when only one-point statistics are considered in the imperfect model.

Statistical Closure Methods in Full Phase Space
The imperfect model prediction skill as well as the improvement through the information–response framework will be compared through checking the models' ability to capture the responses to several different types of perturbed external forcing terms.

The dynamical imperfect model using the closure method offers more precise prediction for the nonlinear responses for both the mean and variance.

Two rows, we show the model outputs for the mean and total variance with closure methods GC1, GC2, and MQG compared with the truth from Monte Carlo simulation.

The model prediction skill increases as more and more detailed calibration about the nonlinear flux are proposed from GC1 to GC2, MQG.

In the signal part, the error in the mean can be minimized to small amount for all three models under optimal parameter; while for dispersion part, MQG and GC2 have much better prediction for the prediction in variance compared with GC1.

Low-Order Models in a Reduced Subspace
Keeping all these shortcomings in mind, we propose the following further corrections to the reduced-order methods and refer the resulting model as the corrected model.

We are interested in checking whether these correction strategies for the reduced methods can actually improve the model prediction skill.

In the full space case, we need to first tune the reduced model parameters in a training phase for optimal responses.

Observing the errors in signal and dispersion part separately for the original model in the second row, it can be found that large inherent information barrier (especially for the mean prediction) exists for improving the model prediction skill no matter how well we tune the parameters in the training phase.

GC2 can even offer better prediction in the reduced-order case than ROMQG model considering that it is also cheaper in computation.

Conclusion and Future Work

Several important points can be concluded from the theoretical analysis and numerical tests using the L-96 test bed: The second-order statistical closure models outperform the linear FDT predictions for capturing responses to external perturbations, especially in regimes with larger perturbations and stronger nonlinearities.

This is an important result showing that higher-order moments can be determined by the lower-order approximations and offers important guideline for designing imperfect closure schemes; Still accurate single-point statistics prediction is not sufficient for the imperfect models to break information barriers (Majda and Gershgorin [53, 56]; Majda and Branicki [57]).

Imperfect model prediction skill can be improved uniformly regardless of the specific perturbation form applied; It is important for practical applications that the information–response framework can also be applied systematically to reduced-order models which focus on capturing the uncertainties in the dominant modes.

Acknowledgement

A machine generated summary based on the work of Majda, Andrew J.; Qi, Di (2015 in Journal of Nonlinear Science).

Book Reading List

An Introduction to the Theory of Climate

By Monin, A.S. *(1986)*.

During the last 20 years the study of, and the prediction of, changes in the climate of our planet have become an urgent social imperative, addressed to scientists the world over. The first principles on which to base such a study were formulated in 1974 in Stockholm, at the international GARP conference on the physical fundamentals of climate theory and climate modeling. In 1979 the World Meteorological Organization and the International Council of Scientific Unions decided to conduct a global program of climate research.

Please see https://www.springer.com/gp/book/9789027719355 for original source.

Non-equilibrium Thermodynamics and the Production of Entropy

By Kleidon, A. (Ed), Lorenz, R. D. (Ed) *(2005)*.

The present volume studies the application of concepts from non-equilibrium thermodynamics to a variety of research topics. Emphasis is on the Maximum Entropy Production (MEP) principle and applications to Geosphere-Biosphere couplings. Written by leading researchers form a wide range of background, the book proposed to give a first coherent account of an emerging field at the interface of thermodynamics, geophysics and life sciences.

Please see https://www.springer.com/gp/book/9783540224952 for original source.

Theory and Practice of Climate Adaptation

By Alves, F. (Ed), Leal Filho, W. (Ed), Azeiteiro, U. (Ed) *(2018).*

Climate change is one of the greatest challenges of our time. As such, both the Fifth Assessment Report (AR5) released by the Intergovernmental Panel on Climate Change (IPCC) and the 25th Conference of the Parties (COP 25) recommendations call for action not only from government, but also from various stakeholders. Apart from the knowledge offered by modeling and forecasts, which allows the readers to understand the problem and how it is likely to develop in the future, the book highlights approaches, methods and tools that can help readers cope with the social, economic and political problems posed by climate change.

Please see https://www.springer.com/gp/book/9783319728735 for original source.

Topics in Geophysical Fluid Dynamics: Atmospheric Dynamics, Dynamo Theory, and Climate Dynamics

By Ghil, M., Childress, S. *(1987).*

The vigorous stirring of a cup of tea gives rise, as we all know, to interesting fluid dynamical phenomena, some of which are very hard to explain. In this book our "cup of tea" contains the currents of the Earth's atmosphere, oceans, mantle, and fluid core. Our goal is to understand the basic physical processes which are most important in describing what we observe, directly or indirectly, in these complex systems. While in many respects our understanding is measured by the ability to predict, the focus here will be on relatively simple models which can aid our physical intuition by suggesting useful mathematical methods of investigation.

Please see https://www.springer.com/gp/book/9780387964751 for original source.

References

1. Rodwell MJ, Palmer TN (2007) Using numerical weather prediction to assess climate models. Q J R Meteorol Soc 133:129–146. https://doi.org/10.1002/qj.23
2. Ma HY, Xie S, Klein S, Williams K, Boyle J, Bony S, Douville H, Fermepin S, Medeiros B, Tyteca S (2014) On the correspondence between mean forecast errors and climate errors in CMIP5 models. J Clim 27(4):1781–1798
3. Karmalkar AV, Sexton David MH, James M, Ben Booth BB, Rostron John MD (2019) Finding plausible and diverse variants of a climate model: part 2 development and validation of methodology. https://doi.org/10.1007/s00382-019-04617-3
4. van den Hurk B, Siegmund P, Klein Tank A et al (2014) KNMI'14: climate change scenarios for the 21st century—A Netherlands perspective. Scientific report WR2014-01, KNMI, Bilt, Netherlands. https://www.climatescenarios.nl115. Accessed 1 Jan 2016
5. CSIRO and Bureau of Meteorology (2015) Climate change in Australia. https://www.climat echangeinaustralia.gov.au/. Accessed 1 Jan 2016
6. Meehl GA et al (2007) The WCRP CMIP3 multimodel dataset: a new era in climate change research. Bull Am Meteorol Soc 88:1383–1394
7. Taylor KE, Stouffer RJ, Meehl GA (2012) An overview of CMIP5 and the experiment design. Bull Am Meteorol Soc 93(4):485–498

8. Scaife AA, Copsey D, Gordon C, Harris C, Hinton T, Keeley S, O'Neill A, Roberts M, Williams K (2011) Improved atlantic winter blocking in a climate model. Geophys Res Lett 38(23):L23703. https://doi.org/10.1029/2011GL049573

9. Hewitt H, Copsey D, Culverwell I, Harris C, Hill R, Keen A, McLaren A, Hunke E (2011) Design and implementation of the infrastructure of HadGEM3: The next-generation Met Office climate modelling system. Geosci Model Dev 4(2):223–253

10. Bony S, Colman R, Kattsov VM, Allan RP, Bretherton CS, Dufresne JL, Hall A, Hallegatte S, Holland MM, Ingram W (2006) How well do we understand and evaluate climate change feedback processes? J Clim 19(15):3445–3482

11. Webb MJ, Lambert FH, Gregory JM (2013) Origins of differences in climate sensitivity, forcing and feedback in climate models. Clim Dyn 40(3–4):677–707

12. Yokohata T, Webb MJ, Collins M, Williams KD, Yoshimori M, Hargreaves JD, Annan JD (2010) Structural similarities and differences in climate responses to CO_2 increase between two perturbed physics ensembles. J Clim 23:1392–1410. https://doi.org/10.1175/2009JCLI2 917.1

13. McKay MD, Beckman RJ, Conover WJ (1979) Comparison of three methods for selecting values of input variables in the analysis of output from a computer code. Technometrics 21(2):239–245

14. Walters D, Williams K, Boutle I, Bushell A, Edwards J, Field P, Lock A, Morcrette C, Stratton R, Wilkinson J et al (2014) The Met Office Unified Model global atmosphere 4.0 and JULES global land 4.0 configurations. Geosci Model Dev 7(1):361–386

15. Gates WL, Boyle JS, Covey C, Dease CG, Doutriaux CM, Drach RS, Fiorino M, Gleckler PJ, Hnilo JJ, Marlais SM (1999) An overview of the results of the atmospheric model intercomparison project (AMIP I). Bull Am Meteorol Soc 80(1):29–55

16. Bony S, Webb M, Stevens B, Bretherton C, Klein S, Tselioudis G (2009) The cloud feedback model intercomparison project: summary of activities and recommendations for advancing assessments of cloud-climate feedbacks. CFMIP Doc

17. Rougier J (2007) Probabilistic inference for future climate using an ensemble of climate model evaluations. Clim Change 81:247–264

18. Collins M, Booth BBB, Harris GR, Murphy JM, Sexton DMH, Webb MJ (2006) Towards quantifying uncertainty in transient climate change. Clim Dyn 27:127–147

19. Murphy JM, Sexton DMH, Barnett DN, Jones GS, Webb MJ, Collins M, Stainforth DA (2004) Quantification of modelling uncertainties in a large ensemble of climate change simulations. Nature 430:768–772

20. Stainforth DA, Aina T, Christensen C, Collins M, Frame DJ, Kettleborough JA, Knight S, Martin A, Murphy J, Piani C, Sexton D, Smith LA, Spicer RA, Thorpe AJ, Allen MR (2005) Uncertainty in predictions of the climate response to rising levels of greenhouse gases. Nature 433:403–406

21. Webb MJ et al (2006) On the contribution of local feedback mechanisms to the range of climate sensitivity in two GCM ensembles. Clim Dyn 27:17–38

22. Barnett DN, Brown SJ, Murphy JM, Sexton DMH, Webb MJ (2006) Quantifying uncertainty in changes in extreme event frequency in response to doubled CO_2 using a large ensemble of GCM simulations. Clim Dyn 26:489–511

23. O'Hagan A, Forster J (2004) Bayesian inference, vol 2b of Kendall's advanced theory of statistics, 2nd edn. Edward Arnold, London

24. Meehl GA, Stocker TF, Collins WD, Friedlingstein P, Gaye AT, Gregory JM, Kitoh A, Knutti R, Murphy JM, Noda A, Raper SCB, Watterson IG, Weaver AJ, Zhao Z (2007) Global climate projections. In: Solomon S, Qin D, Manning M, Chen Z, Marquis M, Averyt KB, Tignor M, Miller HL (eds) Climate change 2007: the physical science basis. Contribution of Working Group I to the fourth assessment report of the Intergovernmental Panel on Climate Change. Cambridge University Press, Cambridge

25. Rougier J, Sexton DMH, Murphy JM, Stainforth DA (2009) Analyzing the climate sensitivity of the HadSM3 climate model using ensembles from different but related experiments. J Clim 22:1327–1353

26. Harris GR, Sexton DMH, Booth BBB, Collins M, Murphy JM, Webb MJ (2006) Frequency distributions of transient regional climate change from perturbed physics ensembles of general circulation model simulations. Clim Dyn 27:357–375
27. Murphy JM, Booth BBB, Collins M, Harris GR, Sexton D, Webb MJ (2007) A methodology for probabilistic predictions of regional climate change from perturbed physics ensembles. Philos Trans R Soc Lond a 365:1993–2028
28. Murphy JM, Sexton DMH, Jenkins G, Boorman P, Booth BBB, Brown K, Clark R, Collins M, Harris GR, Kendon E (2009) Climate change projections. ISBN 978-1-906360-02-3
29. Sanderson BM et al (2008) Constraints on model response to greenhouse gas forcing and the role of subgrid-scale processes. J Clim 21:2384–2400
30. Yokohata T, Webb MJ, Collins M, Williams KD, Yoshimori M, Hargreaves JC, Annan JD (2010) Structural similarities and differences in climate responses to CO_2 increase between two perturbed physics ensembles. J Clim 23(6):1392–1410
31. Houghton J (ed) (2001) IPCC third assessment report: Working Group I report, "The Physical Science Basis." Cambridge University Press, Cambridge
32. Intergovernmental Panel on Climate Change: In: Stocker T et al (eds) The physical science basis IPCC working group I contribution to AR5. Cambridge University Press, Cambridge
33. Knietzsch M-A, Schröder A, Lucarini V, Lunkeit F (2015) The impact of oceanic heat transport on the atmospheric circulation. Earth Syst Dyn 6(2):591–615
34. Intergovernmental Panel on Climate Change (2013) Climate change 2013: the physical science basis. Press, Cambridge University, Cambridge Mass
35. Cox PM, Huntingford C, Williamson MS (2018) Emergent constraint on equilibrium climate sensitivity from global temperature variability. Nature 553(7688):319–322
36. Cox PM (2019) Emergent constraints on climate-carbon cycle feedbacks. Curr Clim Change Rep 5(4):275–281
37. Aengenheyster M, Feng QY, van der Ploeg F, Dijkstra HA (2018) The point of no return for climate action: effects of climate uncertainty and risk tolerance. Earth Syst Dyn 9(3):1085–1095
38. Pillar HR, Heimbach P, Johnson HL, Marshall DP (2016) Dynamical attribution of recent variability in Atlantic overturning. J Clim 29(9):3339–3352
39. Kostov Y et al (2017) Fast and slow responses of Southern Ocean sea surface temperature to SAM in coupled climate models. Clim Dyn 48(5):1595–1609
40. Johnson HL, Cornish SB, Kostov Y, Beer E, Lique C (2018) Arctic Ocean freshwater content and its decadal memory of sea-level pressure. Geophys Res Lett 45(10):4991–5001
41. Cornish SB, Kostov Y, Johnson HL, Lique C (2020) Response of Arctic freshwater to the arctic oscillation in coupled climate models. J Clim 33(7):2533–2555
42. Kuhlbrodt T et al (2007) On the driving processes of the Atlantic meridional overturning circulation. Revi Geophys 45(2):RG2001
43. Hirschi J et al (2003) A monitoring design for the Atlantic meridional overturning circulation. Geophys Res Lett 30:1413
44. Orsi AH, Whitworth T, Nowlin WD (1995) On the meridional extent and fronts of the Antarctic Circumpolar Current. Deep Sea Res Part I: Oceanogr Res Pap 42(5):641–673
45. von Schuckmann K et al (2016) An imperative to monitor Earth's energy imbalance. Nat Clim Chang 6:138–144
46. Hasselmann K, Sausen R, Maier-Reimer E, Voss R (1993) On the cold start problem in transient simulations with coupled atmosphere-ocean models. Clim Dyn 9(2):53–61
47. Held IM et al (2010) Probing the fast and slow components of global warming by returning abruptly to preindustrial forcing. J Clim 23(9):2418–2427
48. Zappa G, Ceppi P, Shepherd TG (2020) Time-evolving sea-surface warming patterns modulate the climate change response of subtropical precipitation over land. Proc Natl Acad Sci 117(9):4539–4545
49. Lucarini V (2009) Evidence of dispersion relations for the nonlinear response of the Lorenz 63 system. J Stat Phys 134(2):381–400
50. Gritsun A, Lucarini V (2017) Fluctuations, response, and resonances in a simple atmospheric model. Physica D 349:62–76

51. Lucarini V, Sarno S (2011) A statistical mechanical approach for the computation of the climatic response to general forcings. Nonlinear Process Geophys 18(1):7–28
52. Ragone F, Lucarini V, Lunkeit F (2016) A new framework for climate sensitivity and prediction: a modelling perspective. Clim Dyn 46(5–6):1459–1471
53. Majda AJ, Gershgorin B (2011) Improving model fidelity and sensitivity for complex systems through empirical information theory. Proc Natl Acad Sci USA 108(25):10044–10049
54. Majda AJ, Gershgorin B (2010) Quantifying uncertainty in climate change science through empirical information theory. Proc Natl Acad Sci USA 107(34):14958–14963
55. Gershgorin B, Majda AJ (2012) Quantifying uncertainty for climate change and long-range forecasting scenarios with model errors. Part I: Gaussian models. J Clim 25(13):4523–4548
56. Majda AJ, Gershgorin B (2011) Link between statistical equilibrium fidelity and forecasting skill for complex systems with model error. Proc Natl Acad Sci USA 108(31):12599–12604
57. Majda AJ, Branicki M (2012) Lessons in uncertainty quantification for turbulent dynamical systems. Discrete Cont Dyn Syst 32(9):3133–3221

Chapter 4
Stochastic Weather and Climate Models

Introduction by Guido Visconti

Stochastic climate modeling was invented in 1976 by Klaus Hasselmann. He proposed to deal with the climate system in the same way the Brownian motion was solved where the fast component (the molecular motion) is the forcing term for the pollen particles. In the case of climate, the fast component is the weather that forces the slow components (ocean, cryosphere, land vegetation). The application of such an approach did explain the natural variability of the atmosphere, with the observed red spectra. The forcing term can be assimilated to a white noise whose high frequency components are attenuated.

In this new perspective the deterministic equations used the climate model need to be substituted by stochastic differential equations (SDE). The numerical integration of such equations was pioneered by Ruslan Stratonovich in the late fifties in Russia and by Kiyoshi Ito in Japan that published his method in 1942. A SDE is made up of a deterministic part and a stochastic term. The difference is that Ito calculus has uncorrelated noise forcing while Stratonovich allows for finite correlations between noise increments and for some time there was a debate of which of the two methods was more correct. Actually it can be shown that the two are equivalent. A SDE does not give a unique solution as the ordinary differential equations and each integration has a different realization so that it is necessary a method to describe the results.

One of the first applications of SDE to a climate problem was published in the early eighties by Alfonso Sutera and coworkers studying the ice age problem. In that case a simple energy balance model was used with a white noise stochastic forcing superimposed to a deterministic periodic signal. The results showed transitions from glacial to interglacial, happening predominantly at the fixed period. This result was the invention of the stochastic resonance theory which had mostly a heuristic value but not strong connection with reality. This first attempt showed that a system based on two potential wells could undergo transitions under the noise action. Later on it was shown that bimodal probability function (PDF) could also be obtained on a single well if forced by a multiplicative (state-dependent) noise.

G. Visconti (ed.), *Climate, Planetary and Evolutionary Sciences*,
https://doi.org/10.1007/978-3-030-74713-8_4

One of the main applications of the stochastic methods to climate and weather models is the parameterization of subgrid-scale processes. The limited time and space resolution of these models does not allow an explicit representation of some process (like convection) so that a parameterization is used when bulk formulas are used based on the assumption that the subgrid-scale state is in equilibrium with the resolved state. It has been shown that errors may arise that could affect not only the variability predicted by the model but also the mean state. For weather forecast models one of the major uncertainties is related to the initial conditions and this problem is usually addressed with the so-called ensemble prediction when the same model is initialized with slightly different initial state. The other problem is related to non-resolved subgrid processes and this is much harder to solve. In a chaotic system the error related to initial conditions and other model errors are really entangled. Different techniques have been suggested and implemented in both climate and forecast models. The results are encouraging with an improvement in the prediction of blocking episodes in the Northern Hemisphere while the weather forecast shows a general improvement in the forecast skill.

We have not treated explicitly the stochastic weather models but are actually much older than the corresponding climate or numerical forecast model. As a matter of fact, the first attempts to use these models can be traced to the end of the nineteen century where sequences of days' weather were produced based on local climatology. The application of these "weather generators" had a very significant impact on water engineering design, in agriculture and so on. They supply a deficiency in detailed meteorological data for rural regions. In some sense they are the poor cousins of the very sophisticated models we mentioned before.

Finally, a very interesting approach to simulate the uncertainties related to climate predictions should be mentioned. Climate models simulate the general circulations of ocean and atmosphere in the present Earth. However, we could imagine having an ensemble of appropriate numbers of planets Earth and to study the evolution of climate in this ensemble.

Machine-Generated Summaries

Keywords: *precipitation, enso, impact, year, climate model, scheme, system, deterministic, improve, ocean, sst, strong, bias, observation, improvement*

Revisiting Meridional Overturing Bistability Using a Minimal Set of State Variables: Stochastic Theory

https://doi.org/10.1007/s00382-013-1992-5

Abstract-Summary

A stochastic analytical model of the Atlantic meridional overturning circulation (AMOC) is presented and tested against climate model data.

AMOC stability is characterised by an underlying deterministic differential equation describing the evolution of the central state variable of the system, the average Atlantic salinity.

The introduction of climatic noise yields an equation describing the evolution of the probability density function of the state variable, and therefore the AMOC.

The method accurately describes the wandering between AMOC-On and AMOC-Off states in the climate model.

Introduction

Based on earlier work by Gregory et al. [1], Rahmstorf [2], Sijp [3] developed a simple non-linear analytical model that links key variables such as overturning strength, average Atlantic salinity (north of 32°S) and the Atlantic salt convergence in a numerical Earth system model to accurately describe their dynamical behaviour in response to changes in the net air-sea exchange of moisture at the Atlantic surface.

The analytical model of Sijp [3] is deterministic, whereas the real system, and many climate models, exhibit significant variability, warranting a stochastic description.

To the main goal of this paper, which is to study the underlying stability structure of climate models, we will also attempt to find a stochastic description of AMOC bistability as a first step in relating the stability of the observed AMOC to more observable aspects of its behaviour, such as longer-term North Atlantic climate variability.

The Model and Experimental Design

Numerical experiments are run using Version 2.8 of the UVic Earth System Model, described in detail in Weaver et al. [4], where an ocean general circulation model (GFDL MOM Version 2.2, Pacanowski [5]) is coupled to a simplified freely evolving one-layer energy-moisture balance model for the atmosphere and a dynamic-thermodynamic sea-ice model of global domain.

Fully coupled climate models and models of intermediate complexity may not represent all atmospheric and oceanic processes accurately, and the Atlantic salt budget of our numerical model may diverge from that in the real system.

To obtain a standard reference configuration, we apply this correction as a constant flux of value $H0 = -0.12$ Sv (where the sign signifies a fresh water loss from the Atlantic) applied to the Atlantic to bring the total surface flux in agreement with the estimate.

We follow an identical procedure to that described in Sijp [3], where the numerical model is equilibrated under a range of values for a constant anomalous freshwater flux H to obtain sets of steady AMOC-On and AMOC-Off states.

Stability of the AMOC: A Dynamical Systems Analysis

The state of a dynamical system can be characterised by a state variable S and a control parameter H. H is a fixed model parameter that controls qualitative features of the model dynamics.

Rather, we explore a range of models (or range of 'climates'), each set by a control parameter, H and its variability.

This could, for instance, be implemented in a model as an explicit atmospheric freshwater flux into the entire Atlantic Ocean (sometimes referred to as "hosing", although the term can attain both positive and negative values).

Including the correction term or control parameter H, the deterministic evolution of S is described by where V is the volume of the Atlantic Ocean and the minus sign appears for H and F_{atm} as they signify freshwater fluxes (although units are chosen to be identical for all terms).

Each choice of freshwater flux H sets a different model climate.

Stability of the AMOC: A Statistical Approach

To test the stochastic analytical model, we have also run stochastically forced numerical experiments in the same manner as case DRY, but where the anomalous Atlantic FW flux is $H_{WET} = H0 + H_{rnd} + 0.0825$ Sv (with an expected value of -0.0375 Sv) such that the difference in expected values is $H_{WET} - H_{DRY} = 0.0125$ Sv.

We have calculated $T_{on}(S)$ and $T_{off}(S)$ by taking the time series of the experiments constituting the WET and DRY cases, and determining for each value of S encountered in our data (and contained within the bounds of the interval) the time taken to the first system exit from that interval (i.e. where the state is "absorbed" at either boundary, performing a "first passage").

Linking Climates to Control Parameter H

The salient AMOC dynamics of a climate with different boundary conditions can be adequately captured by a single value of H relative to a reference climate (here captured by H0).

We will illustrate that a model simulation with glacial boundary conditions can be characterised using the curve $F_H(S)$ obtained from the Holocene configuration, yet using a different value of H from the reference climate.

The glacial model climates require a lower added fresh water flux H to maintain the same average Atlantic salinity compared to their Holocene counterparts.

We could define a correcting fresh water flux function $H_{corr}(S)$ of S from the difference of the natural Atlantic salt divergence $(F_{ocn}(S) - F_{atm}(S))$, i.e. excluding H, between the glacial and Holocene to obtain the glacial flux terms from the Holocene: Here, the superscripts "glac" and "Hol" respectively indicate the fluxes in the glacial and Holocene experiments.

Discussion and Conclusions

The coldest periods during the glacial interval would correspond to an AMOC-Off state, and the system would vacillate between the two states in response to climatic fluctuations and perhaps ice melt.

In the presence of two equilibria under the same control parameter, the AMOC-On and AMOC-Off states, sufficiently strong climatic noise may allow the system to wander into the basin of attraction of the other state, lifting the system across the "potential barrier" associated with the interval of unoccupied states between the two.

To the use of our framework in understanding the bistable and stochastic AMOC behaviour in numerical climate models, if the colder phases of the glacial climate variability in Greenland correspond to AMOC-Off states, our approach could also serve as a basis for a description of the statistical properties of these climate records based on AMOC bistability.

Summary

We have presented an analytical deterministic model of AMOC bistability behaviour and an associated stochastic analytical model of the AMOC statistics in terms of the basic state variable S, the average Atlantic salinity north of 32°S. This description applies with accuracy to our numerical climate model and required little or no tuning.

The stochastic theory yields a description of a bistable overturning system that wanders between AMOC-On and AMOC-Off states under the perpetual influence of climatic noise.

Acknowledgments

A machine generated summary based on the work of Sijp, Willem P.; Zika, Jan D.; d'Orgeville, Marc; England, Matthew H. (2013 in Climate Dynamics).

Stochastic and Deterministic Multicloud Parameterizations for Tropical Convection

https://doi.org/10.1007/s00382-013-1678-z

Abstract-Summary

The authors have shown, in the context of a paradigm two baroclinic mode system, that a stochastic multicloud convective parameterization based on three cloud types (congestus, deep and stratiform) can be used to improve the variability and the dynamical structure of tropical convection.

The stochastic multicloud model is modified with a lag type stratiform closure and augmented with an explicit mechanism for congestus detrainment moistening.

The new stratiform-lag closure allows for increased robustness of the coherent features of the model with respect to the amount of stochastic noise and leading to a multi-scale organization of slowly moving waves envelopes in which short-lived and chaotic convective events persist.

The simulations with the new closure have a higher amount of stochastic noise and result in a Walker type circulation with realistic mean and coherent variability which surpasses results of previous deterministic and stochastic multicloud models in the same parameter regime.

Further, deterministic mean field limit equations (DMFLE) for the stochastic multicloud model are considered.

Aside from providing a link to the deterministic multicloud parameterization, the DMFLE allow a judicious way of determining the amount of deterministic and stochastic "chaos" in the system.

The simulations with the new stratiform heating closure exhibit a mixture of stochastic and deterministic chaos.

It is shown that, in spatially extended simulations, the stochastic multicloud model can capture qualitatively two local statistical features of the observations: long and short auto-correlation times of moisture and precipitation, respectively and the approximate power-law in the probability density of precipitation event size for large precipitation events.

Extended

The stochastic multicloud model (Frenkel et al. [6]; Khouider et al. [7]) (hereafter KBM10 and FMK12) aims to capture these phenomena with a Markov chain lattice model where each lattice site is either occupied by a cloud of a certain type or it is a clear sky site.

Introduction

Cloud-resolving models on fine computational grids and high-resolution numerical weather prediction models with improved convective parameterizations have succeeded in representing some aspects of organized convection (ECMWF [8]; Moncrieff et al. [9]).

These purely deterministic parameterizations were found to be inadequate for the representation of the highly intermittent and organized tropical convection (Palmer [10]) and many of the improvements in GCMs of the last decade came from the relaxation of the QE assumption, through the addition of a stochastic perturbation (Buizza et al. [11]; Khouider et al. [12]; Lin and Neelin [13]; Majda et al. [14]; Majda and Stechmann [15]).

A novel approach to the problem of missing tropical variability in GCMs has been the development of the multicloud parameterizations (Frenkel et al. [6]; Khouider et al. [7]; Khouider and Majda [16–19]).

The multicloud parameterizations capture the interaction of three cloud types (congestus, deep and stratiform) which characterize tropical convection.

When local interactions between the lattice sites are ignored, the dynamical evolution of the cloud area fractions in the stochastic multicloud model takes the form of a coarse grained stochastic process that is intermediate between the microscopic dynamics and the mean field equations (Katsoulakis et al. [20]; Khouider et al. [12]; Majda et al. [14]).

Summary of the Deterministic Multicloud Parameterization

Both deterministic and stochastic multicloud parameterizations assume three heating profiles associated with the main cloud types that characterize organized tropical convective systems (Johnson et al. [21]): cumulus congestus clouds that heat the lower troposphere and cool the upper troposphere, through radiation and detrainment, deep convective towers that heat the whole tropospheric depth, and the associated lagging-stratiform anvils heat the upper troposphere and cool the lower troposphere, due to evaporation of stratiform rain.

Without the meridional dependency, the equations are given by Here H_d, H_s and H_c are the heating rates for deep, stratiform and cumulus congestus clouds obtained by either the deterministic or the stochastic parameterization.

The same expression is used in the parameterization of deep convective heating and CAPE for the stochastic multicloud model.

The parameterization for bulk energy available for congestus clouds can be formally obtained by integrating over the lower troposphere, the convective buoyancy anomaly of a dilute parcel raised from the boundary layer constantly mixing with the environment (KM08a).

The Stochastic Multicloud Model

We note that, just like in the deterministic multicloud model, the congestus heating is tied to the low level CAPE, while, deep convective heating is tied to overall CAPE.

Unlike the nonlinear switch of the deterministic multicloud model, the cloud fractions (of congestus and deep convective clouds) are dependent on both CAPE and dryness.

To the stratiform heating closure of the deterministic multicloud model, the new stratiform heating closure is given by where τ_s and α_s are the stratiform adjustment time scale and the stratiform fraction of deep convection, respectively.

It is possible to make connections between the DMFLE and the heating closures of the deterministic multicloud model by considering a regime where the dependency of the cloud fractions on CAPE (and low level CAPE) is suppressed.

Under these simplifying assumptions, congestus, deep and (by design) stratiform heating fields of the DMFLE parameterization mimic the behavior of the deterministic multicloud model.

Single Column Simulations

We run simulations with the DMFLE to elucidate the role of the stochastic and intrinsic deterministic chaos in the stochastic multicloud model.

The simulations of FMK12, new stratiform heating and congestus detrainment regimes form three distinct examples, which highlight the role of the stochastic noise in the multicloud model.

In terms of structure and variability, the DMFLE simulation is comparable to the stochastic simulation with 900 convective sites n = 30.

By lowering the number of convective sites, we effectively increase the amount of the stochastic noise in the model and improve upon the intrinsic deterministic variability of the system.

It can be seen (by comparing the two bottom panels) that both stochastic and DMFLE simulations are characterized by prolonged periods of congestus activity, which serve to moisten and precondition the environment for deep convection.

Spatially Extended Simulations

Although the full flow structure for the FMK12 simulation is not shown here, there are however some subtle differences between the stochastic FMK12 and the new-lag closure mean circulation: (1) the first baroclinic convergence at the flanks of the warm pool is stronger in the new lag-closure simulation, (2) while the triple peak is present in the deep convection (H_d) profile in the FMK12 case it is only seen in the precipitation field in the stratiform-lag closure case, (3) the mean stratiform and congestus heating rates are much stronger in the new stratiform-lag closure, and (4) the boundary layer equivalent potential temperature θ_{eb} that has a nice bell shape in the case of the FMK12 simulation now displays a much flatter profile to allow non trivial surface fluxes on average.

Conclusions

The speed of propagation of these wave envelopes goes down as the noise level is increased consistent with previous results obtained in the context of a mass-fluxed based simple one-heating-mode model with a stochastic convective inhibition stochastic trigger (Majda et al. [14]).

The only noticeable differences are related to the triple peak structures of the mean deep convective heating and precipitation; they are both present in the FMK12 simulations but only precipitation has a triple peak in the stratiform-lag model simulation.

The increase in mean congestus heating is a by-product of the persistent congestus cloud decks, within the warm pool region, overlying the suppressed phase of the convective envelope waves.

We note that the statistics obtained with improved stochastic multicloud model, both with lag type stratiform closure and congestus detrainment mechanism, are in a better agreement compared to FMK12 and they show a much longer lag correlation in water vapour and much fatter tails in the precipitation distributions.

Acknowledgments

A machine generated summary based on the work of Frenkel, Yevgeniy; Majda, Andrew J.; Khouider, Boualem (2013 in Climate Dynamics).

The Impact of Stochastic Parametrisations on the Representation of the Asian Summer Monsoon

https://doi.org/10.1007/s00382-017-3749-z

Abstract-Summary

The impact of the stochastic schemes Stochastically Perturbed Parametrisation Tendencies (SPPT) and Stochastic Kinetic Energy Backscatter Scheme (SKEBS) on the representation of interannual variability in the Asian summer monsoon is examined in the coupled climate model CCSM4.

SPPT improves the representation of ENSO and through teleconnections thereby the monsoon, supporting previous work on the benefits of this scheme on the model climate.

SKEBS also improves monsoon variability by way of improving the representation of the IOD, in particular by breaking an overly strong coupling to ENSO.

Extended

The impact of the SPPT scheme on ENSO was examined and documented extensively in Christensen et al. [22], where it was stated that SKEBS did not have any notable impact: SKEBS was therefore not considered in that paper.

This will be examined further in future studies.

Introduction

The main local factor influencing the impact of the ENSO teleconnection is the Indian Ocean Dipole (IOD), an oscillation of sea-surface temperatures between the western and eastern regions in the Indian Ocean associated with a localized Walker cell (Saji et al. [23]).

While IOD events can be triggered by ENSO, they often occur independently, and have been shown to influence the monsoon as a stand-alone phenomenon (Yamagata et al. [24]; Meyers et al. [25]).

Whether or not the IOD is in phase with ENSO can have a significant impact on the observed monsoon intensity in a given year (Ashok et al. [26]).

In order for a GCM to accurately simulate the interannual variability of the monsoon, it is therefore crucial to accurately represent ENSO, the IOD, their interactions, and impacts.

A recent study (Christensen et al. [22]) considered the impact of stochastic schemes on ENSO in a coupled GCM.

Model Set-Up, Diagnostics and Observational Data

We expect that the stochastic schemes will lead to a change in the modelled climate, and that the model will require an additional spin-up time for the atmosphere to equilibrate with the ocean.

The SKEBS scheme aims to represent model uncertainty arising from unresolved sub-grid scale processes by introducing random perturbations to streamfunction and potential temperature tendencies.

Different from the original implementation in the ECMWF model (Shutts [27]; Berner et al. [28]), the total instantaneous dissipation rate, which determines the amplitude of the perturbation, is here assumed to be spatially and temporally constant, resulting in a state-independent (additive) stochastic forcing.

Some recent work suggests measuring the IOD as the leading EOF of detrended Indian Ocean SST anomalies averaged over the period September–October–November (SON) rather than with the DMI, to better capture the potential variations of the IOD spatial pattern in different models (Meyers et al. [25]).

Results
It can be seen that both the SPPT and SKEBS schemes notably improve this bias, though in both cases the power at periods of 3 and 4 years is still unnaturally large.

In the deterministic model, the total variability is notably higher than in observations, and the amount of variability in Control coming from the IOD alone, as measured by the associated eigenvalue, is greater than the total variability in observations.

The deterministic model has an excessively large peak at 4 years, compared to observations which appear as a generic red-noise process with no particular power at any one frequency.

Significance testing is done in the same way as for the WY-index, with the observational DMI modelled as an AR(1) process, and indicates no statistically significant difference between observations and SKEBS.

Discussion
SKEBS has slightly improved the correlation between WY and N34 towards that in observations, while significantly reducing both correlations related to the IOD.

Based on the data presented so far, we hypothesize that the improved monsoon variability due to SPPT (SKEBS) is explained primarily by its improved representation of the ENSO (IOD).

For SKEBS, the improvement is both due to the reduced amplitude of the IOD but also to due to reduced periodicity.

We propose that the main reason SKEBS has improved the IOD is because it has broken the overly strong correlation between ENSO and the IOD.

As Control and SPPT also have significant correlation in the majority of the eastern IOD region, ENSO has a strong modulating effect on the IOD amplitude in these runs, while SKEBS elimination of correlation here reduces this effect considerably.

Conclusions
The power spectrum of the WY-index appears to be closely related to ENSO and the IOD, both in observations and in the model, so the impact on these processes was investigated as well.

A simple linear model was formulated to quantify the relative impact of ENSO and IOD on the WY-index power spectrum.

Interannual monsoon variability, as measured by the power spectrum of both the WY-index and a proxy for All India Rainfall, appears in the model to be driven almost entirely by ENSO and the IOD: ENSO works as the primary driver of the IOD, whence both tend to develop in phase to strongly regulate monsoon intensity.

This suggests that focusing on a more realistic Indian Ocean thermocline may be a route to improving the IOD and the monsoon in future model development.

Acknowledgments
A machine generated summary based on the work of Strømmen, K.; Christensen, H. M.; Berner, J.; Palmer, T. N. (2017 in Climate Dynamics).

The Impact of Stochastic Physics on the El Niño Southern Oscillation in the EC-Earth Coupled Model

https://doi.org/10.1007/s00382-019-04660-0

Abstract-Summary
The impact of stochastic physics on El Niño Southern Oscillation (ENSO) is investigated in the EC-Earth coupled climate model.

By comparing an ensemble of three members of control historical simulations with three ensemble members that include stochastics physics in the atmosphere, we find that in EC-Earth the implementation of stochastic physics improves the excessively weak representation of ENSO.

Based on the analogy with the behaviour of an idealized delayed oscillator model (DO) with stochastic noise, we find that when the atmosphere–ocean coupling is small (large) the amplitude of ENSO increases (decreases) following an amplification of the noise amplitude.

The underestimated ENSO variability in the EC-Earth control runs and the associated amplification due to stochastic physics could be therefore consistent with an excessively weak atmosphere–ocean coupling.

The activation of stochastic physics in the atmosphere increases westerly wind burst (WWB) occurrences (i.e. amplification of noise amplitude) that could trigger more and stronger El Niño events (i.e. increase of ENSO oscillation) in the coupled EC-Earth model.

The same analysis of the ENSO behaviour is carried out in a future scenario experiment (RCP8.5 forcing), highlighting that in a coupled model with an extreme warm SST, characterized by a strong coupling, the effect of stochastic physics on the ENSO representation is opposite.

Extended
The impact of additional multiplicative noise forcing is considered.

The impact of stochastic physics on ENSO behaviour in the EC-Earth RCP8.5 scenario is very different compared to that seen in historical experiments.

The impact of the stochastic physics on the simulation of ENSO in EC-Earth future scenario experiments supports the argument that the mean state bias, associated with the coupled feedbacks bias in the tropics, could be responsible for the deficiency of the ENSO simulation in the climate coupled models.

Based on the discussion above, it appears clear that stochastic physics has important beneficial impacts in the simulation of ENSO in terms of amplitude and spectrum in the EC-Earth climate model.

Introduction

Despite model improvements in the last few decades, the representation of ENSO is still biased in several coupled climate models, including errors in the amplitude, in the spatial structure and the temporal variability as well as in representation of the asymmetry of El Niño and La Niña (Flato et al. [29]; Yang and Giese [30]).

In Christensen et al. [31], the Community Climate System Model version 4 (CCSM4) was used to investigate the impact of stochastic parameterization schemes on the representation of ENSO in coupled climate simulations.

The results show that the use of the multiplicative stochastically perturbed parameterization tendencies (SPPT) scheme in the Community Atmosphere Model (CAM, the atmospheric component of CCSM) improves significantly the ENSO power spectrum by reducing the excessive power at periods of 3–4 years and increasing the power with periods less than 3 years.

Data and Methods

In the STO_PHY runs both the stochastically perturbed parameterization tendencies (SPPT) scheme with multiplicative noise and the Stochastic Kinetic Energy Backscatter (SKEB) scheme (Palmer et al. [32]) are implemented.

Since stochastic physics is only implemented in the atmospheric component of the EC-Earth, we make use of another set of the Climate SPHINX runs which includes a series of atmosphere standalone experiments (carried out with the atmospheric component of EC-Earth, see Davini et al. [33] for more details), aiming at investigating the impact of stochastic physics when the atmospheric model is forced by a fixed SSTs (i.e. when there is no mutual interaction between ocean and atmosphere).

We consider ten ensemble members for control and ten members with stochastic physics with the same horizontal and vertical resolution as the coupled runs; each experiment covers the period 1979–2008.

Results

Based on the discussion above, it appears clear that stochastic physics has important beneficial impacts in the simulation of ENSO in terms of amplitude and spectrum in the EC-Earth climate model.

In the CCSM the application of stochastic physics reduces the strong La Niña bias present in the CTRL experiments, but has little impact on El Niño amplitude contrast in EC-Earth the stochastic physics increases the amplitude of El Niño but has little impact on La Niña.

With stochastic physics, the mean state of EC-Earth is improved, principally in the warm pool and along the coastal regions of the South America in the eastern tropical Pacific where the SST warm biases are reduced.

In these "climate change" simulations, the impact of stochastic physics on ENSO is more in agreement with that found in the CCSM model, where warm SST biases (warm SST anomalies in case of RCP8.5 scenario) were associated with strong El Niño events and a sharp narrow ENSO power spectrum.

Discussions and Conclusions
The coupling between SST and winds strengthens with stochastic physics which, when combined with changes of WWBs, it is likely responsible for the improvement of ENSO representation in EC-Earth climate coupled model.

The implementation of stochastic physics not only improves the representation of ENSO in EC-Earth but also the model mean state.

One possibility is that the improvement of the mean ocean and atmospheric state induced by the stochastic physics schemes is reducing the biases in coupled feedbacks in the tropical Pacific, consequently leading to an improvement in the ENSO representation.

The impact of the stochastic physics on the simulation of ENSO in EC-Earth future scenario experiments supports the argument that the mean state bias, associated with the coupled feedbacks bias in the tropics, could be responsible for the deficiency of the ENSO simulation in the climate coupled models.

Acknowledgments
A machine generated summary based on the work of Yang, Chunxue; Christensen, Hannah M.; Corti, Susanna; von Hardenberg, Jost; Davini, Paolo (2019 in Climate Dynamics).

Development of Climate Change Projections for Small Watersheds Using Multi-model Ensemble Simulation and Stochastic Weather Generation

https://doi.org/10.1007/s00382-012-1490-1

Abstract-Summary
Regional climate models (RCMs) have been increasingly used for climate change studies at the watershed scale.

This study developed a two-step downscaling method to generate climate change projections for small watersheds through combining a weighted multi-RCM ensemble and a stochastic weather generator.

The ensemble was built on a set of five model performance metrics and generated regional patterns of climate change as monthly shift terms.

The stochastic weather generator then incorporated these shift terms into observed climate normals and produced synthetic future weather series at the watershed scale.

The ensemble-derived climate change scenario was well reproduced as local daily weather series by the stochastic weather generator.

The proposed combination of dynamical downscaling and statistical downscaling can improve the reliability and resolution of future climate projection for small prairie watersheds.

It is also an efficient solution to produce alternative series of daily weather conditions that are important inputs for examining watershed responses to climate change and associated uncertainties.

Extended

Regional climate models (RCMs) have been increasingly employed to support climate change impact studies at the watershed scale; however, their performance is strongly dependent upon the driving general circulation models (GCMs), internal parameterizations and domain configurations (Christensen et al. [34]; Sanchez et al. [35]; Deque et al. [36]).

The ensemble was able to maintain the skills of good models (HRM3 and WRFG).

The stochastic weather generator, or the second-step downscaling, is thus required to fill the gaps between the regional pattern and the local condition.

The stochastic weather generator (LARS-WG) was then used to convert the ensemble-derived climate change information into daily weather series at the scale of small prairie watersheds.

The stochastic weather generator is able to well reproduce observed climate normals and incorporate pre-defined perturbations.

The proposed combination of statistical downscaling and dynamical downscaling has several advantages.

It is also true that the combination of metrics may have a variety of possible formulations.

Introduction

Many simulation-based studies have been taken to investigate impacts of climate change on freshwater hydrology and biogeochemistry at the watershed scale (Abler et al. [37]; Borah and Bera [38]; Lehner et al. [39]; Maurer [40]; Palmer et al. [41]; Pham et al. [42]; Zhang et al. [43]).

Regional climate models (RCMs) have been increasingly employed to support climate change impact studies at the watershed scale; however, their performance is strongly dependent upon the driving general circulation models (GCMs), internal parameterizations and domain configurations (Christensen et al. [34]; Sanchez et al. [35]; Deque et al. [36]).

With increasing evidences of climate change in the prairie region, along with the growing water demands for agricultural development, urban sprawl and ecosystem health, watershed management is challenged by potential impacts of climate change on hydrological and water quality processes (Huang et al. [44]; Huang and Liu [45]; Chen et al. [46]; Jing and Chen [47]; Mao et al. [48]).

This study aims to (1) develop a two-step method to generate climate change projections for small watersheds through combining a weighted multi-RCM ensemble and a stochastic weather generator and (2) apply the method to the Assiniboia area in the prairie pothole region in southern Saskatchewan, Canada.

Methods

Indices of interannual variability are calculated as: where STD denotes the inter-annual standard deviation for annual temperature, and CV denotes the interannual coefficient of variation (standard deviation divided by the mean) for annual precip-itation; ε is the measure of natural variability, defined as the difference between the maximum and minimum observed values of the 10-year moving averages over the entire simulation period after linearly detrending the data to remove century-scale trends (Xu et al. [49]).

The skill scores of daily maximum temperature, minimum temperature and precip-itation, e.g. $M_3(T_{min})$, $M_3(T_{max})$ and $M_3(P)$, are calculated separately and then multiplied to yield the value of M_3.

When the multi-RCM ensemble is applied for projecting future climate, esti-mations of temperature-related variables (e.g. daily maximum, minimum and mean values) are generally given as: where T_i is the temperature projection by RCM i, assigned with weight W_i.

Study area and Data Collection

Reference data for both RCM evaluation and SWG validation was the Daily 10 km Gridded Climate Dataset for Canada 1961–2003 (hereafter referred to as CAD10), released by the National Land and Water Information Service of Agriculture and Agri-Food Canada.

It includes daily maximum temperature, minimum temperature, and precipitation for the Canadian landmass south of 60°N. Each daily grid contains interpolated point estimates arranged in a regular grid with a spacing of 10 km.

Error estimates of CAD10 were quite low, with mean absolute errors of 1.1 C and 1.6 °C for daily maximum and minimum temperatures, respectively, and about 9% for annual precipitation; the results for daily precipitation and its extremes also compared well with other studies (Hutchinson et al. [50]).

Compared with other gridded climate data, CAD10 has a better coverage for the Canadian Prairies region, making it a better choice in this study.

Results

The multi-RCM ensemble was able to improve on WRFG's skills for temperature, at the cost of reduced performance for precipitation.

It was thus doubtful whether the ensemble's projection for the future period can be directly used in climate change impact studies in small prairie watersheds.

Based on the ensemble projection of 2041–2070, both maximum and minimum temperatures in March would rise 2 °C approximately, which could result in an earlier onset of the snowmelt period and thus lead to a series of shifts in prairie hydrological cycle.

These synthetic weather series were indeed a further downscaling of ensemble-derived climate change scenarios.

It would be desired that shift terms generated by the multi-RCM ensemble be accurately reproduced in the synthetic weather series.

The results indicated that the ensemble-derived regional patterns of climate change were effectively converted into synthetic daily weather series at the watershed scale.

Discussions

This study does not intend to join the debate on the merits and drawbacks of weighted RCM ensemble, but tries to make a step towards the development of high-resolution climate change projections for impact studies at the scale of small watersheds.

Through identifying the models with relatively good performance and generating a single set of results from multiple model runs, the ensemble approach is helpful in exploring and reducing inherent uncertainty in multi-RCM simulations.

A larger sample size would allow for a comprehensive understanding of uncertainties derived from model internal variability and intermodal variations, and would thus improve the robustness of the multi-RCM ensemble in climate change projections.

It was assumed that if a day was simulated as a rainy day by a model with better performance, it was more likely to be a rainy day through the multi-RCM ensemble.

Conclusions

Due to the integrity of RCMs over different time slices, models with good performance during the baseline period are most likely to perform well in the future period, and the advantage of a multi-RCM ensemble is more than likely to be valid in the future period.

This study developed a two-step downscaling to generate climate change projections for small watersheds through combining a weighted multi-RCM ensemble and a stochastic weather generator.

The weighed multi-RCM ensemble led to reduced biases in projections of temperature and precipitation by properly emphasizing models with good performance.

The stochastic weather generator (LARS-WG) was then used to convert the ensemble-derived climate change information into daily weather series at the scale of small prairie watersheds.

The combination thus takes advantage of the robustness of physically-based multi-RCM simulations and the statistical stability of stochastic weather generation.

Acknowledgments
A machine generated summary based on the work of Zhang, Hua; Huang, Guo H. (2012 in Climate Dynamics).

Assessment of Climate Change Impacts on Watershed in Cold-Arid Region: An Integrated Multi-GCM-Based Stochastic Weather Generator and Stepwise Cluster Analysis Method

https://doi.org/10.1007/s00382-015-2831-7

Abstract-Summary

MGCM-SWG–SCA can investigate uncertainties of projected climate changes as well as create watershed-scale climate projections from large-scale variables.

It can also assess climate change impacts on hydrological processes and capture nonlinear relationship between input variables and outputs in watershed systems.

Results also disclose that: (1) the projected minimum and maximum temperatures and precipitation from MGCM change with seasons in different ways; (2) various climate change projections can reproduce the seasonal variability of watershed-scale climate series; (3) SCHM can simulate daily streamflow with a satisfactory degree, and a significant increasing trend of streamflow is indicated from future (2015–2035) to validation (2006–2011) periods; (4) the streamflow can vary under different climate change projections.

The findings can be explained that, for the Kaidu watershed located in the cold-arid region, glacier melt is mainly related to temperature changes and precipitation changes can directly cause the variability of streamflow.

Introduction

The detailed tasks are: (1) to develop a multi-GCM-based stochastic weather generator (MGCM-SWG) through combining multiple global climate models (MGCM) and stochastic weather generator (SWG), with the purpose of not only reducing the uncertainty of climate change projections (e.g. temperature and precipitation), but also constructing watershed-scale climate change projections from large-scale GCM outputs; (2) to formulate a stepwise-clustered hydrological model (SCHM) for streamflow simulation, through employing a generated cluster tree to represent the nonlinear relationship between predictand (streamflow) and predictors (hydro-meteorological factors) in hydrological processes; (3) to create the MGCM-SWG–SCA method through integrating MGCM-SWG and SCHM into a general framework, in order to quantify the climate changes impacts on hydrological processes (e.g. streamflow) by using watershed-scale climate change projections as inputs to drive the SCHM; (4) to apply the MGCM-SWG–SCA method to the Kaidu watershed of northwestern China, focusing on demonstrating its efficiency in a cold-arid region.

Methodology

A multi-GCM-based stochastic weather generator (MGCM-SWG) is developed for reducing the uncertainty effect of projected climate changes as well as for generating watershed-scale climate change projections from large-scale variables, through coupling multiple global climate models (MGCM) with a stochastic weather generator (SWG).

Based on MGCM–SWG, the simulation outputs are effectively extracted from MGCM to reduce the uncertainty effect of climate projection and, then, the large-scale climate change projections are transferred by SWG to construct watershed-scale climate simulations (e.g. daily maximum and minimum temperatures and precipitation).

There are three components such that (1) the simulation results of climate projection from MGCM (i.e. CCSM, GFDL, HadCM3, MRI-CGCM, and multi-average GCMs) are extracted to conduct various climate change scenarios; (2) the climate change projections are utilized by a SWG to produce local-scale synthetic daily time series of climate variables; and (3) various climate change scenarios are used as inputs to drive the SCHM to examine the impacts of climate changes on streamflow.

Study Area and Data

Under the general background of global warming, the climate in Kaidu watershed has been undergoing significant changes, such as increased temperature and precipitation, added river runoff volumes, accelerated retreat of glaciers, and increased lake water surface elevation and area (Mansuer and Chu [51]; Zhang et al. [52]; Dou et al. [53]).

As the unique high alpine cold climate and topography, the streamflow of Kaidu watershed is principally provided by high mountain precipitation and seasonal snow and glacier melting; and the climate factors directly affecting the recharge of the river are temperature and precipitation (Chen et al. [54]).

Difficulties are confronted in evaluating the potential effects of climate change on the streamflow of Kaidu watershed, due to the uncertainties of climate change scenarios, inherent nonlinearity of hydrological processes, and lack of high-quality data.

The developed MGCM-SWG-SCA method can be employed to simulate hydrological processes (e.g. streamflow) under various climate change projections of the Kaidu watershed.

Result Analysis

The simulation outputs of climate changes (time series of daily minimum and maximum temperatures, and precipitation) from MGCM (i.e. CCSM, GFDL, HadCM3, MRI-CGCM, and multi-average GCMs) were firstly extracted and, then, were transferred by SWG to create watershed-scale climate change projections.

Besides, the simulated results (minimum temperature, maximum temperature, and precipitation) would be projected differently from MGCM.

Using the generated cluster tree, SCHM can simulate streamflow under different climate change projections.

The simulated peak streamflows would be significantly different from CCSM, GFDL, HadCM3, MRI-CGCM and multi-average GCMs, which imply the complex relationship between streamflow and the factors related to climate conditions.

Results disclose that the simulated streamflow would vary differently with the various climate change projections from MGCM (CCSM, GFDL, HadCM3, MRI-CGCM and multi-average GCMs).

Results reveal that different characteristics of multi-year mean values, absolute and relative changes of the climate variables and streamflow exist from MGCM.

Discussion

It can be noticeable that the variability of streamflow is sensitive differently to changes of climate variables (e.g. temperature and precipitation) in different months.

During the spring, streamflow is mostly affected by minimum temperature, rather than precipitation.

Changes in temperature may affect streamflow in spring; while streamflow in winter is mostly affected by changes in precipitation.

In these seasons, streamflows are affected both from changes in precipitation and temperature, since snowmelt and rainfall water jointly contribute important portions of runoff.

The possible reasons for the different performances of streamflow variability correspondingly to changes of climate variables (minimum and maximum temperatures and precipitation) in different months are: (1) precipitation runoff occurs heavily and snowmelt water reduces significantly during the summer and autumn seasons, thereby reducing the proportion of snowmelt runoff; and (2) temperature increases during spring season have a relative more positive effect on streamflow than the other seasons because streamflow is primarily supplied by the snowmelt water.

Conclusions

The advantages of MGCM-SWG–SCA are as follows: (1) it can help investigate the uncertainty in projected climate changes, as well as bridge the spatial resolution between GCM outputs and hydrological models, through employing the developed multi-GCM-based stochastic weather generator (MGCM-SWG) to firstly extract the simulated outputs of climate changes from MGCM and, then, to transfer large-scale climate variables to create watershed-scale climate projections; (2) based on the step-wise cluster analysis (SCA) technique, SCHM can reflect the nonlinear relationship between predictand (e.g. streamflow) and predictors (e.g. precipitation and temperature) in hydrological processes with represented by a cluster tree; and (3) it can also assess the variability of streamflow to potential impacts of climate change, through driving the SCHM with projected climate change scenarios from MGCM-SWG.

Various climate change projections (time series of minimum and maximum temperatures, and precipitation) of the Kaidu watershed from MGCM (CCSM, GFDL, HadCM3, MRI-CGCM and multi-average GCMs) during past (1961–1990), recent (2006–2011) and future (2015–2035) periods were generated.

Acknowledgments

A machine generated summary based on the work of Zhuang, X. W.; Li, Y. P.; Huang, G. H.; Liu, J. (2015 in Climate Dynamics).

Book Reading List

Analysis of Climate Variability

by Storch, H. v. (Ed), Navarra, A. (Ed) *(1999)*.

Various problems in climate research, which require the use of advanced statistical techniques, are considered in this book. The examples emphasize the notion that the knowledge of statistical techniques alone is not sufficient. Instead, good physical understanding of the specific problems in climate research, such as the enormous size of the phase space, the correlation of processes on different time and space scales and the availability of essentially one observational record, is needed to guide the researcher in choosing the right approach to obtain meaningful answers.

Please see https://www.springer.com/gp/book/9783540663157 for original source.

Climate Modelling

by Lloyd, E. (Ed), Winsberg, E. (Ed) *(2018)*.

This edited collection of works by leading climate scientists and philosophers introduces readers to issues in the foundations, evaluation, confirmation, and application of climate models. It engages with important topics directly affecting public policy, including the role of doubt, the use of satellite data, and the robustness of models.

Climate Modelling provides an early and significant contribution to the burgeoning Philosophy of Climate Science field that will help to shape our understanding of these topics in both philosophy and the wider scientific context.

Please see https://www.springer.com/gp/book/9783319650579 for original source.

Stochastic Calculus

by Grigoriu, M. *(2002)*.

This work focuses on analyzing and presenting solutions for a wide range of stochastic problems that are encountered in applied mathematics, probability, physics, engineering, finance, and economics. The approach used reduces the gap between the mathematical and engineering literature. Stochastic problems are defined by algebraic, differential or integral equations with random coefficients and/or input.

Please see https://www.springer.com/gp/book/9780817642426 for original source.

Stochastic Climate Models

by Imkeller, P. (Ed), Storch, J. v. (Ed) *(2001)*.

The proceedings of the summer 1999 Chorin workshop on stochastic climate models captures well the spirit of enthusiasm of the workshop participants engaged in research in this exciting field. It is amazing that nearly 25 years after the formal theory of natural climate variability generated by quasi-white-noise weather forcing was developed, and almost 35 years after J. M. Mitchell first suggested this mechanism as the origin of sea-surface-temperature fluctuations and climate variability, there have arisen so many fresh perspectives and new applications of the theory.

Please see https://www.springer.com/gp/book/9783034895040 for original source.

Stochastic Calculus and Applications

by Cohen, S. N., Elliott, R. J. *(2015)*.

Completely revised and greatly expanded, the new edition of this text takes readers who have been exposed to only basic courses in analysis through the modern general theory of random processes and stochastic integrals as used by systems theorists, electronic engineers and, more recently, those working in quantitative and mathematical finance. Building upon the original release of this title, this text will be of great interest to research mathematicians and graduate students working in those fields, as well as quants in the finance industry.

Please see https://www.springer.com/gp/book/9781493928668 for original source.

Stochastic Climate Theory

by Dobrovolski, S. G. *(2000)*.

The author describes the stochastic (probabilistic) approach to the study of changes in the climate system. Climatic data and theoretical considerations suggest that a large part of climatic variation/variability has a random nature and can be analyzed using the theory of stochastic processes. This work summarizes the results of processing existing records of climatic parameters as well as appropriate theories: from the theory of random processes (based on the results of Kolmogorov and Yaglom) and Hasselmann's "stochastic climate model theory" to recently obtained results.

Please see https://www.springer.com/gp/book/9783540663102 for original source.

Stochastic Partial Differential Equations

by Lototsky, S. V., Rozovsky, B. L. *(2017)*.

Taking readers with a basic knowledge of probability and real analysis to the frontiers of a very active research discipline, this textbook provides all the necessary background from functional analysis and the theory of PDEs. It covers the main types of equations (elliptic, hyperbolic and parabolic) and discusses different types of random forcing. The objective is to give the reader the necessary tools to understand the proofs of existing theorems about SPDEs (from other sources) and perhaps even to formulate and prove a few new ones.

Please see https://www.springer.com/gp/book/9783319586458 for original source.

References

1. Gregory JM, Saenko OA, Weaver AJ (2003) The role of the Atlantic freshwater balance in the hyteresis of the meridional overturning circulation. Clim Dyn 21:707–717. https://doi.org/10.1007/s00382-003-0359-8
2. Rahmstorf S (1996) On the freshwater forcing and transport of the Atlantic thermohaline circulation. Clim Dyn 12:799–811
3. Sijp WP (2012) Characterising meridional overturning bistability using a minimal set of state variables. Clim Dyn. https://doi.org/10.1007/s00382-011-1249-0
4. Weaver AJ, Eby M, Wiebe EC et al (2001) The UVic Earth system climate model: model description, climatology, and applications to past, present and future climates. Atmos Ocean 39:1067–1109
5. Pacanowski R (1995) MOM2 documentation user's guide and reference manual: GFDL Ocean group technical report 3. NOAA, GFDL Princeton, 3rd edn, 232 pp
6. Frenkel Y, Majda AJ, Khouider B (2012) Using the stochastic multicloud model to improve tropical convective parameterization: a paradigm example. J Atmos Sci 69(3):1080–1105
7. Khouider B, Biello J, Majda AJ (2010) A stochastic multicloud model for tropical convection. Commun Math Sci 8(1):187–216
8. ECMWF (2003) Proceedings ECMWF/CLIVAR workshop on simulation and prediction of intraseasonal variability with emphasis on the MJO, pp 3–6
9. Moncrieff M, Shapiro M, Slingo J, Molteni F (2007) Collaborative research at the intersection of weather and climate. WMO Bull 56:204–211
10. Palmer TN (2001) A nonlinear dynamical perspective on model error: a proposal for non-local stochastic-dynamic parametrization in weather and climate prediction models. Q J R Meteorol Soc 127:279–304
11. Buizza R, Miller M, Palmer TN (1999) Stochastic representation of model uncertainties in the ECMWF ensemble prediction system. Q J R Meteorol Soc 125:2887–2908. https://doi.org/10.1002/qj.49712556006
12. Khouider B, Majda AJ, Katsoulakis MA (2003) Coarse-grained stochastic models for tropical convection and climate. Proc Natl Acad Sci 100:11941–11946
13. Lin J, Neelin JD (2003) Towards stochastic deep convective parameterization in general circulation models. Geophys Res Lett 30:1162. https://doi.org/10.1029/2002GL016203
14. Majda AJ, Franzke C, Khouider B (2008) An applied mathematics perspective on stochastic modelling for climate. Philos Trans R Soc a Math Phys Eng Sci 366(1875):2427–2453
15. Majda AJ, Stechmann SN (2008) Stochastic models for convective momentum transport. Proc Natl Acad Sci 105(46):17614–17619
16. Khouider B, Majda AJ (2006) Multicloud convective parametrizations with crude vertical structure. Theor Comp Fluid Dyn 20:351–375
17. Khouider B, Majda AJ (2007) A simple multicloud parametrization for convectively coupled tropical waves. Part II: nonlinear simulations. J Atmos Sci 64:381–400
18. Khouider B, Majda AJ (2008) Equatorial convectively coupled waves in a simple multicloud model. J Atmos Sci 65:3376–3397
19. Khouider B, Majda AJ (2008) Multicloud models for organized tropical convection: enhanced congestus heating. J Atmos Sci 65:897–914
20. Katsoulakis MA, Majda AJ, Vlachos DG (2003) Coarse-grained stochastic processes for microscopic lattice systems. Proc Natl Acad Sci USA 100(3):782–787
21. Johnson RH, Rickenbach TM, Rutledge SA, Ciesielski PE, Schubert WH (1999) Trimodal characteristics of tropical convection. J Clim 12(8):2397–2418
22. Christensen HM, Berner J, Coleman D (2016) Stochastic parametrisation and the El Niño Southern oscillation. Accepted for publication
23. Saji N, Goswami BN, Vinayachandran P, Yamagata T (1999) A dipole mode in the tropical Indian Ocean. Nature 401:360–363

24. Yamagata T et al (2004) Coupled ocean-atmosphere variability in the tropical Indian ocean. In: Wang C, Xie SP, Carton JA (eds) Earth's climate. American Geophysical Union, Washington, DC. https://doi.org/10.1029/147GM12

25. Meyers G, McIntosh P, Pigot L, Pook M (2006) The years of El Niño, La Niña, and interactions with the tropical Indian ocean. J Clim 20:2872–2880

26. Ashok K, Guan Z, Yamagata T (2001) Impact of the Indian Ocean Dipole on the relationship between the Indian Monsoon Rainfall and ENSO. Geophys Res Lett 28(23):4499–4502

27. Shutts GJ (2005) A kinetic energy backscatter algorithm for use in ensemble prediction systems. Q J R Meteorol Soc 612:3079–3102

28. Berner J, Shutts G, Leutbecher M, Palmer T (2009) A spectral stochastic kinetic energy backscatter scheme and its impact on flow-dependent predictability in the ECMWF ensemble prediction system. J Atmos Sci 66:603–626

29. Flato G et al (2013) Evaluation of climate models. In: Stocker TF et al (eds) Climate change 2013: the physical science basis. Cambridge University Press, Cambridge, pp 741–866

30. Yang C, Giese BS (2013) El Niño Southern Oscillation in an ensemble ocean reanalysis and coupled climate models. J Geophys Res Oceans 118:4052–4071. https://doi.org/10.1002/jgrc.20284

31. Christensen HM, Berner J, Coleman D, Palmer TN (2017) Stochastic parameterization and the El Niño-Southern Oscillation. J Clim 30:17–38. https://doi.org/10.1175/JCLI-D-16-0122.1

32. Palmer TN, Buizza R, Doblas-Reyes F, Jung T, Leutbecher M, Shutts GJ, Steinheimer M, Weisheimer A (2009) Stochastic parametrization and model uncertainty. ECMWF tech rep 38. J Clim 30(598):1–44. https://www.ecmwf.int/sites/default/files/elibrary/2009/11577-stochastic-parametrization-and-model-uncertainty.pdf

33. Davini P, von Hardenberg J, Corti S, Christensen HM, Juricke S, Subramanian A, Watson PAG, Weisheimer A, Palmer TN, 2017, Climate SPHINX: evaluating the impact of resolution and stochastic physics parameterisations in climate simulations (submitted)

34. Christensen JH, Raisanen J, Iversen T, Bjorge D, Christensen OB, Rummukainen M (2001) A synthesis of regional climate change simulations—a Scandinavian perspective. Geophys Res Lett 28:1003–1006

35. Sanchez E, Gallardo C, Gaertner MA, Arribas A, Castro M (2004) Future climate extreme events in the mediterranean simulated by a regional climate model: a first approach. Glob Planet Change 44:163–180

36. Deque M, Rowell DP, Luthi D, Giorgi F, Christensen JH, Rockel B, Jacob D, Kjellstrom E, de Castro M, van den Hurk B (2007) An intercomparison of regional climate simulations for Europe: assessing uncertainties in model projections. Clim Change 81:53–70

37. Abler D, Shortle J, Carmichael J, Horan R (2002) Climate change, agriculture, and water quality in the Chesapeake Bay Region. Clim Change 55:339–359

38. Borah DK, Bera M (2004) Watershed-scale hydrologic and nonpoint-source pollution models: review of applications. Trans ASAE 47:789–803

39. Lehner B, Doll P, Alcamo J, Henrichs T, Kaspar F (2006) Estimating the impact of global change on flood and drought risks in Europe: a continental, integrated analysis. Clim Change 75:273–299

40. Maurer EP (2007) Uncertainty in hydrologic impacts of climate change in the Sierra Nevada, California, under two emissions scenarios. Clim Change 82:309–325

41. Palmer MA, Liermann CAR, Nilsson C, Florke M, Alcamo J, Lake PS, Bond N (2008) Climate change and the world's river basins: anticipating management options. Front Ecol Environ 6:81–89

42. Pham SV, Leavitt PR, McGowan S, Peres-Neto P (2008) Spatial variability of climate and land-use effects on lakes of the northern Great Plains. Limnol Oceanogr 53:728–742

43. Zhang H, Huang GH, Wang D, Zhang X (2011) Uncertainty assessment of climate change impacts on the hydrology of small prairie wetlands. J Hydrol 396:94–103

44. Huang GH, Cohen SJ, Yin YY, Bass B (1998) Land resources adaptation planning under changing climate—a study for the Mackenzie Basin. Resour Conserv Recycl 24:95–119

45. Huang YT, Liu L (2008) A hybrid perturbation and morris approach for identifying sensitive parameters in surface water quality models. J Environ Inform 12:150–159
46. Chen B, Jing L, Zhang BY, Liu S (2011) Wetland monitoring, characterization and modelling under changing climate in the Canadian subarctic. J Environ Inform 18:55–64
47. Jing L, Chen B (2011) Field investigation and hydrological modelling of a subarctic wetland— the deer river watershed. J Environ Inform 17:36–45
48. Mao XF, Yang ZF, Chen B (2011) Network analysis and comparative studies on baiyangdian and okefenokee wetland systems in China and US. J Environ Inform 18:46–54
49. Xu Y, Gao XJ, Giorgi F (2010) Upgrades to the reliability ensemble averaging method for producing probabilistic climate-change projections. Clim Res 41:61–81
50. Hutchinson MF, McKenney DW, Lawrence K, Pedlar JH, Hopkinson RF, Milewska E, Papadopol P (2009) Development and testing of Canada-wide interpolated spatial models of daily minimum-maximum temperature and precipitation for 1961–2003. J Appl Meteorol Climatol 48:725–741
51. Mansuer S, Chu XZ (2007) Study on the change of climate and runoff volumes of the Tarim river basin in recent 40 years. Areal Res Dev 26(4):97–101
52. Zhang YC, Li BL, Bao AM, Zhou CH, Chen X, Zhang XR (2007) Study on snowmelt runoff simulation in the Kaidu river basin. Sci China Ser D: Earth Sci 50(1):26–35
53. Dou Y, Chen X, Bao AM, Li LH (2011) The simulation of snowmelt runoff in the ungauged Kaidu river basin of TianShan Mountains, China. Environ Earth Sci 62:1039–1045
54. Chen YN, Yang Q, Luo Y, Shen YJ, Pan XL, Li LH, Li ZQ (2012) Ponder on the issue of water resources in the arid region of northwest China. Arid Land Geogr 35(1):1–9

Chapter 5
Progress in Climate Modeling

Introduction by Guido Visconti

In the development of modeling the climate, two principal paths can be distinguished. On one side the effort to increase the resolution to eliminate as much as possible the parameterization of small-scale processes. This approach is an improvement over the "mechanistic" view of modeling climate mainly through the use of General Circulation Models (GCM). The other approach, which is born out of the theory of system dynamics, has as objective an improvement in the theoretical basis of climate science. A minor but promising development has to do with the direct statistical simulation (DSS) following a suggestion made by Edward Lorenz almost sixty years ago. The improvement in GCM is more oriented to the classical application of predicting the future climate and its impact on the human activities while the other two approaches have explained some specific phenomena like atmospheric jets, or El Nino but their use as predicting tools is not in the near future.

When the resolution of a model goes down to about 1 km it is possible to simulate deep convection, ocean eddies and land–atmosphere interactions in detail. An improvement of this kind eliminates the necessity of parameterize small-scale processes and in turn reduces the biases from which suffer most of the GCM results. Preliminary results obtained at such resolution show that also the reliability of regional climate projection is enhanced. A notable improvement has been obtained by increasing the resolution in numerical weather predictions and a similar effect is expected for the climate projections. However, in this case the financial burden is such that the implementation of such techniques must be multinational using the best technological available infrastructures. The reliance on governmental funding may create some conflict of interest within the scientific community.

The theoretical development of climate dynamics refers exactly to those sub-grid scale processes we mentioned before which are accounted for in a stochastic manner. The complexity of the climate system is such that theoretical studies can be carried out only on highly idealized models (like to the Lorenz equations) or in relatively simple systems like a very elementary model of the ocean current. However numerical

G. Visconti (ed.), *Climate, Planetary and Evolutionary Sciences*,
https://doi.org/10.1007/978-3-030-74713-8_5

experiments on such simple systems can give precious indications on the internal variability (IV) or forced variability (FV) of the climate system. A key concept to study such effects is the so-called pullback/snapshot (PBA) attractor. From the theory of non-linear deterministic systems, the attractor is determined by integrating the systems forward in time and studying where its trajectories lie in the phase space. In the PBA approach the system is studied as an ensemble where the relevant equations are initialized in slightly different conditions at some very distant initial time and integrated up to the present time. Here they reveal the nature of their attractor. The different initial conditions may represent an ensemble of climate systems which obeys the same equations and give the results in terms of a probability distribution. As it was mentioned at the beginning this approach has been applied to very simple systems and hardly could contribute to predict the future climate (or study the climate of the past) but could give important indications on the interpretation of those results.

The last development we would like to mention refers to the statistics of the fluid motions on the Earth (oceans and the atmosphere). The core of any climate model is to integrate in time the relevant equations (thermodynamics, fluid dynamics, chemistry) up to the statistical steady state. Any further integration will reveal the statistics of the system. This approach was not appreciated by Edward Lorenz that in his famous 1963 monograph on the general circulation of the atmosphere affirmed:

> More than any other theoretical procedure, numerical integration is also subject to the criticism that it yields little insight into the problem. The computed numbers are not only processed like data but they look like data, and a study of them may be no more enlightening than a study of real meteorological observations. An alternative procedure which does not suffer this disadvantage consists of deriving a new system of equations whose unknowns are the statistics themselves.

The suggestion by Lorenz was implemented by Brad Marston of Brown University using the Direct Statistical Simulation (DSS) that however revealed a few problems as envisaged by Lorenz in the same monograph.

> [DSS] can be very effective for problems where the original equations are linear, but, in the case of non-linear equations, the new system will inevitably contain more unknowns than equations, and can therefore not be solved, unless additional postulates are introduced

Marston has applied DSS to study zonal jets on Earth reproducing some of their features. As in the case of the development of the theoretical basis of climate science this approach has a long way to go before he could find useful applications. Most of these theoretical attempts to study climate reveal something like an inferiority complex of the scientific community that works on these subjects. The question is that not all sciences must necessarily use very sophisticated mathematical stuff (consider biology!) and yet they must have the same respect. Rephrasing Richard Lewontin: it is more interesting to explain while a mice and an elephant fall with the same acceleration in the vacuum or to explain why they came to such different sizes?

Machine-Generated Summaries

Keywords: *variability, forcing, force, period, state, ensemble, approach, parameter, analysis, feedback, response, climate change, base, enso, consider.*

History Matching for Exploring and Reducing Climate Model Parameter Space Using Observations and a Large Perturbed Physics Ensemble

https://doi.org/10.1007/s00382-013-1896-4

Abstract-Summary
We apply an established statistical methodology called history matching to constrain the parameter space of a coupled non-flux-adjusted climate model (the third Hadley Centre Climate Model; HadCM3) by using a 10,000-member perturbed physics ensemble and observational metrics.

History matching uses emulators (fast statistical representations of climate models that include a measure of uncertainty in the prediction of climate model output) to rule out regions of the parameter space of the climate model that are inconsistent with physical observations given the relevant uncertainties.

Our methods rule out about half of the parameter space of the climate model even though we only use a small number of historical observations.

We explore 2 dimensional projections of the remaining space and observe a region whose shape mainly depends on parameters controlling cloud processes and one ocean mixing parameter.

Constraining parameter space using easy to emulate observational metrics prior to analysis of more complex processes is an important and powerful tool.

It can remove complex and irrelevant behaviour in unrealistic parts of parameter space, allowing the processes in question to be more easily studied or emulated, perhaps as a precursor to the application of further relevant constraints.

Extended
An ideal analysis, and perhaps the only way to truly quantify this uncertainty, must involve expert judgement regarding the deficiencies in the mathematical representations of the physics in the model and the ways they might be improved in order to capture these deficiencies in future generations of models (see Goldstein and Rougier [1], for further discussion).

Introduction
The PPE is used to inform us about the behaviour of the model in this space and, in particular, about regions of the space where model-based projections are not predicted to be inconsistent with current observations.

We illustrate the application of existing methodology from the statistics with computer experiments literature to rule out regions of the parameter space of the UK Met Office's Third Hadley Centre Ocean–Atmosphere General Circulation Model (HadCM3) (Pope et al. [2]; Gordon et al. [3]) containing model runs that are inconsistent with a handful of physically important observational metrics.

We use emulators, (fast statistical representations of climate models that include a measure of uncertainty in the prediction of climate model output), and four pre-industrial global and hemispheric averages of climatic variables to remove over half of the explored space.

We explore parts of HadCM3's parameter space previously unstudied by running models with parameter choices outside the ranges established by Murphy et al. [4].

History Matching

The Bayesian approach uses the PPE to learn about the behaviour of the model throughout its parameter space in order to find regions of the space where model based projections are not inconsistent with current observations.

Bayesian calibration requires a stochastic representation of the climate model, called an emulator (Sacks et al. [5]), to be constructed and to be reliable across the whole of parameter space.

History matching, like Bayesian calibration, requires a statistical model that relates the climate model to reality.

History matching, on the other hand, allows us to use even the most simple outputs of the climate model in order to begin ruling out regions of parameter space.

We can take simple outputs that are relatively easy to model statistically, use them to rule out parameter space via history matching, then focus on emulating more complex model output within NROY space, where they may be easier to model statistically.

History Matching HadCM3

A principal motivation for history matching here, and in any application involving a large ensemble of a climate model, is to make our emulators for key and difficult to model quantities easier to fit and more accurate than they would be if fitted using all data within the unconstrained parameter space.

In an analysis that goes on to provide probabilistic climate predictions such as that in Murphy et al. [6], we would run a second ensemble within NROY space and with twentieth century forcing, and either use history matching with more complex constraints to further reduce NROY space or use Bayesian calibration with more complex constraints to generate probabilistic predictions.

This can vary substantially from observations (with values across the ensemble ranging between -5 and 33 °C) as the parameter perturbations alter the radiative balance at the top of the atmosphere.

History Matching with Multi-model Ensembles

This allows us to derive $Var[z - E[f(x)]]$, the denominator of our implausibility measure (4), under the interpretation that the resulting discrepancy variance represents a tolerance to error which is consistent with using CMIP3 as representative of the judgements of the climate community regarding what represents an informative climate model.

Of adopting a formal statistical model relating CMIP3 and HadCM3 to each other and to the true climate, we are able to obtain 'low' and 'high' tolerance alternatives for the discrepancy variance to be used in a sensitivity analysis.

If we observed a small or negative estimate for Var[U], we may re-visit the second order exchangeability assumption for all models in the collection, or investigate the sensitivity of the estimate to the ensemble size.

NROY Space

In our ensemble, although changes in SAT are a result of different parameter perturbations rather than increasing greenhouse gases, it is likely that the sea ice would respond in a similar way, possibly resulting in the negative correlations found between SAT and SGRAD or SCYC.

SAT is, therefore, the dominant constraint due to the high correlations with the other variables and because of the relatively low discrepancy and observation errors of SAT compared with the other variables in relation to the ensemble ranges for those variables.

Although SAT dominates and that, for SCYC and PRECIP at least, no ensemble members that are not ruled out by SAT alone are ruled out by the additional constraints, that does not mean that the additional variables provide no further constraint on parameter space.

The AMOC and MHT in NROY Space

It reveals a nonlinear relationship between SAT and AMOC across the whole parameter space, but an approximately linear relationship within NROY space.

By history matching on simpler (i.e. univariate and easier to emulate) quantities such as global mean SAT, we have removed many ensemble members with unrealistically weak control AMOC strengths and removed a substantial part of the burden faced in emulating a complex quantity such as the AMOC time series over the whole parameter space by just focussing on that part of the parameter space that we are unable to easily rule out on the basis of observations.

This means that the AMOC in NROY space is, on average, more responsive to CO_2 forcing than in the ruled out space, and that the sensitivity of the AMOC depends on the model parameters through the SAT.

Discussion

This tolerance is set using the error on the observations and discrepancy variance information derived from CMIP3, whilst accounting for the uncertainty in our emulator-based predictions.

The potential impact on our conclusions is assessed using a sensitivity analysis showing that even increasing the estimated discrepancy variance by an order of magnitude only results in around 10% less parameter space removed by history matching.

Complex constraints require sophisticated statistical emulators that are valid throughout NROY space in order to impose them.

AOGCMs of different resolution can be linked statistically (Williamson et al. [7]) so that large PPEs using coarser-resolution models and a small PPE using a very expensive model can be used together to emulate the advanced model and quantify its parametric uncertainty in the ways we have described.

Acknowledgments
A machine generated summary based on the work of Williamson, Daniel; Goldstein, Michael; Allison, Lesley; Blaker, Adam; Challenor, Peter; Jackson, Laura; Yamazaki, Kuniko (2013 in Climate Dynamics).

Advances in Projection of Climate Change Impacts Using Supervised Nonlinear Dimensionality Reduction Techniques

https://doi.org/10.1007/s00382-016-3145-0

Abstract-Summary
Due to the complexity of climate-associated processes, identification of predictor variables from high dimensional atmospheric variables is considered a key factor for improvement of climate change projections in statistical downscaling approaches.

The present paper adopts a new approach of supervised dimensionality reduction, which is called "Supervised Principal Component Analysis (Supervised PCA)" to regression-based statistical downscaling.

To capture the nonlinear variability between hydro-climatic response variables and projectors, a kernelized version of Supervised PCA is also applied for nonlinear dimensionality reduction.

The effectiveness of the Supervised PCA methods in comparison with some state-of-the-art algorithms for dimensionality reduction is evaluated in relation to the statistical downscaling process of precipitation in a specific site using two soft computing nonlinear machine learning methods, Support Vector Regression and Relevance Vector Machine.

Extended
Due to the complexity of climate-associated processes, the two main challenges in developing the stochastic regression-based statistical downscaling approaches for climate change projection are: (1) determination of the functional relationship; and (2) identification of predictor variables from high dimensional atmospheric variables conveying climate change information with respect to the hydro-climate variable of interest.

Due to the complexity and nonlinearity of climate associated processes, and the existence of nonlinear interdependency within atmospheric projectors, a kernelized form of supervised dimensionality reduction is able to efficiently model the nonlinear variability of the data.

To better manage future near and long-term surface water resources for various purposes, especially drinking water use, authorities must be ready to mitigate adverse effects of rainfall shortage and surface water reduction under the impact of climate change.

Future research can focus on improving this characteristic of Supervised PCA.

Introduction

The statistical downscaling approaches relying on developing a statistical and quantitative relationship between large-scale atmospheric variables and fine scale variables at a particular site have gained more popularity among hydrologists wanting to predict climate change impacts on hydro-climate variables.

Due to the complexity of climate-associated processes, the two main challenges in developing the stochastic regression-based statistical downscaling approaches for climate change projection are: (1) determination of the functional relationship; and (2) identification of predictor variables from high dimensional atmospheric variables conveying climate change information with respect to the hydro-climate variable of interest.

To address the first challenge, nonlinear soft-computing data-driven regression modeling techniques such as Artificial Neural Networks (ANN) (Tisseuil et al. [8]; Tomassetti et al. [9]), machine learning methods, including Support Vector Machine (SVM) (Chen et al. [10]; Tripathi et al. [11]), and Sparse Bayesian learning algorithm or Relevance Vector Machine (RVM) (Ghosh and Mujumdar [12]; Joshi et al. [13]), have been applied to improve the downscaling of different hydro-climate variables so as to capture the nonlinearity between hydro-climate predictands and atmospheric predictors.

In statistical downscaling processes, projecting a dependent hydro-climate variable from high-dimensional large-scale atmospheric variables leads to inadequate results in terms of performance accuracy, due to the curse of dimensionality.

Dimensionality Reduction Methods

In a given high-dimensional data set, consider projecting a response stochastic variable using a set of independent high-dimensional explanatory random variables.

A preprocessing step deriving an appropriate low-dimensional manifold encoding of a high-dimension data set is crucial to reaching the best performance on learning.

Unsupervised Methods

Since the relationship between climate variables and transformed explanatory atmospheric projectors is still complex and nonlinear, two soft computing nonlinear machine learning methods Support Vector Regression (SVR), and Relevance Vector Machine (RVM) are employed to capture the nonlinearity and evaluate different dimensionality reduction methods in terms of response variable projection performance.

After selecting the best combined dimensionality-reduction and machine-learning method based on the performance criteria, the credibility of the model should be validated under the impact of changing conditions (non-stationarity) arising from global warming.

The different experiments designed to validate the performance of the combined supervised dimensionality reduction and machine-learning models in the current study are discussed in more detail as follows: I. Base experiment (Tr-RAN-Te-RAN) A random selection of training and validating periods (K-fold cross-validation) is used as a scenario for the validity of the model.

Results and Discussion

Doing so, the best-selected models (Kernel Supervised PCA and RVM) are employed on the different dimension-size of the atmospheric projectors formed based on the six predictor domain states to compare the model performances over the study area.

After selecting the best dimensionality reduction method and demonstrating the sensitivities and corresponding sources of uncertainty in terms of predictor sets, the best combination of the Kernel Supervised PCA and the RVM model formed based on the nine surrounding-grid-cells is employed for projecting precipitation time series for the upcoming decades.

Using the same tuned Kernel Supervised PCA model in the modeling section, the derived transformed atmospheric projectors for the upcoming decades based on different scenarios (in the same reduced-dimension extracted in the modeling) are employed for precipitation projection using the best selected RVM data-mining method.

Conclusions

To improve the performance and the predictive power of the statistical downscaling processes with high-dimensional input data, this study has presented a supervised nonlinear dimensionality reduction technique–Supervised PCA–for extracting principal components in which the dependency between the response hydro-climate variable and large-scale atmospheric projectors is maximized.

Due to the complexity and nonlinearity of climate associated processes, and the existence of nonlinear interdependency within atmospheric projectors, a kernelized form of supervised dimensionality reduction is able to efficiently model the nonlinear variability of the data.

The Supervised PCA method is able to capture the complex nonlinear dependency between target precipitation variable and the atmospheric projectors.

The proposed methodology can be used for other hydro-climate variables and also other regression-based statistical downscaling processes to improve the projection accuracy of target hydro-climate variables in the future.

Acknowledgments

A machine generated summary based on the work of Sarhadi, Ali; Burn, Donald H.; Yang, Ge; Ghodsi, Ali (2016 in Climate Dynamics).

The Theory of Parallel Climate Realizations

https://doi.org/10.1007/s10955-019-02445-7

Abstract-Summary

Based on the theory of "snapshot/pullback attractors", we show that important features of the climate change that we are observing can be understood by imagining many replicas of Earth that are not interacting with each other.

These parallel climate realizations evolving in time can be considered as members of an ensemble.

We argue that the contingency of our Earth's climate system is characterized by the multiplicity of parallel climate realizations rather than by the variability that we experience in a time series of our observed past.

The natural measure of the snapshot attractor enables one to determine averages and other statistical quantifiers of the climate at any instant of time.

We recall that systems undergoing climate change are not ergodic, hence temporal averages are generically not appropriate for the instantaneous characterization of the climate.

This can lead in certain climate-change scenarios to the coexistence of two distinct sub-ensembles representing dramatically different climatic options.

The problem of pollutant spreading during climate change is also discussed in the framework of parallel climate realizations.

Extended

A detailed investigation of these and similar quantities is beyond the scope of our paper, and might be the subject of future studies.

Introductory Comments

The traditional theory of chaos in dissipative systems has taught us that on a chaotic attractor there is a plethora of states, all compatible with the single equation of motion of the problem belonging to a fixed set of parameters [14, 15].

The distribution of the parallel states is not arbitrary: an additional property of attractors with chaotic properties is the existence of a unique probability measure, the natural measure [16], which describes the distribution of the permitted states in the phase space [14–16].

In the next Section we show that a changing climate can be described by an extension of the traditional theory of chaotic attractors: in particular, the theory of snapshot/pullback attractors [17, 18] appears to be an appropriate tool to handle the problem.

Particular emphasis is put on the natural measure on snapshot attractors with respect to which averages and other statistics can be evaluated providing a characterization of the climate at any instant of time.

Changing Climates: Mathematical Tools

Speaking, a snapshot or pullback attractor can be considered as a unique object of the phase space of a dissipative dynamical system with arbitrary forcing to which an ensemble of trajectories converges within a basin of attraction.

Even if the concept of snapshot and pullback attractors is practically the same, it is useful to remark here that a pullback attractor is defined as an object that exists along the entire time axis (provided the dynamics remains well-defined back to the remote past), while a snapshot attractor is a slice of this at a given, finite instant of time (their union over all time instants thus constitutes the entire pullback attractor).

However, that if the dynamics is not defined back to the remote past, then the pullback attractor is also undefined, but from some time after initialization the snapshot attractors can be practically identified.

The Theory of Parallel Climate Realizations

In order to make the unusual concept of pullback/snapshot attractors plausible, which might appear too much mathematically-oriented, while the concept of observed time series is widely used, we proposed the term parallel climate realizations in [19].

The ensemble, representing the natural measure of the snapshot attractor, undergoes a change in time due to the time-dependence of the forcing, and, as a consequence, both the "mean state" (average values) and the internal variability of the climate changes with time.

If the climate state itself is changing markedly within such a time interval, these averages unavoidably yield statistical artefacts that may be misinterpreted as they mix up events of the recent and more remote past.

We can say that the ensemble of parallel climate realizations is the generalization of the Gibbs distributions known from statistical physics for a non-equilibrium system whose parameters are drifting in time.

Illustration of Parallel Climate Realizations

An investigation of the Lorenz84 model with seasonal forcing [20] was carried out by Bódai et al. [21] from the point of view of an ensemble approach and led to the conclusion that the snapshot attractor of the forced system appears to be chaotic in spite of the fact that in extended regions of the forcing parameter F of the time-independent system the attractors are periodic.

Additional issues about initialization may arise from the insufficient spinup time: drifts corresponding to the convergence process from some state off the attractor in deep oceanic variables may appear and are actually documented in the MPI-GE [22], while their importance is mostly unknown in the other ensembles.

If one takes the time evolution of the atmospheric variables in different members of such a hypothetical full ensemble, and then constructs an ocean ensemble with each of these atmospheric realizations applied as a fixed forcing, the result will be an extended set of OCCIPUT-type ocean ensembles.

Nonergodicity and Its Quantification

The original observation of Romeiras et al. [17], according to which a single long trajectory traces out a pattern different from that of an ensemble stopped at a given instant, implies that ergodicity (in the sense of the coincidence of ensemble and temporal averages) is not met in nonautonomous systems.

Traditional chaotic attractors are known to be ergodic [16]: sufficiently long temporal averages coincide with averages taken with respect to sufficiently large ensembles.

The nonergodic mismatch can be evaluated along each single realization of the climate ensemble and depends on the realization.

Teleconnections: Analyzing Spatial Correlations

Investigating the teleconnections through the temporal correlations between a so-called teleconnection index and another variable (e.g., temperature or precipitation) a single correlation coefficient can be obtained.

With a sufficiently large number (N) of realizations an ensemble-based instantaneous correlation coefficient can be defined which provides the appropriate characterization of the strength of teleconnections in the spirit of parallel climate realizations.

The NAO teleconnection index (NAOI) is based on the difference in the normalized sea level pressure between Iceland and the Azores.

At a given time instant, it is also possible to compute an instantaneous teleconnection index in the spirit of parallel climate realizations.

An increasing strength of the teleconnection between a particular ENSO index and the Indian summer precipitation has been detected in the MPI Grand Ensemble in the twentieth century.

Ensembles in Experiments

Such experimental investigations nicely complement research based on numerical general circulation models: the latter can, theoretically, access the full set of parameters but with a limited resolution which hides important subgrid-scale nonlinear phenomena that may affect multiple scales.

Reproducing them with the same boundary conditions (forcing) most naturally provides an ensemble of different realizations of the same process, which represents the multitude of possibilities permitted by turbulence or chaotic-like phenomena, i.e., parallel realizations of the minimal climate system model.

Besides climate-related aspects it is worth noting that the ensemble approach may be the proper way to conduct fluid dynamics experiments in which non-equilibrium (non-ergodic) processes and turbulence are involved, i.e. phenomena characterized by inherent internal variability.

Only an ensemble statistics of these lifespans from a multitude of experiments (that are initiated identically within measurement precision) can provide meaningful information of these interesting intermittent phenomena, as demonstrated in e.g. [23].

Splitting of the Snapshot Attractor

An important property of the climate system is that for some range of fixed parameter values, it also allows two coexisting usual (stationary) attractors.

Even when initializing the ensemble entirely inside one of the basins of attraction (that belongs to the initial parameter value), only a fraction of the ensemble may end up on the usual attractor on which the ensemble was started.

During the returning part of the parameter drift, at the point when this usual attractor reappears, the snapshot attractor (as an extended object) may overlap with the basin of attraction of both of the coexisting usual attractors.

The separation of the snapshot attractor to two unconnected branches, between which transition of trajectories is not possible, stems from the fact that the corresponding stationary system is not ergodic in the sense of the existence of a unique global asymptotic probability measure [24].

Spreading of Pollutants in a Changing Climate

As an additional utilization of an ensemble of parallel climate realizations, the change in the intensity of atmospheric large-scale spreading of pollutants can also be investigated in a changing climate.

The intensity of the spreading can be characterized in general by such stretching rates [25–27].

In [27] in order to explore what the typical spreading behavior is in a changing climate, ensemble simulations of the PlaSim and CESM climate models were used.

Of climate change, spreading simulations showed an overall decreasing trend in the stretching rate in the ensembles of both climate models.

Temporal Aspects: An Emerging Research Direction

This is actually rather intuitive, since the statistical or dynamical relationship between two time instants separated by a given time is not temporally invariant any more: it depends on when within the climate change either of these instants is chosen.

To characterize the relationship between temporally separated values of a given variable, a "workaround" is to compute the correlation coefficient between two time instants with respect to the time-dependent natural probability measure (with respect to temporally evolving ensemble members in practice).

It is meaningful to compute the temporal average or standard deviation of some variable for e.g. a given decade, but then this average or standard deviation will have its own probabilistic description as defined via the time-dependent natural measure.

The ensemble average of this interval-wise taken quantifier should not be confused with the corresponding ensemble quantifier of a time instant within the given time interval: while these two characterizations coincide in a stationary climate, biases are introduced if the climate is changing.

Conclusion

The concepts of the average and the deviation from it also appear in the IPCC report [28], but it also considers averages taken over different climate models relevant.

The different models, however, describe climates of "different physics", the differences of which do not reflect the internal variability of the climate, rather the perhaps significant inaccuracies of the models.

In the spirit of the article, it seems more appropriate to evaluate projections within single models based on parallel climate histories.

We wish to briefly address the characterization of model uncertainties within a single climate model.

Acknowledgments

A machine generated summary based on the work of Tél, T.; Bódai, T.; Drótos, G.; Haszpra, T.; Herein, M.; Kaszás, B.; Vincze, M. (2019 in Journal of Statistical Physics).

Understanding the Links Between Climate Feedbacks, Variability and Change Using a Two-Layer Energy Balance Model

https://doi.org/10.1007/s00382-020-05189-3

Abstract-Summary

A simple, two-layer energy balance model (EBM) is used to investigate climate variability in Coupled Model Intercomparison Project Phase 5 (CMIP5) models and examine possible links between variability and climate sensitivity, and the roles of stochastic variability, radiative feedbacks and ocean mixing.

The EBM represents global variability that, while somewhat stronger than the CMIP5 models, simulates reasonable ratios between shorter and longer timescales.

Variability in the EBM to the range of parameters from the Global Climate Models is found to be particularly sensitive to stochastic variability, especially on interannual time-scales.

The EBM results suggests that spread in stochastic forcing across the CMIP5 models is the single greatest factor degrading the correlation between variability and climate sensitivity, although model to model differences in radiative forcing and mixing into the deep ocean are also important.

They also suggest that normalizing variability in general circulation models by stochastic forcing, uptake into the deep ocean and radiative forcing are all important first steps to reduce factors that will otherwise confound the correlations.

Introduction

The approach taken is to develop and utilize a two-layer energy balance/feedback model (EBM) for the climate system, to explore and understand its variability on a range of timescales and to relate these to variability and climate change sensitivity found in the CMIP5 GCMs.

We will use the EBM approach to explore four questions: (1) How well can important aspects of global scale variability on timescales from interannual to multi-decades in CMIP5 models be understood and quantitatively described using a simple two-layer EBM? (2) What relative role do radiative feedbacks play in determining the magnitude of global variability, especially on longer timescales? (3) What parameters control potential relationships between the magnitude of variability and transient climate response (TCR) (Collins et al. [29]) and/or ECS. (4) What do differences across GCMs in their magnitude of stochastic forcing, the strength of radiative feedbacks and in other parameters therefore imply for the potential for constraining ECS or TCR through observations of variability.

Model Description and Analysis Methodology

Estimates of temperature variance and stochastic forcing from the CMIP5 models were calculated by first detrending annual mean temperatures and TOA radiation (to remove any residual drift), then removing the annual cycle by subtracting off mean January, mean February etc. For temperature, annual, monthly, decadal and 30-year variances we calculated after first averaging the monthly temperature fluctuations into annual means, then passing 10 year and 30 year running means through these timeseries prior to the calculation of variances.

Observational estimates based on CERES (Clouds and the Earth's Radiant Energy System) satellite data indicate that global scale total TOA variability has a standard deviation of around 0.62 W m^{-2} on monthly timescales (Trenberth et al. [30]), a value comparable to the multi model mean (although, as with the models, some of the observed value will likely represent the response, i.e. feedback, from surface temperature changes).

Sensitivity of Variability in the EBM

The purpose of this is to understand how changes in these parameters affect temperature variability on different timescales before then considering how these parameters affect correlation between variability and sensitivity.

It is important to collate the climate variability computed numerically via the two-layer model with those represented in GCMs.

This gives us some confidence that the simple two-layer EBM both qualitatively and quantitatively reproduces overall features of variability from interannual to multi-decades in comparison with CMIP5 models.

In calculations, we applied the monothetic OFAT (one-factor-at-a-time) analysis, varying each parameter over its range and holding others at their base (i.e. CMIP5 model average) values.

Section we evaluate how the EBM and CMIP5 variances range across the ensemble of models, and what the EBM implies for the relationship between climate variability across different timescales and climate sensitivity.

Analysis of Climate Variability and Change of CMIP5 Models

The EBM predicts a high degree of correlation (i.e. high explained variance across the models) between variability and ECS with an R^2 of 0.58 at interannual timescales, and up to 0.68 for 30-year.

It can be easily understood in the case of the ECS/variability correlation: the only factor producing spread in the ECS remains F, which plays no role in variability in the EBM.

The EBM predicts that without corresponding feedbacks operating these variables do not produce any significant correlation.

Removing the spread in γ, however, has the counter intuitive effect of decreasing the correlation between variability and ECS.

The impact of F is easily understood: it causes spread in ECS but does not affect variability, so reducing its spread produces greater correlation.

The puzzle is why eliminating the spread in γ reduces the correlation between ECS and variability.

Summary and Conclusions

We use a simple 2-layer energy balance model (EBM) to ask what factors might contribute to the spread in variability, and which factors might provide (or indeed limit) the degree of correlation between the magnitude of unforced variability and climate sensitivity (both ECS and TCR) across timescales from interannual to multi-decadal.

The correlation across CMIP5 models between the GCM variances and those simulated by the EBM are modest, with around 25% variability explained for longer timescale (decadal and 30-year).

The EBM predicts that the correlations between sensitivity and variability should be higher at longer timescales in the GCMs.

The EBM predicts lower correlations between variability and TCR than with ECS, consistent with there ocean heat uptake factors affecting TCR, whereas ECS is dependent on forcing and feedback alone.

The role of stochastic forcing in the current results is striking, as the EBM suggests that it could be a key 'spoiler' of cross GCM climate change/variability correlations.

Acknowledgments

A machine generated summary based on the work of Colman, Robert; Soldatenko, Sergei (2020 in Climate Dynamics).

A Voyage Through Scales, a Missing Quadrillion and Why the Climate is not What You Expect

https://doi.org/10.1007/s00382-014-2324-0

Abstract-Summary

Using modern climate data and paleodata, we voyage through 17 orders of magnitude in scale explicitly displaying the astounding temporal variability of the atmosphere from fractions of a second to hundreds of millions of years.

We identify five of these: weather, macroweather, climate, macroclimate and megaclimate, with rough transition scales of 10 days, 50 years, 80 kyears, 0.5 Myear, and we quantify each with scaling exponents.

Mean temperature fluctuations increase up to about 5 K at 10 days (the lifetime of planetary structures), then decrease to about 0.2 K at 50 years, and then increase again to about 5 K at glacial-interglacial scales.

Both deterministic General Circulation Models (GCM's) with fixed forcings ("control runs") and stochastic turbulence-based models reproduce weather and macroweather, but not the climate; for this we require "climate forcings" and/or new slow climate processes.

Averaging macroweather over periods increasing to ≈30–50 years yields apparently converging values: macroweather is "what you expect".

Macroweather averages over ≈30–50 years have the lowest variability, they yield well defined climate states and justify the otherwise ad hoc "climate normal" period.

Moving to longer periods, these states increasingly fluctuate: just as with the weather, the climate changes in an apparently unstable manner; the climate is not what you expect.

Moving to time scales beyond 100 kyears, to the macroclimate regime, we find that averaging the varying climate increasingly converges, but ultimately—at scales beyond ≈0.5 Myear in the megaclimate, we discover that the apparent point of convergence itself starts to "wander", presumably representing shifts from one climate to another.

Introduction: Foreground or Background, Signal or Noise?
If we attempt to extend Mitchell's picture to the dissipation scales at frequencies 6 or 7 orders of magnitude higher (for millimetric spatial scale variability), the spectral range would increase by an additional ten or so orders of magnitude.

In Mitchell's time, this scale bound view had already led to an atmospheric dynamics framework that emphasized the importance of numerous processes occurring at well defined time scales, the quasi periodic "foreground" processes illustrated as bumps—the signals—on Mitchell's nearly flat background.

The purpose of this paper is therefore to stand Mitchell on his head, to invert the roles of foreground and background—of signal and noise—to treat the spectral continuum with its challenging and nontrivial multifractal scaling, as the fundamental signal and to relegate the residual quasiperiodic processes to the role of background processes where they belong.

Standing Mitchell on His Head: The Scaling Paradigm
By the early 1980s, following the explosion of scaling (fractal) ideas it was realized that scale invariance was a very general symmetry principle often respected by nonlinear dynamics, including many geophysical processes and turbulence.

In nonlinear dynamical systems, power laws arise when over a range of scales there are no processes strong enough to break the scaling symmetry.

Another way of putting this is to say that the dominant dynamical processes occur in synergy over a wide range of scales, with the resulting behaviour displaying no characteristic size or duration.

We can express this in yet another way in terms of systems theory: $H < 0$ indicates negative feedbacks occurring over a wide range of scales in a scale invariant way whereas $H > 0$, indicates positive feedbacks occurring over a wide range (this should not be confused with persistence and antipersistence which for Gaussian processes refer to fluctuations growing more or less quickly than Brownian motion).

Scaling in the Weather, Macroweather and Climate Regimes

Starting with the climate regime, numerous paleo temperature series (mostly from ice and ocean cores) have been analyzed and there is broad agreement on their scaling nature with spectral exponents estimated in the range $\beta_c \approx 1.3$ to 2.1 over the range from hundreds to tens of thousands of years, (Lovejoy and Schertzer [31, 32]; Schmitt et al. [33]; Ditlevsen et al. [34]; Pelletier [35]; Ashkenazy et al. [36]; Wunsch [37]; Huybers and Curry [38]; Blender et al. [39]; Lovejoy [40]; Rypdal and Rypdal [41]).

A seductive feature of the (anisotropic) scaling framework is that it fairly accurately predicts the weather to macroweather transition scale $\tau_w \approx 10$ days.

The analogous calculation for the ocean using the empirical (near surface) ocean turbulent flux $\varepsilon \approx 10^{-8}$ W/kg, yields a lifetime of ≈ 1 year which is indeed the scale separating a high frequency "ocean weather" (with $\beta > 1$) from a low frequency "macro-ocean weather" with $\beta < 1$ (Lovejoy and Schertzer [32]; at depth, ε is much lower and the corresponding lifetimes are much longer).

Real Space Fluctuations and Analyses

The behavior of the mean fluctuation is thus $<\Delta T> \approx \Delta t^H$ so that if $H > 0$, on average fluctuations tend to grow with scale whereas if $H < 0$, they tend to decrease.

While the latter is adequate for fluctuations increasing with scale (i.e. $H > 0$), mean absolute differences generally increase and so when $H < 0$, they do not correctly estimate fluctuations.

Once estimated, the variation of the fluctuations with scale can be quantified by using their statistics; the qth order structure function $S_q(\Delta t)$ is particularly convenient: where " $<$.

For periods longer than this, the statistics are dominated by averages of many planetary scale structures, and these fluctuations tend to cancel out: for example large temperature increases are typically followed (and partially cancelled) by corresponding decreases.

The consequence is that in this macroweather regime, the average fluctuations diminish as the time scale increases.

Discussion

Avoiding anthropogenic effects by considering the pre-1900 epoch, for GCM climate models, the key question is whether solar, volcanic, orbital or other climate forcings are sufficient to arrest the $H < 0$ decline in macroweather fluctuations and to create an $H > 0$ regime with sufficiently strong centennial, millennial variability to account for the background variability out to glacial-interglacial scales.

Whatever the ultimate source of the growing fluctuations in the $H > 0$ climate regime, a careful and complete characterization of the scaling in space as well as in time (including possible space–time anisotropies) allows for new stochastic methods for predicting the climate.

the conventional but ad hoc "climate normal" period—this not only justifies the normal but allows averages of relevant variables over it to define "climate states" and the changes at scales $\Delta t > \tau_c$ to define climate change (again, in the recent period, this defines the scale at which anthropogenic variability starts to dominate natural variability).

Conclusions

A far more realistic picture of atmospheric variability is obtained by standing this scale bound picture on its head: placing the continuum processes in the fore, with the perturbing quasiperiodic processes in the background.

The empirically substantiated picture is rather one of "unstable", "wandering", high frequency weather processes (i.e. $H > 0$) tending—at scales beyond 10 days or so—and primarily due to the quenching of spatial degrees of freedom (intermediate frequency, low variability)—to macroweather processes.

True climate processes are "weather-like" ($H > 0$) and only emerge from macroweather at even lower frequencies, due to new slow internal climate processes coupled with external forcings (including in the recent period, anthropogenic forcings).

Whatever the cause, it is an empirical fact that the emergent synergy of new processes yields fluctuations that on average again grow with scale in at least a roughly scaling manner and become dominant typically on time scales of 30–100 years (somewhat less in the recent period) up to ≈ 100 kyears.

Acknowledgments

A machine generated summary based on the work of Lovejoy, S. (2014 in Climate Dynamics).

A New Framework for Climate Sensitivity and Prediction: A Modelling Perspective

https://doi.org/10.1007/s00382-015-2657-3

Abstract-Summary

The sensitivity of climate models to increasing CO_2 concentration and the climate response at decadal time-scales are still major factors of uncertainty for the assessment of the long and short term effects of anthropogenic climate change.

While the relative slow progress on these issues is partly due to the inherent inaccuracies of numerical climate models, this also hints at the need for stronger theoretical foundations to the problem of studying climate sensitivity and performing climate change predictions with numerical models.

Response theory puts the concept of climate sensitivity on firm theoretical grounds, and addresses rigorously the problem of predictability at different time-scales.

These results show that performing climate change experiments with general circulation models is a well defined problem from a physical and mathematical point of view.

These results show that considering one single CO_2 forcing scenario is enough to construct operators able to predict the response of climatic observables to any other CO_2 forcing scenario, without the need to perform additional numerical simulations.

We also introduce a general relationship between climate sensitivity and climate response at different time scales, thus providing an explicit definition of the inertia of the system at different time scales.

Introduction

We follow a complementary approach to define a robust theoretical framework for the use of GCMs in addressing the problem of climate response, sensitivity and prediction, based upon Lucarini and Sarno [42].

The standard approach to the problem of computing the response of a climate model to the forcing due to a increasing CO_2 concentration is the following: take a model, run it to a stationary state, increase the CO_2 concentration following one specific CO_2 increase scenario, measure the increase of global surface temperature, define on it operational measures of the sensitivity of the system.

The approach suggested here makes it possible to compare models in a new way, showing how the equilibrium climate sensitivity is just one point of a function that contain much more informations about the properties of the response to components of the forcing at different time-scales.

Methods and Materials

Axiom A systems possess a Sinai-Ruelle-Bowen (SRB) invariant measure, which guarantees (a) the asymptotic equivalence of time and ensemble averages of observables (that it is not, despite intuition, a general property of nonequilibrium systems) and (b) the stability of the statistical properties when a weak stochastic forcing is applied.

The use of response formulas in most cases of physical interest is justified thanks to the Chaotic Hypothesis (Gallavotti [43]), which states that chaotic systems with many degrees of freedom effectively behave as Axiom A systems in terms of properties (a) and (b) even if they do not satisfy rigorously requirements (1) and (2), at least when considering the statistical properties of coarse-grained observables (e.g. globally or regionally integrated quantities).

When we compute the expectation value of an observable in a numerical model as the long-term average on a stationary state, we are in fact implicitly assuming that the system is Axiom A-like.

Results

The long-term increase of the surface temperature for the doubling scenario (the equilibrium climate sensitivity) is rather high if compared with what is typically obtained with standard IPCC models, being 8.1 K against typical estimates between 1.5 and 4.5 K (IPCC [44, 45]).

Another interesting application the possibility to estimate the time horizon on which the mean climate change signal is distinguishable from the natural variability of the climate system for different rates of change of the forcing.

Given the Green function of the system, we can compute the expected mean climate change signal for forcing corresponding to different rates of change of the CO_2 concentration, and check after how many years the mean signal is larger than a chosen number of standard deviations of the observable in the unperturbed system.

Summary and Discussion

The approach proposed here bypasses some of these mathematical issues by exploiting formal properties of the response and allows for constructing rigorous definitions of climate sensitivity at different time scales through the susceptibility function.

We have provided a framework for relating the difference between transient and equilibrium climate sensitivity to the inertia of the CS, and have shown how these properties depend of the response of the system on all time scales.

Inaccuracies in representing specific spectral features have serious impacts on our ability to predict climate response on the corresponding time scales, and our findings could help understanding why, e.g., climate response at decadal time scales may be hard to capture.

RRT provides a well defined theoretical framework and tools that allows to diagnose rigorously discrepancies in the properties of the frequency dependent response of different models and to guide the design of the climate change experiments.

Acknowledgments

A machine generated summary based on the work of Ragone, Francesco; Lucarini, Valerio; Lunkeit, Frank (2015 in Climate Dynamics).

Effect of AMOC Collapse on ENSO in a High Resolution General Circulation Model

https://doi.org/10.1007/s00382-017-3756-0

Abstract-Summary

We look at changes in the El Niño Southern Oscillation (ENSO) in a high-resolution eddy-permitting climate model experiment in which the Atlantic Meridional Circulation (AMOC) is switched off using freshwater hosing.

Convergence of this transport deepens the thermocline in the eastern tropical Pacific and increases the temperature anomaly relaxation time, causing increased ENSO period.

The anomalous Ekman transport is caused by a surface northerly wind anomaly in response to the meridional sea surface temperature dipole that results from switching the AMOC off.

To a previous study with an earlier version of the model, which showed an increase in ENSO amplitude in an AMOC off experiment, here the amplitude remains the same as in the AMOC on control state.

Extended

Yu et al. [46] have suggested a link between AMO and El Niño location, stronger AMOC leading to more central El Niño events and conversely weaker AMOC to more eastern El Niño events, very similar to our findings.

Introduction

The AMOC off state in this simulation is stable over the 450 years duration of the model integration (Mecking et al. [47]) which is then compared with a control run making the study a comparatively clean assessment of the impacts of AMOC shutdown.

We look in particular at the differences in ENSO resulting from the global climatic changes that collapse of the AMOC can induce in the model.

There have also been studies of ENSO in hosed, weakened AMOC runs of CMIP3 era models (Timmerman et al. [48]; Dong and Sutton [49]) and in most models there was a substantial weakening of the annual cycle in the eastern equatorial Pacific and an increase in ENSO amplitude.

Using a stochastically forced damped oscillator model of slow ENSO dynamics introduced by Jin [50] to qualitatively understand the response of the much more complicated HadGEM3, we suggest the difference in ENSO amplitude between the different models is due to the balance of changes in ENSO damping and the magnitude of stochastic forcing.

Model setup and Experiment Design

To repeat pertinent details, the model is the Global Coupled 2.0 model (GC2) configuration of the HadGEM3 model (Hewitt et al. [51]).

This consists of an atmosphere, ocean, sea-ice and land-surface models.

The atmosphere model is Global Atmosphere vn6.0 (GA6) (Demory et al. [52]) of the Met Office unified model at N216 horizontal resolution and 85 levels in the vertical.

Two runs of the model are compared, a steady state control run (the AMOC is in its usual on state in this run) and an AMOC off steady state run.

The AMOC off run is initialised after 42 years of the control run.

The AMOC off run is integrated for a total of 450 years from the start of the salinity perturbations.

Analysis we use all 150 years of the control run and the last 300 years of the AMOC off run to determine ENSO properties.

Results

The leading EOF in both the control and the AMOC off run is representative of the ENSO mode.

Relative to the control run, the AMOC off ENSO EOF accounts for proportionally slightly less of the total variance in SST (60 vs. 63%) although this difference, both proportional and absolute, is within estimated error bounds.

The variance differences at each spatial location were tested for significance at the 99% confidence level by performing a two sample f-test on the deseasonalized spatial SST anomaly fields in both control and AMOC off runs.

Time varying properties are analyzed by projecting the leading EOF onto the time ordered fields of deseasonlized monthly SST for both control and AMOC off runs.

There is also a shift to longer ENSO periods in the AMOC off relative to the control run which does appear to be significantly different.

Mechanisms for Differences in ENSO

Having established that ENSO in the AMOC off run relative to the control has (1) a similar amplitude and distribution of SST anomalies, (2) a spatial pattern shifted eastward and (3) a longer, more regular period, we discuss mechanisms that could result in these differences.

One sees a deepening of the thermocline in the east Pacific concentrated in the region of large ENSO variability as expected from the changes in the wind fields and the Ekman transport divergence.

Apart from the eastward shift in ENSO and the mean state, the other change is the mean depth of the thermocline in the eastern equatorial Pacific.

b is difficult to estimate from regressions of east to west thermocline depth difference versus east to west temperature difference using the HadGEM3 simulations as the model never reaches a true equilibrium between the thermocline depth difference and the wind stress at a given time due to the different adjustment time scales (and therefore lags) of the SST and thermocline depths in the east and west Pacific.

Discussion

All 5 models in Timmerman et al. [48] had significantly increased ENSO amplitudes as measured by the power spectra of the SST anomalies in the Niño 3 region.

The increase of peaks without broadening in the CMIP3 model's power spectra suggest their ENSOs become more periodic and less damped.

CMIP5-class models show an improvement in terms of representing both the properties of ENSO (amplitude, frequency, spatial pattern) and the physical processes and feedbacks which are responsible for generating and maintaining the oscillation (Bellenger et al. [53]).

From the power spectra in Timmerman et al. [48] it appears that there is no change in ENSO period for most CMIP3 models.

As argued above, from the CMIP3 model's power spectra it seems likely that their ENSOs become less damped and more periodic.

Conclusion

The increase in ENSO period is backed up using a simple model of ENSO as a stochastically forced damped oscillator.

Using the simple model, one can potentially understand the differing responses in the slow ENSO dynamics as a competition between the decrease in damping tending to increase amplitude and the decrease in forcing tending to decrease the amplitude.

Acknowledgments

A machine generated summary based on the work of Williamson, Mark S.; Collins, Mat; Drijfhout, Sybren S.; Kahana, Ron; Mecking, Jennifer V.; Lenton, Timothy M. (2017 in Climate Dynamics).

On the Relationship Between Atlantic Niño Variability and Ocean Dynamics

https://doi.org/10.1007/s00382-017-3943-z

Abstract-Summary

We address the question whether the equatorial SST bias affects the ability of a coupled global climate model to produce realistic dynamical SST variability.

We assess this by decomposing SST variability into dynamical and stochastic components.

To compare our model results with observations, we employ empirical linear models of dynamical SST that, based on the Bjerknes feedback, use the two predictors sea surface height and zonal surface wind.

We find that observed dynamical SST variance shows a pronounced seasonal cycle.

This indicates that the Atlantic Niño is a dynamical phenomenon that is related to the Bjerknes feedback.

In the coupled model, the SST bias suppresses the summer peak in dynamical SST variance.

Bias reduction, however, improves the representation of the seasonal cold tongue and enhances dynamical SST variability by supplying a background state that allows key feedbacks of the tropical ocean–atmosphere system to operate in the model.

Due to the small zonal extent of the equatorial Atlantic, the observed Bjerknes feedback acts quasi-instantaneously during the dynamically active periods of boreal summer and early boreal winter.

Extended

To compare our results with the evolution of the observed climate system, we use the ERA-Interim (Dee [54]) and the Archiving, Validation, and Interpretation of Satellite Oceanographic (AVISO) datasets.

We find that differences between ERA-Interim SST and other SST datasets are negligibly small. (Analysis results for alternative validation datasets such as the HadISST dataset (Rayner et al. [55]) are not shown.

We find that the Atlantic Bjerknes feedback is near-instantaneous during the dynamically active phases of the year.

In the coupled equatorial system, the atmosphere would react to the ocean-induced SST variability and our empirical models would pick up a statistical co-variability between SST and u10 that would be partly reflected in our SST decomposition— even though u10 in this idealized example was not fundamental in causing the SST variability in the first place.

Future study will help to further our understanding of the Atlantic Niño and its predictability.

Introduction

In a positive feedback, it relates SST and thermocline variability in the eastern ocean basin to zonal surface wind variability in the western ocean basin (u10) and lends growth to the Pacific (Bjerknes [56]) and Atlantic Niños (e.g. Keenlyside and Latif [57]; Burls et al. [58]; Lübbecke and McPhaden [59]; Deppenmeier et al. [60]).

The Pacific Niño generally is the result of a free mode of interannual variability that is driven by the Bjerknes feedback; interactions with the seasonal cycle occur, but do not dominate ENSO SST variability.

Burls et al. [58] argue that the Atlantic Niño hence reflects a modulation of the seasonally active Bjerknes feedback instead of an independent mode of interannual variability.

In contrast to numerous studies that have provided evidence for a relationship between Atlantic Niño variability and the Atlantic Bjerknes feedback, Nnamchi et al. [61, 62] have proposed that the Atlantic Niño is essentially driven by stochastic processes in the atmosphere rather than by dynamical ocean processes that are potentially predictable.

Model and Methods

All ensemble members use the same wind stress forcing, but differ in their initial conditions, which are taken from a control run at a time when the model is close to equilibrium.

In a partially coupled model the ocean and sea ice components are forced with observed wind stress anomalies that are added to the model's monthly mean wind stress climatology.

To diagnose the heat flux correction, we use the same methodology as Ding et al. [63]: During a control integration, we nudge the first ocean level of the model towards the monthly climatology of observed SST with a restoring time scale of 10 days.

We note that Ding et al. [63] showed a substantial improvement in the ability of the partially coupled model runs to reproduce observed SST variability in boreal summer in FLX compared to STD.

Impact of the Coupled Bias on the Equatorial Atlantic

We assess SST and zonal wind biases in the tropical Atlantic for our KCM experiments.

Richter and Xie [64] and Richter et al. [65] have shown in different CGCMs that the equatorial Atlantic SST bias is related to a bias in zonal surface wind in the western equatorial Atlantic, which in turn can be traced back to precipitation deficiencies of the models.

In agreement with Richter et al. [66], a similar process could be at work in CGCMs: Spring zonal winds that are systematically too weak in the western equatorial Atlantic could inhibit seasonal thermocline shoaling in the eastern ocean basin and hence intense surface cooling during early boreal summer.

This behaviour is hardly altered qualitatively in the FLX experiment, indicating that the zonal wind bias depends only weakly on eastern basin SST in the model.

SST Variance Decomposition Method

To diagnose the observed dynamical SST variance, we use observed thermocline depth in the eastern equatorial Atlantic to model ERA-Interim SST variability in the same region.

That our decomposition approach heavily relies on empirical linear models, but that the resulting decomposition of the SST variance is not linear, i.e., the full SST variance is not the sum of the stochastic and dynamical SST variance. (The basic decomposition of the SST anomaly, however, is.) Here, we use the Bjerknes feedback as the dynamical framework for our empirical models of dynamical SST.

This indicates either that the co-variability between our predictors is strong during the respective month and that using either of them provides sufficient information to produce reasonable dynamical SST; or that the removed predictor does not have a strong impact on SST variability during this month. (2) Model adjustment keeps both predictors in a linear combination. (3) Model adjustment increases the complexity of the model by adding a non-linear predictor term, i.e. a quadratic term or a product of SSH and u10.

Seasonality of Dynamical SST Variance in the tropical Atlantic

The overall similarity between the total and dynamical SST variance suggests that the seasonal cycle of total SST variability in the tropical Atlantic is largely shaped by the variable dynamical contribution.

The dynamical and stochastic SST variances for the model experiment FLX are comparable to observations (blue).

The absolute minimum of dynamical SST variance occurs in May—when observed dynamical SST variance is already high and contributes substantially to the overall boreal summer peak.

Once the cold tongue is established, the feedbacks set in and contribute to dynamical SST variability.

The STD SST bias decreases and our empirical models operate on comparable conditions, resulting in dynamical SST variances in the STD experiment that are similar to observations in boreal fall and early boreal winter.

In boreal winter, the single u10 model does not contribute to dynamical SST variance.

Feedback Strengths in the Tropical Atlantic

Recall that the three relationships that make up the closed Bjerknes feedback in our framework are (1) Atl3 SST produces WAtl u10 variability, (2) WAtl u10 variability is translated into Atl3 thermocline—here: SSH—variability via equatorial wave dynamics, and (3) Atl3 SSH positively feeds back to Atl3 SST and lends growth to the initial SST anomaly.

We assess the degree of lag for each of the three Bjerknes relationships via a cross-correlation analysis for each month.

In our cross-correlation analysis for each calendar month and Bjerknes feedback element, we fix the response agent to the calendar month and correlate it sequentially with the forcing agent of all relevant lags.

Black crosses indicate the lag for which the relationship in terms of the ACC is strongest for the considered calendar month and Bjerknes feedback element.

Summary and Discussion

In agreement with numerous previous studies on the dynamics of the Atlantic Niño (e.g. Zebiak [67]; Carton et al. [68]; Ding et al. [69]), we find that dynamical SST variance contributes substantially to equatorial Atlantic SST variability in boreal summer (May–July), the peak phase of the Atlantic Niño.

That, in contrast to May and June, the December peak of enhanced dynamical SST variance is captured by both the FLX and STD experiments, indicating that the KCM appears to be able to reproduce the variability associated with Okumura and Xie [70])'s Atlantic Niño II.

While we have provided further evidence for a dynamically driven Atlantic Niño, research is not yet clear on what exactly these dynamics are: If the Bjerknes feedback is involved in establishing the seasonal cold tongue, which processes govern the feedback modulation that produces the interannual variability of the Atlantic Niño?

Acknowledgments

A machine generated summary based on the work of Dippe, Tina; Greatbatch, Richard J.; Ding, Hui (2017 in Climate Dynamics).

A Theoretical Model of Strong and Moderate El Niño Regimes

https://doi.org/10.1007/s00382-018-4100-z

Abstract-Summary

The existence of two regimes for El Niño (EN) events, moderate and strong, has been previously shown in the GFDL CM2.1 climate model and also suggested in observations.

Although the recent 2015–16 EN event provides a new data point consistent with the sparse strong EN regime, it is not enough to statistically reject the null hypothesis of a unimodal distribution based on observations alone.

We implemented this nonlinear mechanism in the recharge-discharge (RD) ENSO model and show that it is sufficient to produce the two EN regimes, i.e. a bimodal distribution in peak surface temperature (T) during EN events.

Using the Fokker–Planck equation, we show how the bimodal probability distribution of EN events arises from the nonlinear Bjerknes feedback and also propose that the increase in the net feedback with increasing T is a necessary condition for bimodality in the RD model.

Extended

Despite the strong simplifications, we show that this model reproduces two EN regimes and provides insights into the role of the stochastic forcing in El Niño diversity and predictability.

Introduction

We focus on the latter, particularly on our proposal that strong EN events (e.g. 1982–83 and 1997–98) correspond to a separate dynamical regime associated with nonlinearity in the Bjerknes feedback (TD16).

A theoretical model with nonlinear ocean advection (Timmermann et al. [71]; An and Jin [72]) produces strong EN in the form of "bursts" as part of complex self-sustained nonlinear oscillations, but these only have a weak resemblance to observations.

Although all these nonlinear mechanisms could contribute to ENSO, no study to our knowledge has addressed the origin of strong and moderate dynamical regimes of El Niño (warm) events.

This model focuses only on the strength of El Niño events as a first approximation to ENSO diversity, neglecting the spatial distribution or seasonal effects, or nonlinear processes specific to La Niña.

Despite the strong simplifications, we show that this model reproduces two EN regimes and provides insights into the role of the stochastic forcing in El Niño diversity and predictability.

Recharge-Discharge Model

This also includes the nonlinear radiative cloud feedback that enhances damping in the convective regime (Lloyd et al. [73]; Bellenger et al. [53]).

Fitting the linear RD model to the nonlinear RD model run produces a weaker effective linear damping parameter to the original from Burgers et al. [74], as expected.

This reduced by 45% weaker damping that the Burgers et al. [74].

Fokker–Planck Equation

The Fokker–Planck (FP) equation describes the evolution of the probability distribution function (PDF) of states in a stochastically-forced ("Brownian") dynamical system (Risken [75]), which allows us to address issues of predictability in simple climate models by describing how the PDF evolves from an initial condition under all possible realizations of the stochastic forcing (e.g. Hasselmann [76]).

Using the terminology of the FP equation, the PDF evolution is governed by the "drift", which is the displacement, rotation, and deformation of the PDF by the deterministic dynamics, and the "diffusion", which is the spreading of the PDF due to the random walk associated with the stochastic forcing.

Results

The skewness of the distribution of EN T peaks in the nonlinear model (1.47) is larger than the observational value (1.11), whereas the linear model is lower (0.89).

An important aspect of the onset of these observed strong EN events is that, in contrast to the "pure" (unforced) RD dynamics, in general h does not decrease as sharply when T increases towards its peak as afterwards, during EN decline; in 1982, h even increased right up to two months prior to the peak T. This indicates that, if the RD model is indeed representative of the underlying dynamics, the onset of the 1982 (and probably also the other) strong EN was strongly facilitated by external forcing (TD16).

After the observed EN peaks, the pronounced discharge process leads to large negative h, but the associated La Niña peak T anomalies are not as large as the ones for the EN events in the Niño 3 region.

Discussion

Other empirical methods (e.g. Burgers et al. [74]) could be adapted to consider this nonlinearity, or the proposed model could be used to derive an estimate of the non-linear feedback through assimilation of observations.

This model highlights the key role of the stochastic forcing, particularly the component of the forcing on ENSO time-scales, in the growth of the strong EN events (e.g. Levine and Jin [77]; TD16).

It is often assumed that the low-frequency positive forcing is the result of clustering of short-term westerly wind events, either randomly or modulated by SST (e.g. Gebbie et al. [78]; Zavala-Garay et al. [79]; Gebbie and Tziperman [80]).

The strong La Niña following strong EN events in our nonlinear model is consistent with the strong heat content discharge that is seen in observations, except that in observations the discharge does not necessarily produce strong La Niña events.

Conclusions

In a previous study (Takahashi and Dewitte [81]), the convective SST threshold in the eastern Pacific and the associated nonlinearity in the Bjerknes feedback provides a parsimonious explanation for this, motivating further exploration of this possibility suggestive with a simple theoretical model based on this mechanism.

We show that this nonlinearity is sufficient to produce the bimodal distribution associated with strong and moderate EN regimes.

It is a parsimonious theory for the EN regimes, based on a well-known nonlinear SST-convection relation.

It is a simpler model than, for instance, high-dimensional linear models and does not produce exotic behavior as other nonlinear models or require special assumptions about the forcing, thus providing a better null hypothesis for models that exhibit two EN regimes.

Acknowledgments

A machine generated summary based on the work of Takahashi, Ken; Karamperidou, Christina; Dewitte, Boris (2018 in Climate Dynamics).

Climate Change Impact Assessment on Flow Regime by Incorporating Spatial Correlation and Scenario Uncertainty

https://doi.org/10.1007/s00704-016-1802-1

Abstract-Summary

Flooding risk is increasing in many parts of the world and may worsen under climate change conditions.

The current statistical downscaling approaches face the difficulty of projecting multi-site climate information for future conditions while conserving spatial information.

The results showed different variation trends of annual peak flows (in 2080–2099) based on different climate change scenarios and demonstrated that the hydrological impact would be driven by the interaction between snowmelt and peak flows.

The proposed CLWRS approach is useful where there is a need for projection of potential climate change scenarios.

Introduction

Besides precipitation patterns, the prediction of future floods relies on changes in temperature, snowmelt, land use patterns, etc. Future climate changes can be predicted based on the physics described by the General Circulation Models (GCMs), which simulate the interaction among atmosphere, ocean and sea.

LARS-WG was proposed by Racsko et al. [82] as a stochastic weather generator, intended to model meteorological parameters such as precipitation and solar radiation.

Based on the above-mentioned studies, it is found that LARS-WG is effective in simulating climate change for meteorological variables but is limited to being applied for a single site; multi-site RainSim is capable of spatially addressing rainfall but requires modifications to address future climate change.

The objective of this study is to propose a coupled LARS-WG and RainSim (CLWRS) approach to quantify the changes in flood occurrences under future climate projections for different GCMs for future time periods (i.e. 2080–2099).

Methodology

The initial purpose of LARS-WG is to obtain uncorrelated precipitation patterns which serves as the input for RainSim through the Change Factor approach detailed in the following section.

The output of RainSim in turn serves as the input for the reanalysis by LARS-WG to generate meteorological data, now taking into account the spatial patterns among multiple meteorological sites.

Step 1: Use LARS-WG to predict baseline and future meteorological conditions LARS-WG determines the statistical properties of historical meteorological data, and generates long records of simulated data for either future or baseline condition (1961–1990).

Step 3: Use LARS-WG to reanalyse meteorological data based on correlated precipitation The precipitation pattern generated from RainSim is used as the input for the reanalysis by LARS-WG to generate the rainfall statistics.

The temperature and radiation simulated for different future scenarios by LARS-WG along with the precipitation from RainSim are used as the input for LARS-WG.

Case Study and Data

The meteorological data, including maximum temperature (T_{max}), minimum temperature (T_{min}) and precipitation (P) for three meteorological stations are collected for 1965–2007 from Environment Canada website (Environment Canada [83]).

Other meteorological data required are the dew point temperature (T_{dew}) and global radiation (R).

The hydrometric data for this case study for the period 1965–2007 were obtained from Canadian Water Office (Environment Canada [83]).

This relationship is then used for the data sourced from Environment Canada to obtain T_{dew} for the period 1965–2007.

T_{dew} is calculated from relative humidity (R_h), R, T_{max} and T_{min} which is obtained for the period 1979–2007 CFSR data, using the formula given by Lawrence [84].

The change factor approach described in the earlier section is employed to obtain future precipitation data for the Kootenay Watershed incorporating spatial characteristics.

The second type is the topographic data for the Kootenay Watershed (Kite [85]).

Results and Discussion

For 100-year return period, the A1B scenario predicts an increase of flow around 11.46% (based on baseline values); A2 and B1 scenarios predict decreases of 7.27 and 8.02%, respectively.

Another important observation is the difference in change factors and their influence in predicting peak flows.

The variation of the predicted flood peaks exhibits an increasing trend for different return periods, mirroring the trend of 75th percentile.

For lower return periods of 5 and 10 years, decreases in peak flows at 5.98 and 3.49% are predicted; whereas, the higher return periods suggest increasing trends.

Similar to the B1 scenario, the HADCM3 predicts the highest increases in the flow peaks while INCM3 and GIAOM predictions remain conservative and FGOALS suggest decreases.

A1B scenario which predicts increase in temperatures with lesser magnitudes, predicted increasing flood peaks.

Conclusion

The CLWRS framework combining LARS-WG and RainSim with SLURP model is presented.

LARS-WG is used to facilitate the quantification of the predictions' uncertainty arising from emission scenarios.

Although, this variation could complicate quantifying the relationship between changes in temperature, snowmelt and flood peaks; it would facilitate better understanding of the Pacific Northwest.

The importance of this relationship is further emphasized by the changes in the flood peaks predicted by the different scenarios.

As LARS-WG continues to be updated with the CMIP5 scenarios and as more GCMs are incorporated to augment its predictive capabilities, the subsequent evolution of the trends in scenario uncertainty desires further investigation.

This can be further extended to model and simulate temperature patterns at the different meteorological stations.

Acknowledgments

A machine generated summary based on the work of Vallam, P.; Qin, X. S. (2016 in Theoretical and Applied Climatology).

Book Reading List

Climate Change Modeling Methodology

by Rasch, P. J. (Ed) *(2012)*.

The Earth's average temperature has risen by 1.4°F over the past century, and computer models project that it will rise much more over the next hundred years, with significant impacts on weather, climate, and human society. Many climate scientists attribute these increases to the build up of greenhouse gases produced by the burning of fossil fuels and to the anthropogenic production of short-lived climate pollutants. Climate Change Modeling Methodologies: Selected Entries from the Encyclopaedia of Sustainability Science and Technology provides readers with an introduction to the tools and analysis techniques used by climate change scientists to interpret the role of these forcing agents on climate.

Please see https://www.springer.com/gp/book/9781461457664 for original source.

Climate Modelling

by Lloyd, E. (Ed), Winsberg, E. (Ed) *(2018)*.

This edited collection of works by leading climate scientists and philosophers introduces readers to issues in the foundations, evaluation, confirmation, and application of climate models. It engages with important topics directly affecting public policy, including the role of doubt, the use of satellite data, and the robustness of models.

Please see https://www.springer.com/gp/book/9783319650579 for original source.

Demystifying Climate Models

by Gettelman, A., Rood, R. B. *(2016).*

This book demystifies the models we use to simulate present and future climates, allowing readers to better understand how to use climate model results. In order to predict the future trajectory of the Earth's climate, climate-system simulation models are necessary. When and how do we trust climate model predictions? The book offers a framework for answering this question. It provides readers with a basic primer on climate and climate change, and offers non-technical explanations for how climate models are constructed, why they are uncertain, and what level of confidence we should place in them. It presents current results and the key uncertainties concerning them.

Please see https://www.springer.com/gp/book/9783662489574 for original source.

Development and Evaluation of High Resolution Climate System Models

by Yu, R., Zhou, T., Wu, T., Xue, W., Zhou, G. *(2016).*

This book is based on the project "Development and Validation of High Resolution Climate System Models" with the support of the National Key Basic Research Project under grant No. 2010CB951900. It demonstrates the major advances in the development of new, dynamical Atmospheric General Circulation Model (AGCM) and Ocean General Circulation Model (OGCM) cores that are suitable for high resolution modeling, the improvement of model physics, and the design of a flexible, multi-model ensemble coupling framework.

Please see https://www.springer.com/gp/book/9789811000317 for original source.

Mathematics of Climate Modeling

by Dymnikov, V. P., Filatov, A. N. *(1997).*

The present monograph is dedicated to a new branch of the theory of climate, which is titled by the authors, "Mathematical Theory of Climate." The foundation of this branch is the investigation of climate models by the methods of the qUalitative theory of differential equations. In the Russian edition the book was named "Fundamentals of the Mathematical Theory of Climate." Respecting the recommendations of Wayne Yuhasz (we are truly grateful to him for this advice), we named the English edition of the book "Mathematics of Climate Modelling."

Please see https://www.springer.com/gp/book/9780817639150 for original source.

Modeling Dynamic Climate Systems

by Robinson, W. A. *(2001)*.

The world consists of many complex systems, ranging from our own bodies to ecosystems to economic systems. Despite their diversity, complex systems have many structural and functional features in common that can be effectively modeled using powerful, user-friendly software. As a result, virtually anyone can explore the nature of complex systems and their dynamical behavior under a range of assumptions and conditions. This ability to model dynamic systems is already having a powerful influence on teaching and studying complexity. The books in this series will promote this revolution in "systems thinking" by integrating skills of numeracy and techniques of dynamic modeling into a variety of disciplines.

Please see https://www.springer.com/gp/book/9780387951348 for original source.

Systems Representation of Global Climate Change Models.

by Sreenath, N. *(1993)*.

This book bridges the gap between system theory and global climate change research, and benefits both. A representative set of systems problems is listed indicating how such cross-fertilization would enhance present understanding of global problems while assisting the extension of systems theory. The goal is a comprehensive conceptual model of global change which encompasses atmosphere, lithosphere, ocean, biosphere and cryosphere.

Please see https://www.springer.com/gp/book/9783540198246 for original source.

Stochastic Climate Models

by Imkeller, P. (Ed), Storch, J. v. (Ed) *(2001)*.

The proceedings of the summer 1999 Chorin workshop on stochastic climate models captures well the spirit of enthusiasm of the workshop participants engaged in research in this exciting field. It is amazing that nearly 25 years after the formal theory of natural climate variability generated by quasi-white-noise weather forcing was developed, and almost 35 years after J. M. Mitchell first suggested this mechanism as the origin of sea-surface-temperature fluctuations and climate variability, there have arisen so many fresh perspectives and new applications of the theory.

Please see https://www.springer.com/gp/book/9783034895040 for original source.

Models for Tropical Climate Dynamics

by Khouider, B. *(2019)*.

This book is a survey of the research work done by the author over the last 15 years, in collaboration with various eminent mathematicians and climate scientists on the subject of tropical convection and convectively coupled waves. In the areas of climate modelling and climate change science, tropical dynamics and tropical rainfall are among the biggest uncertainties of future projections. This not only puts at risk billions of human beings who populate the tropical continents but it is also of central importance for climate predictions on the global scale.

Please see https://www.springer.com/gp/book/9783030177744 for original source.

References

1. Goldstein M, Rougier JC (2009) Reified Bayesian modelling and inference for physical systems. J Stat Plan Inference 139:1221–1239
2. Pope VD, Gallani ML, Rowntree PR, Stratton RA (2000) The impact of new physical parametrizations in the Hadley Centre climate model: HadAM3. Clim Dyn 16:123–146
3. Gordon C, Cooper C, Senior CA, Banks H, Gregory JM, Johns TC, Mitchell JFB, Wood RA (2000) The simulation of SST, sea ice extents and ocean heat transports in a version of the Hadley Centre coupled model without flux adjustments. Clim Dyn 16:147–168
4. Murphy JM, Sexton DMH, Barnett DN, Jones GS, Webb MJ, Collins M, Stainforth DA (2004) Quantification of modelling uncertainties in a large ensemble of climate change simulations. Nature 430:768–772
5. Sacks J, Welch WJ, Mitchell TJ, Wynn HP (1989) Design and analysis of computer experiments. Stat Sci 4:409–435
6. Murphy JM, Sexton DMH, Jenkins GJ, Booth BBB, Brown CC, Clark RT, Collins M, Harris GR, Kendon EJ, Betts RA, Brown SJ, Humphrey KA, McCarthy MP, McDonald RE, Stephens A, Wallace C, Warren R, Wilby R, Wood R (2009) UK Climate Projections Science Report: climate change projections. Met Office Hadley Centre, Exeter, UK. https://ukclimateprojec tions.defra.gov.uk/images/stories/projections_pdfs/UKCP09_Projections_V2.pdf
7. Williamson D, Goldstein M, Blaker A (2012) Fast linked analyses for scenario based hierarchies. J R Stat Soc Ser C 61(5):665–692
8. Tisseuil C, Vrac M, Lek S, Wade AJ (2010) Statistical downscaling of river flows. J Hydrol 385:279–291
9. Tomassetti B, Verdecchia M, Giorgi F (2009) NN5: a neural network based approach for the downscaling of precipitation fields—model description and preliminary results. J Hydrol 367:14–26
10. Chen S-T, Yu P-S, Tang Y-H (2010) Statistical downscaling of daily precipitation using support vector machines and multivariate analysis. J Hydrol 385:13–22
11. Tripathi S, Srinivas VV, Nanjundiah RS (2006) Downscaling of precipitation for climate change scenarios: a support vector machine approach. J Hydrol 330:621–640
12. Ghosh S, Mujumdar PP (2008) Statistical downscaling of GCM simulations to streamflow using relevance vector machine. Adv Water Resour 31:132–146
13. Joshi D, St-Hilaire A, Daigle A, Ouarda TBMJ (2013) Databased comparison of Sparse Bayesian learning and multiple linear regression for statistical downscaling of low flow indices. J Hydrol 488:136–149
14. Ott E (1993) Chaos in dynamical systems. Cambridge University Press, Cambridge
15. Gruiz M, Tél T (2006) Chaotic dynamics. Cambridge University Press, Cambridge
16. Eckmann JP, Ruelle D (1985) Ergodic theory of chaos and strange attractors. Rev Mod Phys 57(3):617

17. Romeiras FJ, Grebogi C, Ott E (1990) Multifractal properties of snapshot attractors of random maps. Phys Rev A 41(2):784
18. Ghil M, Chekroun MD, Simonnet E (2008) Climate dynamics and fluid mechanics: natural variability and related uncertainties. Physica D 237(14–17):2111
19. Herein M, Drótos G, Haszpra T, Márfy J, Tél T (2017) The theory of parallel climate realizations as a new framework for teleconnection analysis. Sci. Rep. 7(January):44529
20. Lorenz EN (1990) Can chaos and intransivity lead to interannual variability? Tellus a 42A:378
21. Bódai T, Tél T (2012) Annual variability in a conceptual climate model: snapshot attractors, hysteresis in extreme events, and climate sensitivity. Chaos 22(2):023110
22. Maher N, Milinski S, Suarez-Gutierrez L, Botzet M, Kornblueh L, Takano Y, Kröger J, Ghosh R, Hedemann C, Li C et al (2019) The Max Planck Institute grand ensemble-enabling the exploration of climate system variability. J Adv Model Earth Syst 11:2050
23. Avila M, Mellibovsky F, Roland N, Hof B (2013) Streamwise-localized solutions at the onset of turbulence in pipe flow. Phys Rev Lett 110:224502
24. Gardiner C (2009) Stochastic methods: a handbook for the natural and social sciences. Springer, Berlin
25. Haszpra T, Tél T (2013) Topological entropy: a Lagrangian measure of the state of the free atmosphere. J Atmos Sci 70(12):4030
26. Haszpra T (2017) Intensification of large-scale stretching of atmospheric pollutant clouds due to climate change. J Atmos Sci 74(12):4229
27. Haszpra T, Herein M (2019) Ensemble-based analysis of the pollutant spreading intensity induced by climate change. Sci Rep 9(1):3896
28. Stocker T, Qin D, Plattner GK, Tignor M, Allen S, Boschung J, Nauels A, Xia Y, Bex V, Midgley P (2013) IPCC, climate change 2013: the physical science basis. In: Contribution of Working Group I to the fifth assessment report of the intergovernmental panel on climate change. Cambridge University Press, Cambridge
29. Collins M, Knutti R, Arblaster J, Dufresne J-L, Fichefet T, Friedlingstein P, Gao X, Gutowski WJ, Johns T, Krinner G, Shongwe M, Tebaldi C, Weaver AJ, Wehner M (2013) Long-term climate change: projections, commitments and irreversibility. In: Stocker TF, Qin D, Plattner G-K, Tignor M, Allen SK, Boschung J, Nauels A, Xia Y, Bex V, Midgley PM (eds) Climate Change 2013: the physical science basis. In: Contribution of Working Group I to the fifth assessment report of the intergovernmental panel on climate change. Cambridge University Press, Cambridge
30. Trenberth KE, Fasullo JT, Balmaseda MA (2014) Earth's energy imbalance. J Clim 27:3129–3144
31. Lovejoy S, Schertzer D (1986) Scale invariance in climatological temperatures and the spectral plateau. Ann Geophys 4B:401–410
32. Lovejoy S, Schertzer D (2012) Low frequency weather and the emergence of the Climate. In: Sharma AS, Bunde A, Baker D, Dimri VP (eds) Extreme events and natural hazards: the complexity perspective. AGU monographs, Washington, pp 231–254
33. Schmitt F, Lovejoy S, Schertzer D (1995) Multifractal analysis of the Greenland Ice-core project climate data. Geophys Res Lett 22:1689–1692
34. Ditlevsen PD, Svensmark H, Johson S (1996) Contrasting atmospheric and climate dynamics of the last-glacial and Holocene periods. Nature 379:810–812
35. Pelletier JD (1998) The power spectral density of atmospheric temperature from scales of 10^{-2} to 10^6 yr. EPSL 158:157–164
36. Ashkenazy Y, Baker D, Gildor H, Havlin S (2003) Nonlinearity and multifractality of climate change in the past 420,000 years. Geophys Res Lett 30:2146. https://doi.org/10.1029/2003GL018099
37. Wunsch C (2003) The spectral energy description of climate change including the 100 ky energy. Clim Dyn 20:353–363
38. Huybers P, Curry W (2006) Links between annual, Milankovitch and continuum temperature variability. Nature 441:329–332. https://doi.org/10.1038/nature04745

39. Blender R, Fraedrich K, Hunt B (2006) Millennial climate variability: GCMration of δ18O of cultured benthic. Geophys Res Lett 33:L04710. https://doi.org/10.1029/2005GL024919
40. Lovejoy S (2013) What is climate? EOS 94(1) 1 January, pp 1–2
41. Rypdal M, Rypdal K (2014) Long-memory effects in linear-response models of Earth's temperature and implications for future global warming. Clim Dyn (in press)
42. Lucarini V, Sarno S (2011) A statistical mechanical approach for the computation of the climatic response to general forcings. Nonlin Processes Geophys 18:7–28
43. Gallavotti G (1996) Chaotic hypothesis: onsanger reciprocity and fluctuation-dissipation theorem. J Stat Phys 84:899–926
44. IPCC (2007b) In: Solomon S, Qin D, Manning M, Chen Z, Marquis M, Averyt KB, Tignor M, Miller HL (eds) Climate change 2007: the physical science basis. In: Contribution of Working Group I to the Fourth Assessment Report of the intergovernmental panel on climate change. Cambridge University Press, Cambridge, United Kingdom and New York, NY, USA
45. IPCC (2013) In: Stocker TF, Qin D, Plattner G-K, Tignor M, Allen SK, Boschung J, Nauels A, Xia Y, Bex V, Midgley PM (eds) Climate Change 2013: the physical science basis. Contribution of working group I to the fifth assessment report of the intergovernmental panel on climate change stocker. Cambridge University Press, Cambridge, United Kingdom and New York, NY, USA
46. Yu JY, Kao PK, Paek H, Hsu HH, Hung CW, Lu MM, An SI (2015) Linking emergence of the central Pacific El Niño to the Atlantic multidecadal oscillation. J Clim 28:651–662
47. Mecking JV, Drijfhout SS, Jackson LC, Graham T (2016) Stable AMOC off state in an eddy-permitting coupled climate model. Clim Dyn. https://doi.org/10.1007/s00382-016-2975-0:1-16
48. Timmerman A, Okumura Y, An SI, Clement A, Dong B, Guilyardi E, Hu A, Jungclaus JH, Renold M, Stocker TF, Stouffer RJ, Sutton R, Xie SP, Yin J (2007) The influence of a weakening of the Atlantic meridional overturning circulation on ENSO. J Clim 20:4899–4919
49. Dong B, Sutton RT (2007) Enhancement of ENSO variability by a weakened Atlantic thermohaline circulation in a coupled GCM. J Clim 20:4920–4939
50. Jin FF (1997) An equatorial ocean recharge paradigm for ENSO. Part I: Conceptual model. J Atmos Sci 54:811–829
51. Hewitt HT, Copsey D, Culverwell ID, Harris CM, Hill RSR, Keen AB, McLaren AJ, Hunke EC (2011) Design and implementation of the infrastructure of HadGEM3: the next generation Met Office climate modelling system. Geosci Model Dev 4:223–253
52. Demory ME, Vidale P, Roberts M, Berrisfor P, Strachan J, Schiemann R, Mizielinski M (2013) The role of horizontal resolution in simulating drivers of the global hydrological cycle. Clim Dyn 1:25
53. Bellenger H, Guilyardi E, Leloup J, Lengaigne M, Vialard J (2014) ENSO representation in climate models: from CMIP3 to CMIP5. Clim Dyn. https://doi.org/10.1007/s00382-013-1783-z
54. Dee DP et al (2011) The ERA-Interim reanalysis: configuration and performance of the data assimilation system. Q J R Meteorol Soc 137(656):553–597. https://doi.org/10.1002/qj.828
55. Rayner NA, Parker DE, Horton EB, Folland CK, Alexander LV, Rowell DP, Kent EC, Kaplan A (2003) Global analyses of sea surface temperature, sea ice, and night marine air temperature since the late nineteenth century. J Geophys Res 108(D14):4407. https://doi.org/10.1029/2002JD002670
56. Bjerknes J (1969) Atmospheric teleconnections from the equatorial pacic. Mon Weather Rev 97(3):163–172. https://doi.org/10.1175/1520-0493(1969)097%3c0163:ATFTEP%3e2.3.CO;2
57. Keenlyside NS, Latif M (2007) Understanding equatorial Atlantic Interannual variability. J Clim 20(1):131–142. https://doi.org/10.1175/JCLI3992.1
58. Burls NJ, Reason CJC, Penven P, Philander SG (2012) Energetics of the tropical Atlantic zonal mode. J Clim 25(21):7442–7466. https://doi.org/10.1175/JCLI-D-11-00602.1
59. Lübbecke JF, McPhaden MJ (2013) A comparative stability analysis of Atlantic and Paciffic Niño modes. J Clim 26(16):5965–5980. https://doi.org/10.1175/JCLI-D-12-00758.1
60. Deppenmeier AL, Haarsma RJ, Hazeleger W (2016) The Bjerknes feedback in the tropical Atlantic in CMIP5 models. Clim Dyn 1–17. https://doi.org/10.1007/s00382-016-2992-z

61. Nnamchi HC, Li J, Kucharski F, Kang IS, Keenlyside NS, Chang P, Farneti R (2015) Thermodynamic controls of the Atlantic Niño. Nat Commun 6:8895. https://doi.org/10.1038/ncomms9895

62. Nnamchi HC, Li J, Kucharski F, Kang I-S, Keenlyside NS, Chang P, Farneti R (2016) An equatorial–extratropical dipole structure of the Atlantic Niño. J Clim 29(20):7295–7311. https://doi.org/10.1175/JCLI-D-15-0894.1

63. Ding H, Greatbatch RJ, Latif M, Park W (2015) The impact of sea surface temperature bias on equatorial Atlantic interannual variability in partially coupled model experiments. Geophys Res Lett 42(13):5540–5546. https://doi.org/10.1002/2015GL064799

64. Richter I, Xie S-P (2008) On the origin of equatorial Atlantic biases in coupled general circulation models. Clim Dyn 31(5):587–598. https://doi.org/10.1007/s00382-008-0364-z(English)

65. Richter I, Xie SP, Wittenberg AT, Masumoto Y (2012) Tropical Atlantic biases and their relation to surface wind stress and terrestrial precipitation. Clim Dyn 38(5–6):985–1001. https://doi.org/10.1007/s00382-011-1038-9

66. Richter I, Xie SP, Behera SK, Doi T, Masumoto Y (2014) Equatorial Atlantic variability and its relation to mean state biases in CMIP5. Clim Dyn 42(1–2):171–188. https://doi.org/10.1007/s00382-012-1624-5

67. Zebiak SE (1993) Air–sea interaction in the equatorial Atlantic region. J Clim 6(8):1567–1586. https://doi.org/10.1175/1520-0442(1993)006%3c1567:AIITEA%3e2.0.CO;2

68. Carton JA, Cao X, Giese BS, Da Silva AM (1996) Decadal and interannual SST variability in the tropical. Atlantic Ocean. https://doi.org/10.1175/1520-0485(1996)838026%3c1165:DAISVI%3e2.0.CO;2

69. Ding H, Keenlyside NS, Latif M (2010) Equatorial Atlantic interannual variability: role of heat content. J Geophys Res Oceans 115(C9):C09020. https://doi.org/10.1029/2010JC006304

70. Okumura Y, Xie SP (2006) Some overlooked features of tropical Atlantic climate leading to a new Nino-like phenomenon. J Clim 19(22):5859–5874

71. Timmermann A, Jin FF, Abshagen J (2003) A nonlinear theory for El Niño bursting. J Clim 60:152–165

72. An SI, Jin FF (2004) Nonlinearity and asymmetry of ENSO. J Clim 17:2399–2412

73. Lloyd J, Guilyardi E, Weller H (2012) The role of atmosphere feedbacks during ENSO in the CMIP3 models. Part III. The shortwave flux feedback. J Clim. https://doi.org/10.1175/JCLI-D-11-00178.1

74. Burgers G, Jin FF, van Oldenborgh GJ (2005) The simplest ENSO recharge oscillator. Geophys Res Lett. https://doi.org/10.1029/2005GL02295

75. Risken H (1996) The Fokker–Planck equation. Methods of Solution and applications, 3rd edn. Springer, Berlin

76. Hasselmann K (1976) Stochastic climate models. Part I Theor Tellus 28:473–485

77. Levine AFZ, Jin FF (2010) Noise-induced instability in the ENSO recharge oscillator. J Clim. https://doi.org/10.1175/2009JAS3213.1

78. Gebbie G, Eisenman I, Wittenberg A, Tziperman E (2007) Modulation of westerly wind bursts by sea surface temperature: a semistochastic feedback for ENSO. J Atmos Sci. https://doi.org/10.1175/JAS4029.1

79. Zavala-Garay J, Zhang C, Moore AM, Wittenberg AT, Harrison MJ, Rosati A, Vialard J, Kleeman R (2008) Sensitivity of hybrid ENSO models to unresolved atmospheric variability. J Clim 21(15):3704–3721

80. Gebbie G, Tziperman E (2009) Predictability of SST-modulated westerly wind bursts. J Clim. https://doi.org/10.1175/2009JCLI2516.1

81. Takahashi K, Dewitte B (2016) Strong and moderate nonlinear El Niño regimes. Clim Dyn. https://doi.org/10.1007/s00382-015-2665-3

82. Racsko P, Szeidl L, Semenov M (1991) A serial approach to local stochastic weather models. Ecol Model 57:27–41

83. Environment Canada (2014) [online]. Available from https://climate.weather.gc.c/ and https://wateroffice.ec.gc.ca. Accessed 26 Sept 2014

84. Lawrence MG (2005) The relationship between relative humidity and the dewpoint temperature in moist air: a simple conversion and applications. Bull Am Meteorol Soc 86(2):225–233
85. Kite G (1995) The SLURP model: computer models of watershed hydrology. Water Resources Publications, Littleton CO

Chapter 6
Maximum Entropy Production (MEP) and Climate

Introduction by Guido Visconti

The idea of Maximum Entropy Production (MEP) appeared for the first time in 1975 in a paper by Garth Paltridge. He used a very simple energy balance climate model to evaluate the latitudinal temperature distribution requiring that the net exchange of the entropy for each latitude belt be at a minimum. Notice that MEP was born as a minimum but as remarked in the review paper by Ozawa et al." *...the exchange rate was defined positive inward. This condition is identical to the maximum entropy discharge into the surrounding system, which corresponds to the maximum entropy production due to the turbulent dissipation.*" The same argument was used in a prompt critique by Clive Rodgers that in the same year (1975) showed that the same results could be obtained in a much simpler way and neglecting any internal process within the latitude zone. Rodgers also observed that the method could not be applied to planets devoid or having tiny atmospheres (like Mercury and Mars) but also dense atmospheres like Venus. He also observed that actually Paltridge was maximizing internal entropy production (associated with irreversible processes within the system) by minimizing the external production. At that time, it was known as a theorem due to Prigogine about minimum entropy production in linear systems. About 25 years later the MEP principle was applied to evaluate the pole to equator temperature for Mars and Titan claiming a good agreement with the data. However, a more accurate analysis by Richard Goody showed that the claim was not justified. He also observed that MEP could be applied only to the fluctuating component of the heat flow and the application also to the steady part would lead to error. At the moment MEP cannot be enunciated as a theorem because it has not been proved as such although it seems to work in many applications. Beside the early applications to climate, MEP has been applied to study Earth mantle convection, earthquake dynamics, Jupiter red spot but also biology and ecology. In brief MEP affirms that in a physical system far from the equilibrium state and subject to an irreversible process from one state to another the path chosen by the system is such to maximize the production of entropy. This is equivalent to affirm that the system tends to dissipate as much energy in the form of heat.

The core of the problem can be understood if we refer to our planet. Earth absorbs roughly 240 W/m^2 from a high temperature source (the Sun at about 6000 K) and at equilibrium emits the same power at a much lower temperature (the Earth's upper atmosphere at 255 K). The result is a large increase in entropy that is given by the ratio of the energy flux to temperature. Of the produced entropy 93% is attributed to absorption of shortwave radiation (visible and UV), 2.7% to infrared radiation, 2.6% to the water cycle, 1% to atmospheric circulation, 0.7% to life processes and only 0.04% to humans. We now see how MEP can be connected with almost anything and the discussions have concentrated especially on the smallest fraction of the budget (humans and life). The advantage of MEP (if it is right) is that climate models could be built without adjusting too many parameters. A closer examination however revealed that also Paltridge treated the entire climate system with MEP subject to enough empirical constraints to reduce the problem to a single variable (the horizontal heat flux) subsequent work by other groups on the same problems revealed some instability.

One of the main actors of MEP is Axel Kleidon not only for his immense work on the subject but also for his claim to have found evidence of Gaia theory in the folds of MEP theory. In a 2010 paper the Gaia hypothesis was reformulated where the Earth system has evolved further away from a state of thermodynamic equilibrium mostly by the increased generation rate of chemical free energy by life. It was recognized that, as Gaia, MEP apparently assumes teleology (it knows which entropy has to be maximized). This hypothesis raised some criticism mainly on the claim that a biotic Earth creates much more free energy with respect to an abiotic one and the main mechanisms is photosynthesis. Also the claim that is life which determines the thermodynamic disequilibrium was disputed. An interesting point of view on MEP, was expressed by Tyler Volk and Olivier Pauluis when they observed that the important factor is not how much entropy is produced by the system but rather the way it is produced. They reached this conclusion by observing that the production of entropy in a dry and humid atmosphere is the same while the internal processes of dissipation are quite different. They also showed that life containing systems can either increase or decrease entropy production rate.

The conclusion by Paltridge at the end of a meeting on MEP is still valid "....*it is likely that MEP will remain entirely in the realm of academic discussion until someone—probably someone from outside the field—identifies a complex system whose behavior cannot be predicted by any means other than application of MEP. It is also likely that the "someone" will need to visualize fairly easily the physics of what is going on so that the result can be understood without relying entirely on some esoteric mathematical proof.*"

Machine-Generated Summaries

Keywords: *production, entropy production, mep, state, temperature, climate, principle, equation, flux, study, surface, parameter, steady state, result, due.*

Entropy Production Selects Nonequilibrium States in Multistable Systems

https://doi.org/10.1038/s41598-017-14485-8

Abstract-Summary

Best candidate theories based on the maximum entropy production principle could not be unequivocally proven, in part due to complicated physics, unintuitive stochastic thermodynamics, and the existence of alternative theories such as the minimum entropy production principle.

We use a simple, analytically solvable, one-dimensional bistable chemical system to demonstrate the validity of the maximum entropy production principle.

To generalize to multistable stochastic system, we use the stochastic least-action principle to derive the entropy production and its role in the stability of nonequilibrium steady states.

Introduction

Similar to [1], we distinguish between two types of MaxEPPs: the 'state selection' principle, addressing which steady state is selected in a multistable system, and the 'gradient response' principle, focusing on the response of the average entropy production of a stochastic system to changes in a parameter.

Since the above argument combines the entropy production from the bistable ODE model with the weights from the master equation, which is mono-stable in the infinite volume limit, this mixing of models may have led to the wrong conclusion regarding MaxEPP.

Even without the Gaussian approximation, we can obtain the weights of the states and their entropy production rates using the master equation.

MaxEPP is a principle for multistable systems in which the entropy production biases the evolution of the system towards the highest-entropy producing state.

Discussion

MaxEPP applies in the former because the weights of the low and high states shift in the exact stochastic approach due to a first-order phase transition.

However the discrepancy in how the MaxEPP is achieved in the two approaches: using the master equation we observe a first-order phase transition and state switching at the critical point, while using the Langevin approximation, the high state is selected.

To this local MaxEPP for states of a multistable system for a fixed driving force, there is also a trivial global MaxEPP principle, which simply says that the more a system is driven away from equilibrium the more it produces entropy ('gradient response' principle) [1, 2].

Our interpretation of MaxEPP is in line with the recent finding that the entropy production, by itself, is not a unique descriptor of the steady-state probability distribution [3].

Acknowledgement

A machine generated summary based on the work of Endres, Robert G. (2017 in Scientific Reports).

A Parametric Sensitivity Study of Entropy Production and Kinetic Energy Dissipation Using the FAMOUS AOGCM

https://doi.org/10.1007/s00382-011-0996-2

Abstract-Summary

The possibility of applying either the maximum entropy production conjecture of Paltridge (Q J R Meteorol Soc 101:475–484, [4]) or the conjecture of Lorenz (Generation of available potential energy and the intensity of the general circulation.

The climate response is analysed in terms of its entropy production and the strength of the Lorenz energy cycle.

No maximum is found in the total material entropy production, which is dominated by the hydrological cycle and tends to increase monotonically with global-mean temperature, which is not constant because the parameter variations affect the net input of solar radiation at the top of the atmosphere (TOA).

A peak in the generation of APE is found in association with the maximum baroclinic activity, but no trade-off of the kind shown by simple climate models is found between meridional heat transport and the meridional temperature gradient.

We conclude that the maximum entropy production conjecture does not hold within the climate system when the effects of the hydrological cycle and radiative feedbacks are taken into account, but our experiments provide some evidence in support of the conjecture of maximum APE production (or equivalently maximum dissipation of kinetic energy).

Introduction

The idea has been proposed of using material entropy production and kinetic energy dissipation (or, equivalently, generation of available potential energy) as objective functions for parameter tuning (Kleidon et al. [5, 6]; Kleidon [7]; Kunz et al. [8]).

Without the hydrological cycle in the model the material entropy production corresponds almost entirely to the one associated with heating due to kinetic energy dissipation.

In doing this we want to address the following issues: (i) to test the applicability and validity of MEP and MKED in a complex AOGCM; (ii) to highlight differences between material entropy production and kinetic energy dissipation when the whole hydrological cycle is included; (iii) to understand which parameters and which processes are significant for MEP; (iv) to understand to what extent MEP may provide an inherent principle for parameter optimisation of GCMs (Kunz et al. [8]).

Model and Method

The atmosphere model is run with fixed sea surface temperature and sea-ice conditions through the day and the various fluxes are accumulated after each atmospheric time step.

At the end of the day these fluxes are passed to the ocean model which is then integrated forwards in time and supplies sea surface conditions back to the atmosphere.

The integral is extended over the Earth's surface A and the sums are extended over the vertical model levels z and over all the atmospheric/ocean non-radiative diabatic processes k (radiative processes are treated as an external energy source or sink (Goody, [9]; Volk and Pauluis [10]).

The release in the atmosphere of the heat associated with dissipated kinetic energy leads to the entropy production (Peixoto et al. [11]): All these quantities are obtained as global time means over a sufficiently long period in which the system can be considered in a steady state (typically tens of years).

Experiments with the Full AOGCM

A similar non-monotonic behaviour of mean dry static energy, eddy kinetic energy and near-surface eddy kinetic energy (as a measure of surface storminess) in midlatitude baroclinic zones has been recently found by O'Gorman and Schneider [12] and Schneider et al. [13] in a series of simulations with an idealised moist general circulation model resulting from changes in longwave optical thickness.

Since the eddy kinetic energy, i.e. the kinetic energy not associated with the mean flow but with the midlatitude macroturbulences, tends mostly to be dissipated on a timescale of few days, the result found by O'Gorman and Schneider [12] is likely to be equivalent to a present-day climate with maximum kinetic energy dissipation.

Let us note that in moist coupled GCM a state with maximum eddy activity does not necessarily correspond with a state of maximum meridional heat transport as in a dry atmospheric model (Kleidon et al. [5]), since about half of it is due to latent heat which increases with temperature (Schneider et al. [13]) and since cloud feedback may change the net insolation.

Experiments with Fixed Absorbed Shortwave Radiation

Although of greater generality, the experiments described so far are hard to compare directly with Kleidon et al. [5]; Kleidon et al. [6] and the two-box model because in these earlier studies the amount of absorbed shortwave radiation is prescribed, and this is a key element for explaining the results.

The main effect of a constant planetary albedo and no cloud radiative forcing is the reduction of internal variability.

The meridional temperature gradient decreases with the increase of the poleward heat transport and the rate of generation of APE increases.

The fact that the MEP and MKED states here correspond to the one with the largest atmospheric poleward heat transport and the smallest equator-pole temperature gradient confirms results by Kleidon et al. [5] and Kleidon et al. [6], in which the maximum dissipation state is the most baroclinically active one (and also the one with the maximum entropy production, since in a dry model the entropy production is mostly due to kinetic energy dissipation).

Discussion and Conclusions

This paper represents the first study of parametric variation aimed at examining MEP in a complex atmospheric-ocean general circulation model, extending previous approaches (Kleidon et al. [5, 6]; Kleidon [7]; Kunz et al. [8]).

From the MEP point of view, the total material entropy production does not seem to be a very interesting quantity in this experiment.

This result stands to be a qualitative confirmation of Lorenz's [14] conjecture and, to a certain extent, of Paltridge's [15] hypothesis (based on the entropy production associated with the meridional heat transport only).

Variation of the meridional heat flux does not manage to show a "trade-off" mechanism for MEP as described in Lorenz et al. [16] and Kleidon [17].

This demonstrates that in complex, highly constrained models changes to basic climatic fluxes and gradients deriving from parameter perturbations may not detect MEP, as questioned by Goody [18].

Results seem to be in agreement with Kleidon's explanation of MEP states as states of maximum atmospheric poleward heat transport (Kleidon et al. [5, 6]; Kleidon [7]).

Acknowledgement

A machine generated summary based on the work of Pascale, Salvatore; Gregory, Jonathan M.; Ambaum, Maarten H. P.; Tailleux, Rémi (2011 in Climate Dynamics).

Does Maximal Entropy Production Play a Role in the Evolution of Biological Complexity? A Biological Point of View

https://doi.org/10.1007/s12210-020-00909-7

Abstract-Summary

A considerable literature has developed around the concept that the complexity of biological organisms, and the development of ever increasing complexity during biological evolution, is driven, in some way, by the maximization of entropy production (MEP).

It is concluded that the MEP ideas seem to have little relevance either for the development of biological complexity or biological evolution.

Introduction
This, according to the great majority of biologists, represents the basis of biological evolution, which has led to the generation of increasing structural and functional complexity in both the plant and animal kingdoms over the last 1–2 billion years.

While the theory of evolution by natural selection provides a mechanism for self-organization of complex biological structures, it is largely descriptive and makes little or no reference to underlying physical laws.

As many have noted, the complexity of biological structure and its evolution seem to place it outside the laws of physics, and in particular the second law of thermodynamics.

The attempt to discuss the complexity and evolution of the LS in physical terms seems to have commenced with the little known study of the mathematician Fantappiè [19] and the subsequent publication of the well-known book "What is Life?"

MEP and Biological Evolution
Points (i), (ii) and (iii) are the main postulates on which the concept of MEP as the "vital force" in biological evolution is based. (i)Steady-state assumption applied to the LS In general terms, one may think of the steady state (SS) as representing a stable, or metastable, state for open thermodynamic systems, in much the same way as equilibrium does for closed systems.

If this highly ordered structure of chlorophyll/protein complexes were to be less ordered, with longer charge separation times (leading to a lowering of the electronic energy transfer efficiency), the electronic energy of the excited state would be increasingly dissipated as heat and/or fluorescence and entropy production would increase.

Even assuming that the suggestion of Ulanowicz and Hannon (20) is correct, i.e., the LS does indeed increase the rate of entropy (heat) production with respect to that in its absence, an affirmation which has not been proven as far as we are aware, there is not a logical demonstration that MEP is in some way a "driving force" for the LS and its evolution.

Conclusions
We also reject the idea that it is the entropy production or its rate which drives biochemical and biophysical processes.

Several well-known examples of cellular structures, examples of biological complexity, are shown to function in terms of maximizing efficiency and minimizing entropy production which is "lost work".

Acknowledgement
A machine generated summary based on the work of Jennings, Robert C.; Belgio, Erica; Zucchelli, Giuseppe (2020 in Rendiconti Lincei. Scienze Fisiche e Naturali).

Application of Maximum Entropy Method for Droplet Size Distribution Prediction Using Instability Analysis of Liquid Sheet

https://doi.org/10.1007/s00231-011-0797-5

Abstract-Summary
This paper describes the implementation of the instability analysis of wave growth on liquid jet surface, and maximum entropy principle (MEP) for prediction of droplet diameter distribution in primary breakup region.

The early stage of the primary breakup, which contains the growth of wave on liquid–gas interface, is deterministic; whereas the droplet formation stage at the end of primary breakup is random and stochastic.

The stage of droplet formation after the liquid bulk breakup can be modeled by statistical means based on the maximum entropy principle.

The deterministic aspect considers the instability of wave motion on jet surface before the liquid bulk breakup using the linear instability analysis, which provides information of the maximum growth rate and corresponding wavelength of instabilities in breakup zone.

Extended
The deterministic aspect was done, which involves the determination of the liquid bulk breakup length and the mass mean diameter by means of hydrodynamic instability theory and unstable wave development on the liquid jet surface.

Introduction
The distribution of droplet size and velocity in sprays is a crucial parameter needed for fundamental analysis of practical spray systems.

Since the mid-1980s, the Maximum Entropy Principle (MEP) method has gained popularity in atomization and spray field to predict droplet size and velocity distribution and has obtained reasonable success.

The MEP approach can predict the most likely droplet size and velocity distributions under a set of constraints expressing the available information related to the distribution.

The liquid sheet instability is responsible for spray formation and determines the resultant spray characteristics such as droplet size, velocity and their distributions, and hence must be included in the model for the completeness and success of the model.

Mitra and Li [21, 22] also used the MEP to obtain droplet size distributions of liquid sheets and used instability analysis, which provides information for prior distribution for the droplet sizes corresponding to the unstable wave growth.

Linear Instability Analysis of Annular Liquid Sheet
Inner gas is considered as a forced vortex, as it is restricted by liquid jet' sheet.

Due to the swirling coaxial flow effect, centrifugal forces act on the annular liquid sheet.

The sum of these forces determines whether the annular liquid sheet is going to breakup or remains stable.

The displacement disturbances at the inner and outer interfaces are The linearized disturbed equations for the inner and outer air are written in component form as: where Boundary conditions must be applied at the liquid interface.

The dispersion equation is obtained by substituting the pressure disturbances inside the liquid and the gas phases into the dynamic boundary conditions at the two interfaces.

The result shows that linear theory accurately predicts the wave length and growth rate of the unstable wave responsible for sheet breakup for the low perturbed liquid jet.

Prediction of Droplet Distribution

The droplet formation process deals with the stochastic sub-model, where a probability density function (PDF) is used to describe the distribution of droplets in sprays.

To extract governing equations and to determine size and velocity distribution for particles, a control volume is considered from the outlet of the injector to the droplet formation location.

S_m, S_{mu} and S_e are the source terms for mass, momentum and energy equations, respectively.

The source terms of mass and energy are set to zero indicating that the evaporation and heat transfer during spraying process have been neglected.

If there is any energy conversion whiting the control volume, it is not considered as a source term.

To solve the governing equations, analytical domains for non dimensional diameter and velocity are considered from 0 to 3.

Coupling Two Sub-models

Linear instability is used to obtain the initial mean drop size and breakup length of sheet breakup.

According to the linear instability theory, waves on the liquid bulk grow either with time or with the downstream distance and the wave with the maximum growth rate, whose wavelength is called dominant wavelength, would dominate the liquid bulk breakup and droplet formation process.

The length of liquid sheet can be estimated using the breakup length according to wave analysis sub model.

In their approach the unstable wave motion before the liquid bulk breakup is taken into account through the instability analysis and is coupled to MEP model by various source terms.

The linear instability theory is used to determine the maximum wave growth rate and consequently the jet breakup length, as well as mean droplet size.

Droplet Distribution at Downstream of Sheet Breakup Region

To consider drag force on droplets, the momentum source term should be modified.

Using the sum of the drag forces on individual droplets, the new momentum source term can be extracted.

The final droplet's size and velocity distributions belong to the droplets formed at downstream of jet breakup regime or final stage of primary breakup.

Results and Discussion

It is also observed that the theoretical distribution predicts slightly greater values for small droplets compared to the experimental distribution.

The results shown so far were related to onset of sheet breakup and showed the size distribution for initial droplets formed at the beginning of droplet formation region.

These data have been used to investigate the droplet size distribution downstream of breakup position using a comparison of results for two different positions, 5 and 10 mm from the nozzle exit.

According to the diagram, droplet size distribution converges to the unique result that belongs to the end part of primary breakup, where there is a steady condition for droplet formation and droplet distribution in spray.

The theoretical distribution in 10 mm downstream direction predicts slightly greater values for big droplets compared to the experimental distribution.

Conclusion

The random process of distributing the diameter and velocity of the droplets in the primary breakup region was modeled implementing maximum entropy principle (MEP).

Two sub-models are coupled together by the momentum source term and mass mean diameter of droplets.

A satisfactory agreement is achieved between the predicted droplet size distributions and experimental measurements for two different sprays with and without high speed surrounding gas.

The present model may be applied to obtain the initial droplet size and velocity distributions for sprays.

Acknowledgement

A machine generated summary based on the work of Movahednejad, E.; Ommi, F.; Hosseinalipour, S. M.; Chen, C. P.; Mahdavi, S. A. (2011 in Heat and Mass Transfer).

Climate Entropy Budget of the HadCM3 Atmosphere–Ocean General Circulation Model and of FAMOUS, Its Low-Resolution Version

https://doi.org/10.1007/s00382-009-0718-1

Abstract-Summary

The rate of material entropy production of the climate system is found to be $\sim$50 mW m^{-2} K^{-1}, a value intermediate in the range 30–70 mW m^{-2} K^{-1} previously reported from different models.

Numerical entropy production in the atmosphere dynamical core is found to be about 0.7 mW m^{-2} K^{-1}.

The material entropy production within the ocean due to turbulent mixing is $\sim$1 mW m^{-2} K^{-1}, a very small contribution to the material entropy production of the climate system.

The rate of change of entropy of the model climate system is about 1 mW m^{-2} K^{-1} or less, which is comparable with the typical size of the fluctuations of the entropy sources due to interannual variability, and a more accurate closure of the budget than achieved by previous analyses.

Results are similar for FAMOUS, which has a lower spatial resolution but similar formulation to HadCM3, while more substantial differences are found with respect to other models, suggesting that the formulation of the model has an important influence on the climate entropy budget.

Since this is the first diagnosis of the entropy budget in a climate model of the type and complexity used for projection of twenty-first century climate change, it would be valuable if similar analyses were carried out for other such models.

Extended

The rate of entropy production of the system, due to irreversible processes, must be positive in order to satisfy the second law of thermodynamics.

The rate of change of entropy must be zero if the system is in a steady state.

All this suggests the need for future work to improve our knowledge of the basic thermodynamics of the climate system.

Introduction

The physical quantity which is able to describe both the irreversibility (non-equilibrium) and the steadiness of the climate system at once (and of any non-equilibrium steady state system) is the entropy production rate of the system (Kondepudi and Prigogine [23]; DeGroot and Mazur [24]).

The rate of entropy production of the system, due to irreversible processes, must be positive in order to satisfy the second law of thermodynamics.

General circulation models (GCMs) are generally not designed to diagnose entropy, but at the moment there is increasing interest in entropy production as a diagnostic tool for GCMs.

Entropy seems to have a role in the basic thermodynamics of the climate: for example Pauluis and Held [25] showed that the efficiency of convection is reduced by the entropy production of non-viscous processes (such as water vapour diffusion).

Lucarini et al. [26] used entropy production concepts and other thermodynamic quantities to study a solar constant variation hysteresis experiment.

Models and Experiments
By virtue of the reduced horizontal and vertical resolution and increased timestep FAMOUS runs about ten times faster than HadCM3.

The atmosphere model of HadCM3 has a horizontal grid spacing of $2.5° \times 3.75°$ and 19 vertical levels.

The horizontal atmospheric grid-spacing of FAMOUS is $5° \times 7.5°$, twice that used in HadCM3, and the vertical resolution is reduced to 11 levels.

Hyper-diffusion (sixth-order in HadCM3, eighth in FAMOUS) is applied after the advection to the model prognostics.

The ocean component in HadCM3 is a 20 level version of the Cox [27] model on a $1.25° \times 1.25°$ latitude–longitude grid.

A 1-h time step is used to integrate the model in HadCM3, 12 h in FAMOUS.

At the end of the day these fluxes are passed to the ocean model which is then integrated forwards in time and supplies SSTs and sea-ice conditions back to the atmosphere.

The Climate Entropy Budget
The second term is hard to deal with because in a GCM F is not represented by processes transporting heat down a temperature gradient; many parametrised processes are typically involved, which do not involve fluxes and gradients explicitly.

A heating rate can be written in terms of the convergence of F and, from elementary calculus, over a volume V with boundary A. If we integrate over the whole climate system, by construction there is no boundary term, because heat fluxes through the boundary are purely radiative, not material.

The global integral of this term can be evaluated instead from the rate at which heat is added (removed) to (from) the hydrological cycle, that is from the latent heating associated with evaporation at the surface and phase changes within the atmosphere and from the temperatures at which this occurs.

Radiative Entropy Production Within the Climate System
The radiation generates a large entropy sink (~ -390 mW m^{-2} K^{-1}) in the body of the atmosphere because it is dominated by longwave cooling, the atmosphere being heated mainly non-radiatively from the surface by latent and sensible heat.

The large irreversible radiative entropy production is not relevant to the operation of the material climate system, as can be understood by a thought experiment, in which we imagine there is no radiation, but instead we provide the climate system at every point with heat sources and sinks equal to shortwave absorption and net longwave emission.

In the radiation scheme the solar radiation is represented as a vertically orientated downward beam which is partly absorbed at every level and finally at the surface.

16c) in Ozawa et al. [28] requires knowledge of the upwelling longwave radiation emitted by the surface (say e_{sur}) and partially absorbed at the level i and conversely of the radiation e_i emitted at the level i and absorbed at the surface.

Sensible and Latent Heating
These variables are the total water content $q_t = q + q_L + q_F$ (where q is the water vapour and q_L and q_F the liquid and frozen water contents, respectively) and the liquid-frozen water potential temperature $\theta_L = \theta - (L_C/c_p)q_L - \{(L_C + L_F)/c_p\}q_F$ (where L_C and L_F are the latent heat of condensation and fusion respectively and c_p the specific heat capacity at constant pressure).

An approximate estimate of the entropy produced by the vertical turbulent fluxes of sensible heat within the boundary layer can be obtained when we recall that at the surface (which coincides with the first boundary layer level in the model), $F = H$, because at the surface $q_L = q_F = 0$, i.e. there is no surface flux of cloud water, only water vapour.

The Energy Correction and the Dissipation of Kinetic Energy
There are four explicit processes causing the dissipation of the specific kinetic energy $(v \cdot v)/2$: (a) the turbulent eddy stresses in the boundary layer scheme associated with the vertical diffusion of horizontal momentum; (b) the hyper-diffusion term in the momentum equation for (u, v); (c) the gravity wave drag and (d) the impact on the large scale flow due to the convective eddy-flux stresses.

The total amount of viscous dissipation $D = \int_{V\_at} \rho\varepsilon \, dV$ due to the eddy-stresses is given by the mass integral over the entire atmospheric volume of the specific kinetic energy tendency $\partial_t(v^2/2) = v \cdot \partial_t v$ caused by them.

We describe how HadCM3 deals with the kinematic impact of the boundary layer, convection, gravity wave schemes and horizontal hyperdiffusion. (a) Turbulent stresses in the boundary layer: vertical momentum diffusion.

From (37–40) the kinetic energy tendencies and the dissipation terms are estimated and compared with the energy correction which is about 3.1 W m^{-2} in HadCM3 and 3.4 W m^{-2} in FAMOUS.

Numerical Entropy Production in the Dynamical Core
In numerical models there is a small entropy source of a numerical nature (Johnson [29]; Egger [30]).

In the numerical integration phenomena of pure numerical nature such as the numerical dispersion, Gibbs oscillation and numerical diffusion and filtering yield an entropy source.

The numerical entropy produced by the advection scheme is not associated with a heating term.

Entropy Production in the Ocean
The material entropy production in the ocean is due to three different processes: heat transport, viscous dissipation and molecular diffusion of salt ions.

Numerical entropy production is negligible in the ocean model.

Several irreversible processes are parametrised in the ocean model in order to reproduce the effects of the turbulence due to the sub-grid scale mixing: vertical diffusion, mixed layer physics, convective adjustment and isopycnal diffusion.

They have a non-zero contribution to the ocean entropy budget since they move heat down a temperature gradient.

A further scheme implemented in the HadCM3 ocean model is Visbeck et al. [31] version of the Gent and McWilliams [32] scheme for the adiabatic thickness diffusion.

The mixed layer scheme (Kraus and Turner [33]) works out a balance between the energy required for mixing the water column and the energy available as turbulent kinetic energy from the wind and from the introduction of buoyancy at the ocean surface, leading to convective instability.

Climate Entropy Budget and the Uncertainty in the Material Entropy Production

This difference is due mainly to differences in the shortwave fluxes in the two models, i.e. HadCM3 has a higher value of the shortwave energy radiation absorbed in the atmosphere than FAMOUS (283 vs. 266 W m^{-2}).

This highlights that the maximum uncertainity is associated with the entropy produced by the moist processes leading to latent heating, being the range $\sim$(20–36) mW m^{-2} K^{-1}.

Goody estimates the entropy produced by stresses in the frictional boundary layer to be $\sim$6 mW m^{-2} K^{-1}, due to dry convection in the free atmosphere to be around 4 mW m^{-2} K^{-1} and due to moist convection 1.3 mW m^{-2} K^{-1}.

The order-of-magnitude value provided by Peixoto et al. [11] is 7 mW m^{-2} K^{-1}, estimated from a diabatic frictional heating in the boundary layer of 1.9 W m^{-2} at a temperature of 280 K. HadCM3 and FAMOUS have an entropy production due to the dissipation of kinetic energy which is higher than the other model and observational estimates.

Conclusions

A detailed and comprehensive entropy budget analysis of a complex atmosphere–ocean general circulation model, HadCM3, and its low resolution version, FAMOUS, has been presented in this paper.

HadCM3 and FAMOUS are rather similar as far as entropy terms are concerned although the resolution is different.

Even though the results confirm to a certain extent the order-of-magnitude numbers obtained from (Goody [9]; Fraedrich and Lunkeit [34]), differences between HadCM3 and the other two models are much more substantial and this shows that the differences in the model formulation are very influential on the entropy budget.

The numerical entropy produced by the atmosphere model dynamical core is even smaller, $\sim$0.7 mW m^{-2} K^{-1} in both HadCM3 and FAMOUS, and it is the combined effect of the diffusion (positive) and the filtering-adjustment-advection (negative).

Acknowledgement

A machine generated summary based on the work of Pascale, Salvatore; Gregory, Jonathan M.; Ambaum, Maarten; Tailleux, Rémi (2009 in Climate Dynamics).

Revisiting the Global Surface Energy Budgets with Maximum-Entropy-Production Model of Surface Heat Fluxes

https://doi.org/10.1007/s00382-016-3395-x

Abstract-Summary

The maximum-entropy-production (MEP) model of surface heat fluxes, based on contemporary non-equilibrium thermodynamics, information theory, and atmospheric turbulence theory, is used to re-estimate the global surface heat fluxes.

The new MEP-based global annual mean fluxes over the land surface, using input data of surface radiation, temperature data from National Aeronautics and Space Administration–Clouds and the Earth's Radiant Energy System (NASA CERES) supplemented by surface specific humidity data from the Modern-Era Retrospective Analysis for Research and Applications (MERRA), agree closely with previous estimates.

The MEP model also produces the first global map of ocean surface heat flux that is not available from existing global reanalysis products.

Extended

The MEP model also uses fewer model parameters (than existing models) that are independent of wind speed and SRLs.

The MEP model only uses R_n and T_s data for the case of saturated land surfaces (e.g., saturated soils, irrigated farm lands and canopy under no water stress) where q_s is a function of T_s alone according to the Clausius–Clapeyron equation.

The MEP model predicts heat fluxes without using temperature and humidity gradients, wind speed, and surface roughness data.

The MEP model estimated H is 28 ± 3 W m^{-2}, which is higher than the previous estimates of ~10–20 W m^{-2}.

The MEP model produces new estimates of global annual mean evaporation rate of 667 mm year^{-1} (± 76 mm year^{-1}), sensible heat flux of 30 W m^{-2} (± 4 W m^{-2}), ground (conductive) heat flux (over land) of 12 W m^{-2} (± 10 W m^{-2}) and ocean surface heat flux of 139 W m^{-2} (± 10 W m^{-2}) (through conductive cool-skin).

The MEP model may serve as an effective physical parameterization of the land–ocean–atmosphere interaction in regional and global numerical weather prediction and climate models, contributing to the study of changes of water–energy–carbon cycles in response to radiative forcing perturbations of both natural and anthropogenic origins.

The uncertainties of the MEP estimated heat fluxes can be further reduced with the improved accuracy of radiation measurements in the future.

New results will be reported in the near future.

Introduction

These studies conclude that the existing estimates of the surface (radiative and turbulent) fluxes have substantial uncertainties and do not close the surface energy budget.

Parameterization of transfer coefficients in terms of wind speed and surface roughness length(s) (SRL) may cause substantial errors in the bulk fluxes.

The estimation errors of bulk fluxes are often unbounded without the constraint of surface energy balance.

In the MEP model, the surface turbulent and/or conductive energy fluxes effectively result from the partitioning of radiation fluxes automatically balancing the energy budget.

The global climatology of the surface heat fluxes together with the corresponding uncertainties is re-estimated using the MEP model utilizing the input data of radiation and temperature from the National Aeronautics and Space Administration (NASA) Clouds and the Earth's Radiant Energy System (CERES) products supplemented by the (land) surface specific humidity data from the NASA Modern-Era Retrospective analysis for Research and Applications (MERRA) products.

Method and Data

The MEP theory (Wang and Bras [35]) allows the turbulent latent heat E, sensible heat H and conductive heat flux Q over the Earth–atmosphere interface to be simultaneously solved in terms of analytical functions of surface radiation fluxes, temperature and/or humidity as the most probable partitioning of radiation fluxes while closing the surface energy budgets (satisfying the conservation of energy) by seeking an answer to the question "What is the best prediction of energy partitioning of surface radiation fluxes into surface heat fluxes based on the available surface energy and moisture states?".

The MEP model predicts heat fluxes without using temperature and humidity gradients, wind speed, and surface roughness data.

The difference is that the MEP method makes more effective use of the information most relevant to the heat fluxes provided by the surface variables (radiation, temperature, and/or humidity) than conventional methods.

Results

The MEP modeled annual mean E over oceans is 733 ± 88 mm year^{-1} (58 ± 7 W m^{-2}), which is lower than the previous estimates in the range of 1130–1370 mm year^{-1} ($90 - 109$ W m^{-2}).

The MEP model estimated H is 28 ± 3 W m^{-2}, which is higher than the previous estimates of ~10–20 W m^{-2}.

The uncertainties of the trends of MEP E and H were estimated as 0.06 and 0.04 W m^{-2} year^{-1}, respectively.

The newly estimated global annual mean E, 53 ± 6 W m^{-2}, is lower than previous estimates of 80–90 W m^{-2} largely due to the lower MEP E over oceans.

The MEP estimate of global annual net surface heat flux is ~31 ± 5 W m^{-2}, while previous reanalysis products all have nearly zero annual ground/water–snow–ice heat flux.

Conclusion

The MEP model produces new estimates of global annual mean evaporation rate of 667 mm year^{-1} (± 76 mm year^{-1}), sensible heat flux of 30 W m^{-2} (± 4 W m^{-2}), ground (conductive) heat flux (over land) of 12 W m^{-2} (± 10 W m^{-2}) and ocean surface heat flux of 139 W m^{-2} (± 10 W m^{-2}) (through conductive cool-skin).

The MEP model produces the first estimate of global ocean surface heat flux that is not available from existing data products.

The MEP modeled surface heat fluxes not only (by definition) close surface energy budgets at all space–time scales, but also avoid explicit uses of temperature/moisture gradients, wind speed and surface roughness as model inputs and parameters.

Constructing a new global surface energy budget and putting the MEP model results in a broader perspective of the climate system's energy cycle is the first step of our ongoing efforts.

Acknowledgement

A machine generated summary based on the work of Huang, Shih-Yu; Deng, Yi; Wang, Jingfeng (2016 in Climate Dynamics).

A Simplified Climate Model and Maximum Entropy Production

https://doi.org/10.1140/epjp/s13360-020-00879-7

Abstract-Summary

A simplified climate model based on maximum entropy production, described by a variational principle, is revisited and an analytical solution to its Euler–Lagrange equation is found.

Introduction

The principle that open thermodynamical systems in nature tend to maximize entropy production (MEP) has been debated extensively in the earth sciences, and in atmospheric science in particular.

Although, in numerical studies of these models, one can check easily whether entropy production is maximized, minimized, or has a saddle point, in these studies one is always committed to particular choices of parameters and initial conditions, and numerical confirmation does not constitute mathematical proof.

We revisit a simple one-dimensional model with energy transport in the meridional direction, based on MEP, which was proposed in [36].

A recurrent puzzle in MEP-based system is that sometimes there is a maximum and sometimes a minimum of entropy production [4, 6, 8–16, 17–37, 38, 39, 40, 41, 42, 43, 44, 45, 46, , 47–51]: here we prove (as opposed to checking numerically for special configurations) that a maximum always occurs in this model.

The Model
The Murakami and Kitoh one-dimensional climate model in the meridional direction [36] is based on a very idealized radiative formulation.
Maximum or Minimum?

Conclusions
We have revisited the simplified climate model of [36] and have simplified and solved its main equation, which is derived from an action integral corresponding to the entropy production rate and involves radiative absorbed and radiated fluxes and poleward transport of energy from tropical regions.

Although MEP is routinely verified numerically in this kind of model, it is not understood.

There have often been surprises in natural processes associated with open thermodynamical systems, in which entropy production is sometimes maximized and other times minimized, and it is useful to set MEP on a firm footing with rigorous statements before proceeding with numerical studies.

We have shown that, in the one-dimensional climate model of [36], the entropy production rate is indeed maximized.

Acknowledgement
A machine generated summary based on the work of Faraoni, Valerio (2020 in The European Physical Journal Plus).

Book Reading List

Bayesian Inference and Maximum Entropy Methods in Science and Engineering

by Polpo, A. (Ed), Stern, J. (Ed), Louzada, F. (Ed), Izbicki, R. (Ed), Takada, H. (Ed) (2018).

These proceedings from the 37th International Workshop on Bayesian Inference and Maximum Entropy Methods in Science and Engineering (MaxEnt 2017), held in São Carlos, Brazil, aim to expand the available research on Bayesian methods and promote their application in the scientific community. They gather research from scholars in many different fields who use inductive statistics methods and focus on the foundations of the Bayesian paradigm, their comparison to objectivistic or frequentist statistics counterparts, and their appropriate applications.

Please see https://www.springer.com/gp/book/9783319911427 for original source.

Entropy and Entropy Generation

by Shiner, J. (Ed) (2002).

Entropy and entropy generation play essential roles in our understanding of many diverse phenomena ranging from cosmology to biology. Their importance is manifest in areas of immediate practical interest such as the provision of global energy as well as in others of a more fundamental flavour such as the source of order and complexity in nature. They also form the basis of most modern formulations of both equilibrium and nonequilibrium thermodynamics.

Please see https://www.springer.com/gp/book/9781402004162 for original source.

Entropy Measures, Maximum Entropy Principle and Emerging Applications

by Karmeshu, () (Ed) (2003).

This book is dedicated to Prof. J. Kapur and his contributions to the field of entropy measures and maximum entropy applications. Eminent scholars in various fields of applied information theory have been invited to contribute to this Festschrift, collected on the occasion of his 75th birthday. The articles cover topics in the areas of physical, biological, engineering and social sciences such as information technology, soft computing, nonlinear systems or molecular biology with a thematic coherence.

Please see https://www.springer.com/gp/book/9783540002420 for original source.

Maximum-Entropy Networks

by Squartini, T., Garlaschelli, D. (2017).

This book is an introduction to maximum-entropy models of random graphs with given topological properties and their applications. Its original contribution is the reformulation of many seemingly different problems in the study of both real networks and graph theory within the unified framework of maximum entropy. Particular emphasis is put on the detection of structural patterns in real networks, on the reconstruction of the properties of networks from partial information, and on the enumeration and sampling of graphs with given properties.

Please see https://www.springer.com/gp/book/9783319694368 for original source.

Maximum Entropy and Bayesian Methods

by Smith, C. (Ed), Erickson, G. (Ed), Neudorfer, P. O. (Ed) (1992).

Bayesian probability theory and maximum entropy methods are at the core of a new view of scientific inference. These 'new' ideas, along with the revolution in computational methods afforded by modern computers, allow astronomers, electrical engineers, image processors of any type, NMR chemists and physicists, and anyone at all who has to deal with incomplete and noisy data, to take advantage of methods that, in the past, have been applied only in some areas of theoretical physics.

This volume records the Proceedings of Eleventh Annual 'Maximum Entropy' Workshop, held at Seattle University in June, 1991.

Please see https://www.springer.com/gp/book/9780792320319 for original source.

Maximum Entropy and Bayesian Methods

by Skilling, J. (Ed), Sibisi, S. (Ed) (1996).

This volume records papers given at the fourteenth international maximum entropy conference, held at St John's College Cambridge, England. It seems hard to believe that just thirteen years have passed since the first in the series, held at the University of Wyoming in 1981, and six years have passed since the meeting last took place here in Cambridge. So much has happened. There are two major themes at these meetings, inference and physics.

Please see https://www.springer.com/gp/book/9780792334521 for original source.

Non-equilibrium Thermodynamics and the Production of Entropy

by Kleidon, A. (Ed), Lorenz, R. D. (Ed) (2005).

The present volume studies the application of concepts from non-equilibrium thermodynamics to a variety of research topics. Emphasis is on the Maximum Entropy Production (MEP) principle and applications to Geosphere-Biosphere couplings. Written by leading researchers form a wide range of background, the book proposed to give a first coherent account of an emerging field at the interface of thermodynamics, geophysics and life sciences.

Please see https://www.springer.com/gp/book/9783540224952for original source.

The Maximum Entropy Method

by Wu, N. (1997).

The Maximum Entropy Method addresses the principle and applications of the powerful maximum entropy method (MEM), which has its roots in the principle of maximum entropy introduced into the field of statistical mechanics almost 40 years ago. This method has since been adopted in many areas of science and technology, such as spectral analysis, image restoration, mathematics, and physics.

Please see https://www.springer.com/gp/book/9783642644849 for original source.

References

1. Meysman FJ, Bruers S (2010) Ecosystem functioning and maximum entropy production: a quantitative test of hypotheses. Phil Trans R Soc B 365:1405–1416
2. Schneider ED, Kay JJ (1994) Life as a manifestation of the second law of thermodynamics. Math Comput Model 19:25–48
3. Zia RKP, Schmittmann B (2006) A possible classification of nonequilibrium steady states. J Phys A Math Gen 39:L407–L413
4. Paltridge G (1975) Global dynamics and climate-a system of minimum entropy exchange. Quart J Roy Meteorol Soc 101:475–484
5. Kleidon A, Fraedrich K, Kunz T, Lunkeit F, (2003) The atmospheric circulation and the states of maximum entropy production. Geophys Res Lett 30(23)
6. Kleidon A, Fraedrich K, Kirk E, Lunkeit F (2006) Maximum entropy production and the strenght of boundary layer exchange in an atmospheric general circulation model. Geophys Res Lett 33:L08709
7. Kleidon A (2005) Hyperdiffusion, maximum entropy production, and the simulated equator-pole temperature gradient in an atmospheric general circulation model. https://hdl.handle.net/1903/1927
8. Kunz T, Fraedrich K, Kirk E (2008) Optimisation of simplified GCMS using circulation indices and maximum entropy production. Clim Dyn 30:803–813
9. Goody R (2000) Sources and sinks of climate entropy. Q J R Meteorol Soc 126:1953–1970
10. Volk T, Pauluis O (2010) It is not the entropy you produce, rather, how you produce it. Philos Trans R Soc B 365:1317–1322. https://doi.org/10.1098/rstb.2010.0019
11. Peixoto J, Oort A, de Almeida M, Tomé A (1991) Entropy budget of the atmosphere. J Geophys Res 96:10981–10988
12. O'Gorman PA, Schneider T (2008) Energy of midlatitude transient eddies in idealized simulations of changed climates. J Clim 21:5797–5808. https://doi.org/10.1175/2008JCLI2099.1
13. Schneider T, O'Gorman PA, Levine XJ (2010) Water vapor and the dynamics of the climate changes. Rev Geophys 48(RG3001):22. https://doi.org/10.1029/2009RG000302
14. Lorenz EN (1960) Generation of available potential energy and the intensity of the general circulation. Pergamon, Tarrytown
15. Paltridge GW (1978) The steady-state format of global climate. Quart J Royal Meteorol Soc 104:927–945
16. Lorenz RD, Lunine JI, Withers PG (2001) Titan, Mars and Earth: entropy production by latitudinal heat transport. Geophys Res Lett 28:415–418
17. Kleidon A (2009) Nonequilibrium thermodynamics and maximum entropy production in the earth system. Naturwissenschaften 96:653–677
18. Goody R (2007) Maximum entropy production in climate theory. J Atmos Sci 64:2735–2739
19. Fantappiè L (1943) Sull'interpretazione dei potenziali anticipati della meccanica ondulatoria e su un principio di finalità che ne discende, In: Rend. d. Accad. d'Italia, cl. di sc. fis., mat. e natur., s. 7, IV, pp 81–86
20. Ulanowicz RE, Hannon BM (1987) Life and the production of entropy. Proc R Soc Lond 232:181–192
21. Mitra SK, Li X (1999) A predictive model for droplet size distribution in sprays. at Sprays 9:29–50
22. Kim WT, Mitra SK, Li X (2003) A predictive model for the initial droplet size and velocity distributions in sprays and comparison with experiments. Part Part Syst Charact 20:135–149
23. Kondepudi D, Prigogine I (1998) Modern thermodynamics: from heat engines to dissipative structure. Wiley, Hoboken
24. DeGroot S, Mazur P (1984) Non-equilibrium thermodynamics. Dover, Kluwar
25. Pauluis O, Held M (2002) Entropy budget of an atmosphere in radiative–convective equilibrium. Part I: Maximum work and frictional dissipation. J Atmos Sci 59:125–139

26. Lucarini V, Fraedrich K, Lunkeit F (2009) Thermodynamic analysis of snowball earth hysteresis experiment: efficiency, entropy production and irreversibility. QJR Meterol Soc (in press)
27. Cox MD (1984) A primitive equation, three dimensional model of the ocean. GFDL ocean group technnical report no. 1, Princeton NJ, USA, 143 p
28. Ozawa H, Ohmura A, Lorenz RD, Pujol T (2003) The second law of thermodynamics and the global climate system: a review of the maximum entropy production principle. Rev Geophys 41(4):1018. https://doi.org/10.1029/2002RG000113
29. Johnson D (1997) "General coldness of climate" and the second law: implications for modelling the earth system. J Clim 10:2826–2846
30. Egger J (1999) Numerical generation of entropies. Mon Weather Rev 127:2211–2216
31. Visbeck M, Marshall J, Haine T, Spall M (1997) On the specification of eddy transfer coefficients in coarse resolution ocean circulation models. J Phys Oceanogr 27:381–402
32. Gent P, McWilliams J (1990) Isopycnal mixing in ocean circulations models. J Phys Oceanogr 20:150–155
33. Kraus E, Turner J (1967) A one dimensional model of the seasonal thermocline. Part II. Tellus 19:98–105
34. Fraedrich K, Lunkeit F (2008) Diagnosing the entropy budget of a climate model. Tellus a 60:921–931
35. Wang J, Bras R (2009) A model of surface heat fluxes based on the theory of maximum entropy production. Water Resour Res 45:W11422. https://doi.org/10.1029/2009WR007900
36. Murakami S, Kitoh A (1953) Euler-Lagrange equation of the most simple 1-d climate model based on the maximum entropy production hypothesis. Quart J Roy Meteorol Soc 131:1529–1538
37. Paltridge G (1981) Thermodynamic dissipation and the global climate system. Quart J Roy Meteorol Soc 107:531–547
38. Grassl H (1981) The climate at the maximum-entropy production by meridional atmospheric and oceanic heat fluxes. Quart J Roy Meteorol Soc 107:153–166
39. Noda A, Tokioka T (1983) Climates at minima of the entropy exchange rate. J Meteorol Soc Jpn 61:894–908
40. Pujol T, Fort J (2002) States of maximum entropy production in a one-dimensional vertical model with convective adjustments. Tellus a 54:363–369
41. Pujol T (2003) Eddy heat diffusivity at maximum dissipation in a radiative-convective one-dimensional climate model. J Meteorol Soc Jpn 81:305–315
42. Kleidon A (2004) Beyond Gaia: thermodynamics of life and earth system functioning. Clim Change 66:271–319
43. Kleidon A (2010) A basic introduction to the thermodynamics of the earth system far from equilibrium and maximum entropy production. Philos T Roy Soc B 365:1303–1315
44. Jupp T, Cox P (2010) MEP and planetary climates: insights from a two-box climate model containing atmospheric dynamics. Philos T Roy Soc B 365:1355–1365
45. Herbert C, Paillard D, Dubrulle B (2011) Entropy production and multiple equilibria: the case of the ice-albedo feedback. Earth Syst Dyn 2:13–23
46. Mobbs SD (1982) Extremal principles for global climate models. Quart J Roy Met Soc 108:535–550
47. Kleidon A, Fraedrich K, Kunz T, Lunkeit F (2003) The atmospheric circulation and the states of maximum entropy production. Geophys Res Lett 30:2223
48. Pascale S, Gregory J, Ambaum M, Tailleux R (2012) A parametric sensitivity study of entropy production and kinetic energy dissipation using the FAMOUS AOGCM. Clim Dyn 38:1211–1227
49. Maldonado S (2020), Do beach profiles under non-breaking waves minimize energy dissipation? J. Geophys. Res Oceans 125, e2019JC015876
50. Maldonado S, Uchasara M (2019) On the thermodynamics-based equilibrium beach profile derived by Jenkins and Inman (2006). arXiv:1908.07825 [physics.geo-ph]
51. Faraoni V (2020) On the extremization of wave energy dissipation rates in equilibrium beach profiles. J Oceanogr. https://doi.org/10.1007/s10872-020-00556-4

Chapter 7
Astrobiology and Development of Human Civilization

Introduction by Guido Visconti

Astrobiology is a young science interested in the origin, evolution and the distribution of forms of life in the universe and in particular aims to discover other forms of life outside the Earth. It is interesting that astrobiology looks in particular at life on Earth in the most extreme conditions like at some depth in the crust or in the deepest of the oceans. Astrobiology has received official attention from all space agencies like ESA or NASA. Earth itself could be a subject for astrobiology because as a planet it went through different stages. From a water planet (before the rise of the continents) to an ice planet (snowball Earth) to a hothouse planet (during the carboniferous) and all these stages have seen co-evolving life forms. Life has contributed very much to change the atmospheric composition. Think about the cyanobacteria that probably gave us the initial oxygen in the atmosphere. Now the most evolved life form (the human being) apparently has reached a changing capacity that can be compared to a geological force so that the Holocene ended to give way to the Anthropocene. Not all the scientific community agrees with such classification and some people like Adam Frank likes to talk about the Astrobiology of the Anthropocene.

There are a few connections between Astrobiology and the future of the Earth and human beings. Looking at what happened to the other terrestrial planets of the solar system, Venus and Mars, we see the first with an exasperated greenhouse (runaway) effect while the second now a very cold and barren desert. Although we know that the physical Earth's climate will never evolve to a runaway greenhouse, both planets are an indication that climate can change to the point of wiping out all the life forms. The Earth itself was subject to five major extinctions since the end of snowball Earth about half a billion of years ago and so Astrobiology may give us indications about our future intended not in the IPCC horizon of a hundred years but rather in the range of million years. A further connection of Astrobiology with human future is the so-called Great Filter, which is discussed also in the Fermi Paradox chapter. The Great Filter was introduced by Robin Hanson in 1998 and refers to some kind of barrier produced during the evolving trajectory of an advanced civilization that

G. Visconti (ed.), *Climate, Planetary and Evolutionary Sciences*,
https://doi.org/10.1007/978-3-030-74713-8_7

215

blocks further development to the point to prevent the same galactic civilization to communicate with others. This could be the explanation of the Great Silence we observe in the universe (Fermi paradox).

On the other hand, Astrobiology is also the study of other planets "habitability" and among other things has developed maps of habitability zones in our Galaxy. The habitability is based on those physical conditions that favor the thriving of life as we know it. However, parallel to habitability there is the sustainability to consider. Sustainability is related to the energy or resource consumption and how these can change the environment. There are several reasons to believe that most of the major extinctions we mentioned before may be related to the rapid accumulation in the atmosphere of greenhouse gases. Astrobiology may add further knowledge about these past episodes happening on Earth and other planets.

There are a few papers that deal with some philosophical aspects like the development of universal biology that not necessarily is constrained by what we know on Earth or how important is the definition of life also for the design of future experiments. In this sense we like to conclude with an excerpt from the book by Richard Lewontin and Richard Levins, *Biology under the influence*, that after criticizing the criteria used to search for life on other planets conclude. *The importance of a correct formulation of the problem is, of course, not in finding life on other planets, a project whose probability of success is exceedingly small. Rather a proper model for its solution is a model for the management of terrestrial life. Things in the future cannot be exactly as they were in the past. Ecosystems will change and species will go extinct. Life "as we know it" cannot be maintained. But neither is a future possible that is bound only by imagination and desire. Our methodological problem is to develop an approach to planning and agitation that takes into account both historical contingency and the limits to is possibility.* Although the book was published in 2007 this particular piece (*Life on other worlds*) was printed in 1998 when Astrobiology was in its infancy.

Machine-Generated Summaries

Keywords: *pace, temperature, theory, planetary, solar system, science, organic, universe, star, extraterrestrial, high, surface, molecule, energy, scientist.*

What is Astrobiology?

https://doi.org/10.1007/978-3-030-46087-7_1

Abstract-Summary
Astrobiology is an inherently multidisciplinary field that is focused on the origins, evolution, and distribution of life throughout the Universe.

For scientists, this ambitious endeavor begins with Earth, as it represents the only known example of life in the Universe.

Understanding Earth is, therefore, the first step to understanding the requirements for life to emerge and make a habitable world.

With the collaboration of scientists from many disciplines, we gather the knowledge about the requirements, diversification, and characteristics of terrestrial life, as well as the characteristics of potentially habitable worlds in our Solar System and beyond.

Aliens Everywhere

Our mythology about aliens often includes wrong ideas about scientists and their search for extraterrestrial life.

The scientific community has created a science to understand life on Earth and to better define how to find it elsewhere.

NASA's of astrobiology is the study of the origins, evolution, distribution, and future of life in the Universe (Hubbart [1]; NASA [2]).

Astrobiology articulates the knowledge about origins, the characteristics and evolution of life, (its geology and chemistry), planetary bodies in our, and planets around other stars to understand life as a universal phenomenon.

What Astrobiologists Are Looking for?

We can say that life on Earth has a chemistry based on carbon and requires liquid water.

When astrobiologists say "we are looking for life as we know it," we are not thinking on eyes, antennas, or gray humanoids, but in carbon-based molecules and liquid water.

Scientists assume that the chemistry of carbon and the availability of liquid water are our best choices in the current working definition to search for life in other planets.

Despite being a single example of life (based on carbon and liquid water), the study of life on Earth is relevant for astrobiology because it (1) gives us the tools to understand the requirements for the origins of life and (2) allows us to understand the most extreme conditions where life can survive.

The Limits of Life

Extremophiles provide good models for the study of biodiversity on Earth; they are relevant for the recognition of the boundaries of life, the formulation of theories about the origin of life, to sustain the search for life in extraterrestrial scenarios, and to get insights into the potential of other worlds to support life, all of which are important issues in astrobiology.

While considering the possibility for life to originate and exist on other planetary bodies, it is important to understand the variability of Earth's local conditions when compared to the planetary mean.

Whether or not other planetary bodies such as Mars, Enceladus, or Europa could or did support life, the search for Earth's life true limits will inform our exploration of space and could provide insight into processes that have led to the origin of life on our planet (Merino et al. [3]).

Extreme Sites on Earth

There are some terrestrial sites with extreme ranges of temperature, salinity, mineral content, pH, pressure, and water availability that have been classified as analogues to any of the planetary bodies of astrobiological importance, such as the rocky planets Venus or Mars, the Jovian moons Europa or Ganymede, and the Saturn satellites Titan and Enceladus (Marlow et al. [4]; Preston and Dartnell [5]).

Among the most representative analogue sites relevant to astrobiology are Rio Tinto in Spain and the cold and hyperarid Atacama Desert in Chile as analogue sites for the present planet Mars, as well as the black smokers in the deep sea or the Antarctic lake system in Antarctica as analogues sites for the icy satellites, or the Alaskan oil fields for Titan's environment.

Where Are Astrobiologists Looking for Life?

Potentially habitable planets are those that have liquid water.

The geological record on Mars includes minerals and structures that required liquid water, although this compound is not liquid anymore on the surface of the Red Planet (e.g., Carr and Bell [6]; Golombek and McSween [7]).

The only habitable planet we know now is Earth, and we use it to understand how a potentially habitable planet may work and change in time.

Unlike the case of life, where there is only one example, the case of potentially habitable planets has many examples.

The Closest Neighbors: Planetary Bodies in the Solar System

The first experiment designed and performed to search life on other planet traveled in two of the four spacecrafts from the Viking mission; both landers arrived on Mars surface on 1976, Viking 1 at Chryse Planitia and Viking 2 at Utopia Planitia (NASA Viking missions).

From the observation with telescopes, scientists knew there was frozen water in the Martian poles, and missions have found indirect and direct evidence of frozen water beneath the surface of the Red Planet.

Putting together the pieces collected by a century of observations, spacecraft and rover missions, we are certain that liquid water was present on Mars surface since 3800 billion years ago.

The Red Planet was potentially habitable just by the time life on Earth was arising; even more, both planets shared similar conditions: active volcanoes, liquid water on the surface, an atmosphere of carbon dioxide and nitrogen, and availability of elements such as sulfur and phosphorus, fundamental for making life on Earth.

Habitable Planets Around Other Stars

Numerical models calculate the habitable zone for a star with a given luminosity (amount of stellar energy emitted per unit of time), assuming planets with atmospheres made of water, CO_2, and N_2 (Domagal-Goldman and Segura [8]).

For a planet in the habitable zone of a red dwarf, XUV radiation can heat the planet's upper atmosphere as a result; the atmosphere will flow away from the planet so that there will be no liquid water on the surface without an atmosphere (Luger and Barnes [9]).

For Proxima the calculated inner limit of the habitable zone started beyond 0.2 AU (Barnes et al. [10]); by the time we suppose Proxima b was formed, the planet was in the zone where it received too much energy for having liquid water at the surface.

If Proxima b started with a hydrogen atmosphere equivalent to one thousandth Earth masses, then some of that atmosphere may have survived, preserving the planet's habitability potential (Barnes et al. [10]).

Signatures of Life

Under pervasive extreme conditions (such as high salinity, temperature, or pH), methanogens can further face bioenergetic constraints to use less or non-competitive substrates to produce methane (with different carbon isotopic composition).

In hypersaline environments, such as the ones located at Guerrero Negro's salt flats in Baja California or at the sulfate-rich sediments of Tirez Lagoon in Spain, the diversity of substrates and phylogenetic types (phylotypes) producing methane is considerably reduced and bioenergetically constrained within methanogenic archaea (Lozada-Chávez et al. [11]; Montoya et al. [12]).

Current research in biosignatures to answer this question is focused on (1) understanding false positives for life, that is, planetary processes that may produce detectable amounts of compounds that can be confused with those produced by living things, and (2) generating observational strategies to distinguish between both scenarios (Meadows [13]; Catling et al. [14]; Kiang et al. [15]; Fujii et al. [16]; Walker et al. [17]).

What About Intelligent Life?

Drake had the idea of searching for signals of extraterrestrial (ET) civilizations with the telescopes at Green Bank that were under construction.

Just before starting the project OZMA, the first campaign to search for ET signals, two high energy physicists at Cornell University published a paper proposing the use of radiofrequencies for interstellar communications with other civilizations (Cocconi and Morrison [18]).

They started the manuscript by stating that there were no theories to calculate if life may emerge on other planets, but they made a good point at the end by saying: "The probability of success is difficult to estimate; but if we never search, the chance of success is zero."

Frank Drake proposed an equation that has been used to calculate the number of technological civilizations in our galaxy.

The efforts for searching extraterrestrial civilizations were called: Search for Extraterrestrial Intelligence.

Our Habitable Planet

In the history of humanity, we can perform experiments that may lead us to find life in other planets or moons.

For now, there is only one habitable planet we know of: Earth.

As the Sun increases its luminosity as part of its natural evolution, Earth will lose its oceans and become not habitable for any kind of life; that future is millions of years ahead.

A revision of Earth's history shows us that life survived when our planet was totally frozen and after the impact of large rocks from the space.

Global warming and pollution do influence human life, and we do not have any other place to go.

Acknowledgments

A machine generated summary based on the work of Segura, Antígona; Ramírez Jiménez, Sandra Ignacia; Lozada-Chávez, Irma (2020 in Cuatro Ciénegas Basin: An Endangered Hyperdiverse Oasis).

Astrobiology: Basic Concepts

https://doi.org/10.1007/978-1-4614-8730-2_9

Abstract-Summary

Astrobiology is a challenging interdisciplinary field of contemporary science which appeared in the second half of the last century and stimulated a better understanding of the frontiers of biology.

Astrobiology aims to answer the fundamental questions: Is there life beyond Earth?

This field is rooted in the synergy between astrophysics and biology and is intimately related with planetary sciences, in particular with planetary systems formation and evolution.

Some Historical Highlights

The main point is that, unlike prokaryotes, which used biomineral cycles outside a cell (making them vulnerable to the environmental conditions), metabolic processes inside the eukaryote's cells made the skeleton formation independent of the environment, which changed the cyclic processing of key elements pertinent to life, such as calcium, silicon, etc. This concept gives us a better understanding of the peculiarities of life emergence and the way in which organisms act on their environment, and also allows us to formulate the conditions necessary for life to appear.

They are the functions of a single, indivisible set of organisms, a set comprising the numerous morphologically diverse forms that cause the complexity of life and the fundamental distinction between living matter and inert matter, as well as those reflecting the progress of life through time, which is directly linked with irreversible thermodynamics and Prigogine's notable principle of "the arrow of time."

Prerequisites and Constraints

Earth perfectly fits the position inside the solar system in terms of natural conditions necessary for the origin of life, at least for the well-known form based on carbon and liquid water.

The existence of water, carbon, and other volatiles on an Earth-like planet (or their delivery through migration mechanisms) is key for the origin of life.

They showed that complex organic compounds, including many important to life on Earth, could be readily produced under conditions that likely prevailed in the turbulent environment of the primordial solar system and probably other planetary systems as well.

As we discussed, given the key role of water in the origin of life, we should note that modeling has indicated that the Earth could have received a large influx of volatile matter from comet and asteroid bombardment: a quantity of water that could be comparable to the volume of our planet's oceans.

Tentative Models of the Origin of Life

An alternative to the conception of an ancient RNA world is that of a sequential ordering of the processes of the origin and early evolution of organic matter as the chemical basis of life.

As part of this concept, in which the basic functions of RNA molecules also play an important role, the origin of life is conceived of as a continuous ordering process in an open stationary system, which, in contrast to a conservative (Hamiltonian) system, which conserves energy, is an open dissipative system that exchanges energy with the environment.

The increasing ordering of the original (chaotic) system takes the form of a sequence of bifurcations, from the appearance of primitive polymer structures and the development of the universal catalytic function of peptides, to the emergence of the nucleotide sequences involved in protein synthesis, and ultimately, to the genetic code in which the general plan of organism development as well as its numerous individual peculiarities are recorded.

Evolution

The emergence of the genetic code completes the stage of prebiotic evolution, and biological evolution itself (the evolution of life) begins.

We emphasize again the important role of Darwinism in biological evolution, but not at the early stages of the establishment of life and the development of the molecular self-organization mechanisms of biological systems.

The emergence of intellectual life can then be regarded as an even much less probable and more vulnerable pathway in the overall problem of life's evolution.

The probability of finding life features is certainly non-zero, because, bearing in mind matter transport in space, one may assume that once seeds are planted on a planet with a not too extreme natural environment, primitive life forms can survive.

This relation became more accurate after extrasolar planetary systems including Earth-like planets (Terrans) were discovered, but it is still somewhat ambiguous, especially in an attempt to assess a lifetime of technologically developed and well-advanced civilization.

Acknowledgments
A machine generated summary based on the work of Marov, Mikhail Ya. (2014 in).

Astrobiology and Planetary Sciences in Mexico

https://doi.org/10.1007/978-3-030-46087-7_2

Abstract-Summary
A small community of scientists in Mexico has been contributing to the study of planetary bodies in our Solar System and around other stars, including their potential for habitability.

Particular aspects of this research told as a journey: from the first attempts to reproduce cells and the laboratories where the first Mexican astrobiologists were educated to the sites in Mexico where scientists are studying the extremes of life and likely environments of other planets.

Our journey continues toward other stars where we search for planets beyond our Solar System, known as exoplanets, that have shown a surprising diversity more familiar to science fiction with hot Jupiters, lava worlds, mini-Neptunes, super-Earths, and potentially habitable worlds.

Astrobiology and Planetary Sciences in Mexico
She graduated as chemical pharmacobiologist at the Universidad Nacional Autónoma de México (UNAM) in 1968; by then, she was interested in the chemical evolution of life and decided to study a PhD in the laboratory of Cyril Ponnamperuma, an expert of the chemical evolution of life, that was involved with NASA in the Viking and Voyager programs and was one of the founders of the International Society for the Study of the Origin of Life (ISSOL) that became The International Astrobiology Society in 2005.

A new strategy for the last EMA—held last September 2019 in Puebla at the National Institute of Astrophysics, Optics and Electronics (INAOE)—was enforced toward teaching the scientific background and research being developed for two complementary topics in astrobiology (e.g., and extremophiles), through an intense schedule of courses, research talks, outreach activities, training on field sites, computational analysis, laboratory experiments, and telescope observation.

The Rocks That Fell from the Sky
In the context of the class Geology and Planetary Atmospheres taught at the School of Science (UNAM), the first subject that we teach is the difference between the terms: asteroid, meteoroid, micrometeoroids, meteor, and meteorite.

The group also studies the formation of structures within meteorites, such as chondrules (Cervantes-de la Cruz et al. [19]) and chondrites (Reyes-Salas et al. [20]; Corona-Chávez et al. [21]).

Magnetism of planetary bodies that is found as a remanent in meteorites has also been studied in Mexico (Flores-Gutiérrez et al. [22, 23]; Urrutia-Fucugauchi et al. [24]).

Karina Cervantes de la Cruz, a geologist and Antígona Segura, an astronomer and both authors of this chapter, formed an interdisciplinary group to study the evolution of minerals in meteorites, since their formation until the years that they spend on Earth.

Study the evolution of minerals in chondritic meteorites from their formation and their transformation due to heating processes in the parental bodies.

Planetary Collisions

They used information about nuclear weapons, scaling laws, empirical data, and numerical simulations to study several topics related to Curuça event: a possible impact crater related to this event, height of explosion, energy of explosion, seismic records, sounds, and its possible relation with the Perseids meteor shower.

The objectives of are to (a) study the fragmentation of meteoroids and small bodies in the Earth's atmosphere, (b) determine values of important parameters used to model the dynamics of objects through the atmosphere, (c) understand the interaction between the atmospheric shock wave and the surface, and what information this process provides about the height and energy of the explosion, (d) determine trajectories and orbital parameters of cosmic objects, (e) recover meteorites to study them, and (f) give information to population and Civil Protection to get its help and to avoid fear among inhabitants.

Mars: The History of the Search for Organics

Several scientific controversies have resulted from such research; in this section, we will focus on one that changed NASA's strategy to study organic matter on Mars and was starred by the Mexican scientist, Rafael Navarro-González.

For the Viking landers, analysis to detect organic matter was linked to the search for life.

Although the experiments showed contradictory results that seem to indicate the presence of life, organics were not detected, and the conclusion was that there was no life at the Viking's landing sites.

Higher temperature allowed to detect organics in several sites, indicating that an oxidizing reaction during the analysis was responsible for not detecting organics on Mars (Navarro-González et al. [25]).

In 2014, the first results that hinted to the presence of organic compounds in Mars, as predicted by Navarro-González and collaborators in 2010, were published.

Icy Worlds: Europa and Enceladus

The dissipation of deformation energy is a valuable source of heat in satellites from outer Solar System and help to explain, for example, Io's volcanic activity and the plausible hydrothermal systems on Europa and Enceladus.

This approach underlined the likely of hydrothermal systems to be a source of chemicals and energy to organisms (by hydrocarbon cleavage, hydrogen oxidation, and methane synthesis) in Europa seafloor.

The obviousness of compatible solutes, i.e., low molecular weight and organic molecules responsible for cell internal stability, came from the fact that they are widespread in those linked to icy satellites studies (halophiles, psychrophiles, and hyperthermophiles), including methanogens and sulfate reducers, and these are present in all domains of life.

Titan

Promote the chemical transformation of the dense and mild-reducing atmosphere of Titan originating gaseous hydrocarbons and nitriles, which after different polymerization processes yield solid aerosol particles that grow by a variety of mechanisms and fall to the surface.

The study of the organic chemistry happening on Titan requires direct observations, generation of models, in situ sampling, as well as laboratory simulations and analysis of the synthesized products, all of these from atmospheric and surface processes.

This solid can be considered an analog of the aerosols that form the haze layers detected in Titan's stratosphere, also known as tholins in the jargon of experimentalists.

Seven amino acids and urea were detected demonstrating an interesting fate for the aerosols gravitationally deposited on Titan's surface, where in spite of the low temperatures, chemistry can lead to the abiotic synthesis of organic and biologically attractive compounds.

Planets Around Other Stars

The first known exoplanet system (PSR + 1257 12) was discovered in 1992 with three Earth-mass planets orbiting around a rapidly rotating neutron star or pulsar (remnant of a supernova explosion; Wolszczan and Frail [26]; Wolszczan [27]).

This new and unexpected type of planet is named hot Jupiter as planets in this class have a mass like Jupiter, but orbit once around their star every few days receiving from it up to 10,000 times the irradiation that the Earth receives from the Sun (e.g., Demory and Seager [28]).

Any new exoplanet discoveries from SAINT-EX will provide optimal targets to observe with the next generation of space telescopes (e.g., NASA's James Webb Space Telescope) and explore the atmospheres of rocky planets outside of our own Solar System.

Habitable Worlds and Signatures of Life

Results published in 2003 (Segura et al. [29]) showed which compounds of the simulated earths would be detectable in planets around those stars if chemicals were produced by life at the same rates than on Earth.

This kind of radiation may make a planet around an M dwarf an unsuitable place for life, so non-active stars would be better, but there were no measurements available of UV emission for those stars classified as non-active.

After the 2005 work by Segura and collaborators demonstrated the importance of UV to predict the biosignatures for planets around M dwarfs, there was a renewed interest on these stars (Scalo et al. [30]; Tarter et al. [31]; Billings [32]; Shields et al. [33]) strengthened by the detection of several planets in the habitable zone of M dwarfs including our closest neighbor, Proxima Centauri (Anglada-Escudé et al. [34]).

Since Antígona moved to Mexico, along with students and collaborators, they have been exploring some of the problems for habitability and life detection on planets around M dwarfs.

Acknowledgments
A machine generated summary based on the work of Cervantes de la Cruz, Karina; Cordero-Tercero, Guadalupe; Gómez Maqueo Chew, Yilen; Lozada-Chávez, Irma; Montoya, Lilia; Ramírez Jiménez, Sandra Ignacia; Segura, Antígona (2020 in Cuatro Ciénegas Basin: An Endangered Hyperdiverse Oasis).

Astrobiology and the Possibility of Life on Earth and Elsewhere…

https://doi.org/10.1007/s11214-015-0196-1

Abstract-Summary
Astrobiology is an interdisciplinary scientific field not only focused on the search of extraterrestrial life, but also on deciphering the key environmental parameters that have enabled the emergence of life on Earth.

Astrobiology pushes us to combine different perspectives such as the conditions on the primitive Earth, the physicochemical limits of life, exploration of habitable environments in the Solar System, and the search for signatures of life in exoplanets.

From 2011 to 2014, the European Space Agency (ESA) had the initiative to gather a Topical Team of interdisciplinary scientists focused on astrobiology to review the profound transformations in the field that have occurred since the beginning of the new century.

The present paper is an interdisciplinary review of current research in astrobiology, covering the major advances and main outlooks in the field.

The following subjects will be reviewed and most recent discoveries will be highlighted: the new understanding of planetary system formation including the specificity of the Earth among the diversity of planets, the origin of water on Earth and its unique combined properties among solvents for the emergence of life, the idea that the Earth could have been habitable during the Hadean Era, the inventory of endogenous and exogenous sources of organic matter and new concepts about how chemistry could evolve towards biological molecules and biological systems.

Introduction

Chaired by Dr. Julia Michelle Kotler and Prof. Hervé Cottin, the team was asked to produce an update more than 10 years after the previous topical team focusing on exobiology (note ESA's switch between the use of the terms exobiology and astrobiology) which produced the ESA Special Report entitled Exobiology in the Solar System & the Search for Life on Mars (Brack et al. [35]).

The goal of the new team was to address the profound transformations in the field of astrobiology that have occurred since the 1999 exobiology topical team report, and, using new scientific questions and discoveries, to focus specifically on experimental studies either in the field (i.e. using Earth as a tool for astrobiology) or in space, (i.e. using space as a tool for astrobiology).

What Were Conditions on the Primitive Earth Like?

The origin of water on terrestrial planets is a dynamic research topic focused on determining how much water was accreted during planet formation or delivered later via asteroid and comet impacts early in the history of our solar system (Elkins-Tanton [36]; Morbidelli et al. [37]).

To add confusion to the discussion of the origin of Earth's water, some authors claim that the Earth was formed from relatively dry material and that our planet's volatile components (including water) were brought in later (Albarède [38]; Albarède et al. [39]), while other scenarios indicate that rocky planets were all built of material that contained enough water to form the oceans and atmospheres, and that later impact sources were relatively insignificant (Elkins-Tanton [36]; Hamano et al. [40]; Wood et al. [41]).

The presence of life on Earth depends on the availability of liquid water.

It may even be highly probable that if any other form of life exists in the Universe, it is also based on liquid water.

What Was the Prebiotic Soup?

Besides extraterrestrial sources for these organic compounds, which will be described below, two important terrestrial sources merit discussion: atmospheric and deep geochemical syntheses.

Terrestrial surface conditions may also have been important factors when considering the efficiency of atmospherically-mediated organic synthesis on the primitive Earth.

For about a decade, it has been shown that the argument that a reducing atmosphere was required to produce organic compounds may not be strictly true, and various oxidation state atmospheres could have provided similar types of organic molecules, albeit in yields varying over many orders of magnitude.

The degree to which such synthesis could have competed in quantity and quality with atmospheric organic synthesis and extraterrestrial input on global and local scales on the primitive Earth is open to debate.

All of the major organic compound classes important in terrestrial biochemistry have been identified as extra-terrestrial components in carbonaceous meteorites, indicating that exogenous delivery of organic compounds may have been important for prebiotic chemistry on the early Earth.

How Did Chemistry Turn into Biology?

Since the Miller-Urey experiment, prebiotic chemistry has mostly been devoted to the study of the formation of building blocks of life and how these can sequentially be linked to each other.

This is based on the idea that replication in chemical systems may have preceded autonomous life, i.e. the first protocells may have had a degree of autonomy similar to that of present day free-living microorganisms.

The self-organization process which gave rise to life can be considered as one aspect of a new scientific field called Systems Chemistry in which chemical processes involving a large number of intermediates and reactant steps are studied rather than simply analyzing simple chemical transformations (Kindermann et al. [42]; Ludlow and Otto [43]).

Other forms of energy (chemical gradients in hydrothermal systems for instance) may be useful, though this utility would be limited to driving the formation of organic matter without direct connection with the self-organization process.

What Are the Limits of Life?

The extremes of life are defined by the conditions under which the organisms can no longer harvest sufficient energy to repair or overcome the detrimental effects of stress and background mutation (Hoehler et al. [44]).

By studying the different defense and adaptational mechanisms of organisms in response to their environments, we expand our knowledge about the physical and chemical limits of life, which will enable a target-oriented search for life on other planets.

The exploration of extreme environments on Earth may increase our knowledge of the limits of life, and inform the search for extraterrestrial life Lake Vostok, in Antarctica, is buried under almost 4000 m of ice and may serve as a model for the hypothetical sub-surface ocean on Europa (Marion et al. [45]; Rothschild and Mancinelli [46]).

These different habitability scales should inform our understanding of the ability of life to inhabit different environments on other planets and satellites.

Is There Life Beyond Earth?

The detection of perchlorates in the Martian soil by instruments on the Phoenix lander (Hecht et al. [47]) and the reports of methane in the Martian atmosphere (Formisano et al. [48]; Mumma et al. [49]; Webster et al. [50]) suggest that it may be time to reconsider the presence of organic compounds on the red planet.

Results from previous missions suggest that while in general Mars may be classified as a habitable planet (Jakosky et al. [51]; Stoker et al. [52]; Ulrich et al. [53]) and some terrestrial microorganisms have the potential to survive on Mars (de la

Vega et al. [54]; de Vera et al. [55]; Nicholson et al. [56]; Schirmack et al. [57]; Wassmann et al. [58]), the surface radiation environment and lack of water may limit the possibility of life to subsurface niches.

Europa, Enceladus, and Titan are central objects in astrobiology research representing modern habitats in the outer solar system with conditions that may favor complex organic chemistry and possible life, since liquid water and organic compounds both occur.

Saturn's moon Titan reveals the presence of liquid hydrocarbon oceans and river deltas, apart from complex organic molecules on the surface and in its dense atmosphere.

Conclusion

The study of the origin and limits of life on Earth are informing the search strategy for life elsewhere in the Solar System and beyond.

Scientists have still not managed to reach a consensus about a definition of life (see for instance Bersini and Reisse [59]; Gayon et al. [60]), thus a consensus about an announcement regarding the discovery of extraterrestrial life may be similarly difficult.

The words of Alexander Ivanovitch Oparin regarding the study of the origin of life (Oparin [61]) are still relevant and invite us to exercise great caution and realism in our research: "We are faced with a colossal problem of investigating each separate stage of the evolutionary process as it was sketched here. (...) The road ahead of us is hard and long but without doubt it leads to the ultimate knowledge of the nature of life.

The same can be said about the search for extraterrestrial life.

Acknowledgments

A machine generated summary based on the work of Cottin, Hervé; Kotler, Julia Michelle; Bartik, Kristin; Cleaves, H. James; Cockell, Charles S.; de Vera, Jean-Pierre P.; Ehrenfreund, Pascale; Leuko, Stefan; Ten Kate, Inge Loes; Martins, Zita; Pascal, Robert; Quinn, Richard; Rettberg, Petra; Westall, Frances (2015 in Space Science Reviews).

Earth as a Tool for Astrobiology—A European Perspective

https://doi.org/10.1007/s11214-017-0369-1

Abstract-Summary

Scientists use the Earth as a tool for astrobiology by analyzing planetary field analogues (i.e. terrestrial samples and field sites that resemble planetary bodies in our Solar System).

They expose the selected planetary field analogues in simulation chambers to conditions that mimic the ones of planets, moons and Low Earth Orbit (LEO) space conditions, as well as the chemistry occurring in interstellar and cometary ices.

This paper reviews the ways the Earth is used by astrobiologists: (i) by conducting planetary field analogue studies to investigate extant life from extreme environments, its metabolisms, adaptation strategies and modern biosignatures; (ii) by conducting planetary field analogue studies to investigate extinct life from the oldest rocks on our planet and its biosignatures; (iii) by exposing terrestrial samples to simulated space or planetary environments and producing a sample analogue to investigate changes in minerals, biosignatures and microorganisms.

The European Space Agency (ESA) created a topical team in 2011 to investigate recent activities using the Earth as a tool for astrobiology and to formulate recommendations and scientific needs to improve ground-based astrobiological research.

The major conclusions of the topical team and suggestions for the future include more scientifically qualified calls for field campaigns with planetary analogy, and a centralized point of contact at ESA or the EU for the organization of a survey of such expeditions.

An improvement of the coordinated logistics, infrastructures and funding system supporting the combination of field work with planetary simulation investigations, as well as an optimization of the scientific return and data processing, data storage and data distribution is also needed.

Extended

The European Space Agency (ESA) funded a Topical Team in 2011 to summarize the most recent achievements in the field of astrobiology (Cottin et al. [62]).

A Ground Reference Sample Storage, where samples collected on the field testing campaigns will be stored for future reference, and an Electronic Database, where simulated planetary environment parameters and environmental field parameters are logged, field campaign data and laboratory data stored, and all these can be interconnected with existing databases.

Introduction

This is achieved by partly analyzing materials from specific environments on Earth that exhibit similar conditions as planets and moons in our Solar System (i.e. planetary field analogues), and by using laboratory set-ups that mimic planetary and deep space conditions (i.e. laboratory analogues).

Planetary field analogues are used to (i) perform in-situ measurements during field campaigns, allowing to test methodologies, protocols, and technologies, and (ii) collect samples, both biotic (extant and extinct or fossilized) and abiotic for analyses by state-of-the-art methods in the laboratory, e.g. to investigate the relationship between biosignatures and their rock/mineral context, and to identify microorganisms from extreme environments.

Laboratory analogues on Earth provide laboratory access to selected parameters of extra-terrestrial environments for a wide variety of experiments, e.g. the formation and preservation of biosignatures, the alteration of minerals and rocks, physical and chemical processes or the survival and adaptation mechanisms of microorganisms, as well as their interaction under simulated atmospheric composition, pressure, temperature and radiation of the space environment of interest.

Planetary Field Analogue Environments on Earth
Environmental conditions, such as extremes of dry, cold, with relatively high radiation background and weak seasonal fluctuations (e.g. in the planetary field analogue environment of permafrost in Antarctica), extreme dry and relatively hot or cold (e.g. in the planetary field analogue environments of the Atacama Desert in Chile, and the Antarctic Dry Valleys), extreme pH conditions (e.g. in the planetary field analogue environment of the Río Tinto), and extreme salinity (e.g. in the planetary field analogue environment of the saturated salt areas of the Dead Sea), make these environments similar to space- and planetary environments.

Prokaryotes (e.g. bacteria, bacterial spores, cyanobacteria) have been studied in such extreme planetary field analogue environments due to the fact that early life forms on Earth were prokaryotes and the assumption that if any extra-terrestrial life exists in the Solar System, it must be simple cellular organisms (Hansen [63]; Westall et al. [64], 65).

Field Test Campaigns
Field research at Mars analogue sites, such as desert environments can provide important constraints for instrument calibration and landing site strategies of robotic exploration missions to Mars that will investigate habitability and life beyond Earth during the next decade.

The Mars Desert Research Station (MDRS) in southern Utah (USA) is an analogue site where human space mission simulations are performed in order to investigate all possible factors that may interact to affect mission success, such as human factors and geological in-situ resource exploration (Schlacht et al. [66]; Direito et al. [67]; Ehrenfreund et al. [68]; Foing et al. [69–71]; Kotler et al. [72]; Martins et al. [73]; Orzechowska et al. [74]; Thiel et al. [75]).

EuroGeoMars 2009 was an example of a Moon-Mars field research campaign dedicated to the demonstration of astrobiology instruments and a specific methodology of comprehensive measurements from selected sampling sites (Foing et al. [69]).

In order to help in the interpretation of Mars missions' measurements from orbit (Mars Express, MRO) or from the surface (MSL Curiosity) at Gale crater, several field research campaigns (e.g. ILEWG EuroMoonMars) were performed in the extreme conditions of the Utah desert.

Laboratory Analogues
Planetary and Low Earth Orbit (LEO) space simulation facilities are designed to reproduce a wide range of conditions of space, as the ones accessible in LEO (i.e. on satellites or the International Space Station (ISS)), or of a specific planet or moon, including vacuum or the relevant atmospheric pressure and composition, UV radiation, and (surface) temperature.

While simulation of space and Mars surface environment was the main focus in the past and is adapted for anaerobic organisms in preparation of the ESA MEXEM experiment, the facilities are upgraded now also for the simulation of the icy moon Europa ocean conditions in preparation of the ESA IceCold active exposure experiment featuring in situ life measurements.

Conclusions and Recommendations

The role of the Earth as a tool for Astrobiology has been emphasized by establishing a wide range of planetary analogue field sites and simulation facilities in the laboratory.

The first activities, such as the EU supported EUROPLANET or CAFE, have just started to coordinate systematic interactions and common projects combing planetary field site research with planetary simulation experiments.

Establish an EU or ESA supported international survey of all expeditions performing scientific work in planetary field analogue sites, which could provide logistics and infrastructure, enabling an interdisciplinary scientific cooperation.

Support the use of simulation facilities, which is currently covered in the ESA-member states by national or international cooperation agreements.

The planetary field analogue projects should be supported by ESA like it is accomplished by NASA in the United States.

An ESA supported User Analytical Facility Network should be established to facilitate uniformity in analyses and therefore allowing the different campaigns to be compared to each other.

Acknowledgments

A machine generated summary based on the work of Martins, Zita; Cottin, Hervé; Kotler, Julia Michelle; Carrasco, Nathalie; Cockell, Charles S.; de la Torre Noetzel, Rosa; Demets, René; de Vera, Jean-Pierre; d'Hendecourt, Louis; Ehrenfreund, Pascale; Elsaesser, Andreas; Foing, Bernard; Onofri, Silvano; Quinn, Richard; Rabbow, Elke; Rettberg, Petra; Ricco, Antonio J.; Slenzka, Klaus; Stalport, Fabien; ten Kate, Inge L.; van Loon, Jack J. W. A.; Westall, Frances (2017 in Space Science Reviews).

Microbial Diversity and Biosignatures: An Icy Moons Perspective

https://doi.org/10.1007/s11214-019-0620-z

Abstract-Summary

The icy moons of the outer Solar System harbor potentially habitable environments for life, however, compared to the terrestrial biosphere, these environments are characterized by extremes in temperature, pressure, pH, and other physico-chemical conditions.

The search for life on these icy worlds is anchored on the study of terrestrial extreme environments (termed "analogue sites"), which harbor microorganisms at the frontiers of polyextremophily.

Such model systems and communities in extreme terrestrial environments may provide important information relevant to the astrobiology of icy bodies, including the composition of potential biological communities and the identification of biosignatures that they may produce.

The phylogenetic diversity of extremophiles is very high, leading to their broad dispersal across the phylogenetic tree of life together with a wide variety in metabolic diversity.

Some metabolisms are specific to archaea, for example, methanogenesis, an anaerobic respiration during which methane (CH_4) is produced.

Methanogenesis and sulfur reduction are of specific interest for icy moon research as it might be one of the few known terrestrial metabolisms possible on these celestial bodies.

Introduction

The issue of the origin of life is usually addressed from two different perspectives: the first examines the conditions under which the basic building blocks and macro-molecules significant for life may have emerged on the early Earth, the second explores the origin of functional subsystems (metabolism, replication) and basic structural organization (i.e., the cell) of what is recognized as "alive".

Early reports on the diversity of life (Woese [76]) also failed to consider the incredible metabolic sophistication that microbial life has shown in acquiring energy from its environment, even under the most inhospitable and extreme conditions.

Understanding the uncommon properties of extremophiles has led to questions about their origin (have these organisms recently adapted to the extreme conditions of their environment or are they relics of organisms that existed on the Early Earth and that had to face even harsher environmental conditions?).

It is therefore not surprising that astrobiology studies the properties of life in Earth's extreme environments.

Extremophiles: Diversity, Adaptation and Biosignatures

Discoveries of abundant life in diverse high-pressure environments, including the deep oceans, hydrothermal vents, and crustal rocks, supports the existence of an adaptation of life to HHP, and is consistent with the significance of HHP in the prebiotic synthesis of key biomolecules and the origin of life on Earth (Hazen et al. [77]).

Homeoviscous adjustment should also be regarded as a manner of adapting the composition, and therefore the functionality, of the membrane to abrupt shifts in the environment, or to stresses, including those of temperature, salinity, osmotic stress, pressure and pH. Under optimal physiological conditions, membranes are rather fluid and formed of disordered liquid crystalline phases.

Deep sea hydrothermal vents are among the ecosystems on Earth where polyex-tremophilic conditions or multi-stress situations are encountered by living organisms, such as high or low temperature, high salinity, high hydrostatic pressure and nutrient starvation within the same environment.

Methanogens as Model Organisms for Icy Moon Related Cultivation: Adaptation to Extreme Conditions

A genome comparison of the psychrophilic methanogens, M. burtonii and Methanogenium frigidum was performed to identify characteristics which distinguish cold adaptation mechanisms in these organisms from other archaea.

The cultivation of methanogens under high-pressure conditions offers an opportunity for astrobiological studies, for example, the investigation of physiological responses and metabolic adaptations, and for investigating the ecology of hydrothermal vent systems proposed for ocean worlds.

M. Jannaschii was used to investigate growth and CH_4 production kinetics at high-pressure conditions and at different temperatures, in the presence of He or Ar in addition to H_2/CO_2.

The above described characteristics and adaptations towards low- and high-pressure conditions, psychrophily and (hyper)thermophily, acidiphily and alkaliphily and osmolarity reveal that methanogens thrive under a variety of extreme growth conditions, but also during multi-factorial stress conditions (e.g. simultaneous multivariate concentrations of gaseous and liquid inhibitors, low pH, and pressure influences) and they respond to environmental disturbances in numerous ways (Kral et al. [78]; Taubner et al. [79, 80]).

Origins of Life and Biosignatures on Icy Worlds

It is necessary that the geological environment of the origin of life was able to naturally produce or directly receive a wide range of organic molecules.

The case for syngenicity in carbonaceous microfossils on Earth is often strengthened by Raman spectroscopy demonstrating that the carbonaceous material and its host rock have equivalent thermal histories (e.g. van Zuilen et al. [81]; Marshall et al. [82]) The third, governing consideration in biogenicity is the environment of formation, i.e., does the purported biosignature occur in a geological context consistent microbial habitability?

One can state that deducing the geology and geochemistry of putative hydrothermal vent deposits on Europa or Enceladus would open up the possibility to appraise the habitable niches of ocean worlds and consider the likelihood of a fossilised biosphere of purely chemotrophic life.

Conclusions

Organisms living in extreme environments and in particular microorganisms have, over the evolutionary process, developed a large variety of adaptive strategies.

Metabolic markers such as membrane lipids (saturated and polyunsaturated fatty acids, archaeol and caldarcheaol, etc.), compatible solutes (amino acids and derivatives, sugars and derivatives, polyols) or gas production (e. g. methane), witnesses of biological activity, have been detected in increasingly improbable environments previously considered sterile: thermal springs, hydrothermal vents, acidic lakes, alkaline lakes, hypersalines, deep marine sediments, oil reservoirs, glaciers, etc. The physico-chemical and energetic characteristics of some extreme terrestrial environments are analogous to those of other planets and icy moons in the Solar System, which raises the question of the past or present existence of life on these planets and icy moons, or the fulfilment of all the conditions for another origin of life.

Acknowledgments
A machine generated summary based on the work of Jebbar, Mohamed; Hickman-Lewis, Keyron; Cavalazzi, Barbara; Taubner, Ruth-Sophie; Rittmann, Simon K.-M. R.; Antunes, Andre (2020 in Space Science Reviews).

Review Komatiites: From Earth's Geological Settings to Planetary and Astrobiological Contexts

https://doi.org/10.1007/s11038-007-9135-9

Abstract-Summary
They are among the most ancient lavas of the Earth following the 3.8 Ga pillow basalts at Isua and they represent some of the oldest ultramafic magmatic rocks preserved in the Earth's crust at 3.5 Ga. This fact, linked to their particular features (high magnesium content, high melting temperatures, low dynamic viscosities, etc.), has attracted the community of geoscientists since their discovery in the early sixties, who have tried to determine their origin and understand their meaning in the context of terrestrial mantle evolution.

Komatiites may be extremely significant in the study of the origins and evolution of Life on Earth.

Besides reviewing the main geodynamic, petrological and geochemical characteristics of komatiites, this paper also aims to widen their investigation beyond the classical geological prospect, calling attention to them as attractive rocks for research in Planetology and Astrobiology.

Introduction

It was thought that komatiite eruptions appeared mainly in the Archean.

It is hypothesized that, in the late Archean, komatiite volcanoes would have built edifices on the surface forming oceanic plateaux.

In the Archean (4.0–4.2 Ga), mid-ocean ridges themselves may have been komatiitic.

Although their origin is still controversial, it is generally accepted that komatiites were generated most probably at depth of 150–200 km by massive partial melting of the Archean mantle (Takahashi and Scarfe [83]) and that the ancient komatiitic lava flows erupted at high temperatures of 1400–1700 °C (Arndt et al. [84]; Huppert et al. [85]).

The origin of komatiitic Archean volcanism has also been discussed, considering, among the various hypothesis, a possible association with meteorites and meteoritic impacts (Green [86]; Jones [87]; Jones et al. [88]).

The study of komatiites has also been extended to planetary geology, as they could be of a great help in understanding igneous processes outside the Earth on other planets and moons, where analog materials could exist (Venus, Mars, Io and lunar lavas).

Age

Komatiites were produced most commonly during the Archean (>2.7 Ga) and late Archean.

The prevalence of komatiite almost entirely in the Archean indicates fundamental differences between ancient and modern mantle conditions.

There is a prevalence of komatiites around 2.7 Ga, most of them located in Canada and Australia.

Komatiites are very rare in the Phanerozoic geological record.

The most prominent representatives are the Mesozoic komatiites from Gorgona Island in Colombia associated with the Caribbean oceanic plateau (Arndt et al. [89]; Echeverria [90]).

They are the youngest known komatiitic lavas (89 Ma) and their existence has been interpreted by some authors (Brandom et al. [91]) as unusually high temperatures in the mantle.

In this assumption, the Mesozoic appearance of komatiites may be due to local hydration of the upper mantle after dehydration of hydrous phases (Inoue et al. [92]).

Origin

Campbell et al. [93] have argued from fluid-dynamics calculations that both basalts and komatiites could have been produced by a starting thermal plume rising in a warmer Archean mantle.

Komatiitic magmas generated by mantle plume activities could contribute to the formation of Archean oceanic plateaux which, in some cases, could be later on buried in the mantle via subduction (Kerrich et al. [94]; Polat and Kerrich [95]; Polat et al. [96]; Puchtel et al. [97]).

With such geochemical resemblances, komatiites are strongly supposed to be produced by similar melting processes as for modern boninites but under hotter mantle temperature since one assumed that the Archean mantle was 100–500 °C hotter than the modern mantle (Parman et al. [98]).

Experimental data indicates that the Archean sub-arc mantle needs only to be 1500–1600 °C to produce hydrous komatiitic melts (Parman et al. [98]).

Petrology and Geochemical Characteristics

Some komatiites are enriched in ^{186}Os and ^{187}Os (Brandon et al. [91]; Gangopadhyay et al. [99]).

One of the last attractive discoveries about komatiites has been the report of diamonds (Capdevila et al. [100]).

Capdevila et al. [100] proposed that a primary, anhydrous komatiite magma formed by deep melting, then penetrated hydrated lithosphere beneath the ancient island arc where it collected both water and diamonds.

Melt inclusions found in nearly fresh Belingwe komatiites contain up to 1.1 wt.% of water, detectable CO_2 (up to 200 ppm), sulfur (500–750 ppm), chlorine (400–700 ppm) and Cl with Cl/K_2O ratios ~0.8–1.5 (Woodhead et al. [101]).

It is unclear whether the volatile content of mineralised sequences plays a role in the genesis of komatiite-hosted NiS mineralisation, either in the control of sulphur solubility or/and in the concentration of sulphide blebs (Fiorentini et al. [102]).

Komatiites in the Solar System

Lunar basalts, beside the resemblance in composition to terrestrial komatiite lavas, may have flowed in a similar way as komatiites.

Those extrusive shergotites are ultramafic pyroxene-rich lavas with coexisting pigeonite and augite, like komatiitic basalts, making early Precambrian basaltic komatiites on Earth a possible candidate for an analog of Martian lavas (Reyes and Christensen [103]).

Huppert et al. [85] demonstrate that terrestrial Precambrian komatiitic lavas, erupting as very hot, highly fluid, turbulent flows, may be capable of eroding deep channels, melting and assimilating rocks over which they flowed.

Because of its orthopyroxene spinifex and high MgO content resulting in a high liquidus temperature of 1611 °C, the 3.3 Ga komatiite in the Commondale greenstone belt of South Africa has been proposed (Williams et al. [104]) as a useful terrestrial analog for the Ionian lavas.

Komatiites and Implications in Astrobiology

Although basalts are major extrusive rocks in the Archean (De Witt and Ashwal [105]), komatiites may also represent another compositionally rocky component of the primitive environments where possibly early life on Earth could emerge and evolve, and it's plausible that they may have played a role in the evolution of the primitive atmosphere and maybe in some changes in the compositions of the hydrosphere.

Reasoning, early life on the Archean Earth may have colonized a wide variety of hydrothermal environments which would have included hydrothermal systems on mid-ocean ridges, around komatiite plume volcanoes and in shallow water settings.

A wide variety of trace elements like Zn, Cu, Se and other rare elements also used in essential processes of life, not easily accessible to the early unskilled biosphere, would have been introduced to living organisms most likely by hydrothermal systems and possibly around komatiite plume volcanoes (Nisbet and Fowler [106]).

Summary

The unique character and origin of komatiites make them excellent indicators of the early composition and development of the Earth's mantle.

Considered as "primitive" lavas, a significant constituent of the Hadean crust, and beside basalt another component of Archean greenstone belts, komatiites are important for understanding the evolution of the early Earth.

Komatiites are not exclusively related to the early Earth, as they may resemble some geological features of other planetary bodies of the Solar System, particularly on Io, and could help us to understand some extraterrestrial volcanisms.

It's probable that komatiites may have participated in the production of hydrocarbons and fixed nitrogen in the early Earth.

Volcanic rocks such as komatiites could be a habitat for early microbial life, since at the beginning of Earth's history komatiite plume volcanoes could provide possible sites for the evolution of various biological processes.

Acknowledgments
A machine generated summary based on the work of Nna-Mvondo, Delphine; Martinez-Frias, Jesus (2007 in Earth, Moon, and Planets).

Why We Should Care About Universal Biology

https://doi.org/10.1007/s13752-017-0280-8

Abstract-Summary
The scientific perspective of life-as-it-could-be has expanded in part by research in astrobiology, synthetic biology, and artificial life.

In the face of such scientific developments, we argue there is an ever-growing need for universal biology, life-as-it-must-be, the multidisciplinary study of non-contingent aspects of life as guided by biological theory and constrained by the universe.

Three distinct but connected ways of universalizing biology—with respect to characterizing aspects of life everywhere, with respect to the explanatory scope of biological theory, and with respect to extending biological insights to the structure of nonbiological entities.

Introduction
The accelerating pace of research in astrobiology spurs a need for a unified, universal approach to biological thought.

Such research is approaching or already engaging in the study of universal biology, creating a need for unification, and philosophical and theoretical input.

Simply put, universal biology is the multidisciplinary study of the noncontingent properties of life as guided by biological theory and constrained by the universe.

Despite the diversity of approaches to such a study, we claim that research in universal biology can be characterized in three distinct but related ways: (1) with respect to generalizations about life everywhere, (2) with respect to the explanatory scope of biological theory, and (3) with respect to biological theory applied to the organization of nonbiological phenomena.

We explore the questions that motivate the study of universal biology, and the research programs that consider those questions.

Life

By exploring universal biology in the sense of universal expectations for biology, many believe we can help define or develop a theory of life.

As the study of noncontingent aspects of life, universal biology is approachable independently of clearly distinguishing life from nonlife, since features of the universe that also apply to living phenomena may be relevant to such research.

This project is one of discovering the necessary (and thus universal) aspects of life, or universal biology in the sense described above.

To address life as it must be elsewhere, to do universal biology, minimal genome research must (appropriately) also favor a minimal conception of its own approach.

Although minimal genome concepts may yet discover necessary features of biological systems, the underlying assumptions about the nature of life must be seriously considered in a study of universal biology.

Biological Theory

Biological theories may be universal in scope and thus not limited to life's history on Earth (Powell and Mariscal [107]).

This notion of universal biology is complementary to the approach discussed above, of course, although it focuses on biological theory rather than on potential life-forms elsewhere in the universe.

By including non-universal theoretical tools, the EES is a provincial theory of biology.

To the aforementioned approach to universal biology, which relies on the context of evolutionary theory, there is another common approach to universal biology focused on entirely different theoretical aspects of biology: complexity theory and self-organization.

Stuart Kauffman's theory of the self-organization of life is a good example of how a framework of universal biology not only improves the status of biology as a science, but also clearly demonstrates how such a theory can incorporate physics and chemistry, yet still be a discussion about biology.

The Universe

Such inquiry needs to be free of teleological thinking; but, as illustrated with CNS, it is not easy to apply biological theories to the structure of the universe without bringing along the unfortunate biological assumptions of design and purpose.

McShea, recognizing the problem with teleological thinking in biology, explains that (the appearance of) teleological or is merely a result of structural hierarchy in a system.

Although it may seem like a step backward to reintroduce some teleological language into how we discuss the structure of the universe, McShea's naturalized teleology is a big improvement upon CNS in terms of its range of explanation, as well as providing a way to explain and investigate the area of overlap between the structure of life and the structure of the universe—opening the door for a more coherent universal biology that may serve to link the two.

Conclusion

Universal biology is the multidisciplinary study of the noncontingent properties of life as guided by biological theory and constrained by the universe.

We've presented three separate but compatible ways of characterizing the study of universal biology: (1) generalizing from our understanding of life on Earth and our knowledge of the universe to life everywhere, (2) assessing the explanatory scope of biological theory and the extent to which its elements are universal, and (3) using biological principles to explain the structure and organization of other phenomena in the universe and perhaps the universe itself.

Rather than define life, we advocated an approach to biology that begins with well-accepted universal principles, such as those derived from physics, chemistry, and probability theory.

We explored the potential universal scope of biological theory, especially with respect to evolutionary theory and broader principles of self-organizing complexity.

Acknowledgments

A machine generated summary based on the work of Mariscal, Carlos; Fleming, Leonore (2017 in Biological Theory).

From Earth to the Universe: Life, Intelligence, and Evolution

https://doi.org/10.1007/s13752-017-0266-6

Abstract-Summary

Most astrobiologists assume that "first contact" with extraterrestrial life, if it is ever to occur, will likely be the discovery of microbial life elsewhere in our solar system.

This article will touch on theory and research relating to the origins and evolution of life and intelligence on Earth and speculation about extraterrestrial life and intelligence.

An Introduction

Many take pains to distinguish between the scientific search for evidence of extraterrestrial (henceforth ET) microbial life in our solar system and the search for evidence of extraterrestrial intelligent (henceforth ETI) life in the universe.

In its early years, from the 1960s through the 1980s, NASA's scientific search for evidence of ET life focused on looking for evidence of microbial life on Mars and of intelligent life beyond our solar system.

By the mid-1990s, NASA's search for ET life steered away from the search for extraterrestrial intelligence (SETI) to focus on looking for evidence of microbial life in our solar system.

The search for evidence of ET life in the solar system appeared, and still appears, to be a more promising scientific endeavor than the search for ETI life.

While the scientific discourse on astrobiology leans toward optimism about the possibility of ET life, optimistic thinking is tempered by the limits of evidence and observations gathered thus far.

Theories, Findings, and Dogmas
In this document, the science community has identified six major topics of research in the field today: identifying abiotic sources of organic compounds; synthesis and function of macromolecules in the origin of life; early life and increasing complexity; co-evolution of life and the physical environment; identifying, exploring, and characterizing environments for habitability and biosignatures; and constructing habitable worlds.

The demonstration of ribose formation under some prebiotic conditions does not necessarily mean that we have to punt to Mars, but rather that a problem once thought intractable is now yielding to broader scientific inquiry. (p. 259) As to what "life" is, and is not, an international group of scientists engaged in origins of life research—many of whom identify as astrobiologists, including NASA's senior scientist for astrobiology—considered the question at a meeting in Japan in 2015 and decided that a consensus definition of "life" is not necessary to advance this field of research: A definition of life is notoriously difficult, although many putative ones exist....

Public Perceptions
The religious prophecy is one of apocalypse.) The so-called "Drake equation" $-N = R^* f_p n_e f_l f_i f_c L$—has been widely popularized by the SETI community—and consequently widely repeated in news and entertainment media—to promote the idea that ETI life is widespread in the universe: $N =$ The number of civilizations in the Milky Way galaxy whose electromagnetic emissions are detectable.

In a rather extreme example of taking leaps of faith about life in the universe, a group of UFOlogists (Deardoff et al. [108])—implicitly if not explicitly SETI believers—constructed an argument that recent advances in scientific knowledge provide a scientific justification for taking claims of the existence of UFOs—and, consequently, ETI life—seriously: Anthropic reasoning applied to inflation theory reinforces the prediction that we should find ourselves part of a large, galaxy-sized civilization....

Conclusions
That said, astrobiology is growing rapidly, its goals and objectives are evolving and expanding in response to advances in biology and astronomy, the pace of discovery is brisk, and the possibility of ET life is a serious scientific question.

Astrobiology has largely abandoned the search for ETI life in favor of more promising lines of research, and SETI as a scientific endeavor has veered closer to the fringes of legitimacy with its dependence on a dogma that is more than 50 years old and essentially unexamined—belief that the origin of life is easy, that extraterrestrial life is common, that extraterrestrial intelligent life is inevitable and technological, and that ETI life will save us from ourselves.

While he has endorsed the idea of listening for radio signals of ETI origin, cosmologist Stephen Hawking has also warned against efforts to make contact with extraterrestrial intelligence, speculating that ETI life would more likely raid Earth for its resources than share its resources with us.

Acknowledgments
A machine generated summary based on the work of Billings, Linda (2017 in Biological Theory).

Is Defining Life Pointless? Operational Definitions at the Frontiers of Biology

https://doi.org/10.1007/s11229-017-1397-9

Abstract-Summary
To this pessimistic conclusion, we argue that critically rethinking the nature and uses of definitions can provide new insights into the epistemic roles of definitions of life for different research practices.

This paper examines the possible contributions of definitions of life in scientific domains where such definitions are used most (e.g., Synthetic Biology, Origins of Life, Alife, and Astrobiology).

We examine contexts where definitions integrate criteria for life into theoretical models that involve or enable observable operations.

We show how these definitions of life play important roles in influencing research agendas and evaluating results, and we argue that to discard the project of defining life is neither sufficiently motivated, nor possible without dismissing important theoretical and practical research.

Introduction
Synthetic Biology, Origins of Life, Astrobiology, and computational Artificial Life are examples of fields in which definitions of life have become an established research topic connected to the characterisation, detection, demarcation, and synthesis of life (Luisi [109]; Ruiz-Mirazo et al. [110]).

It is our aim in this paper to examine the arguments of these criticisms in comparison to the roles that definitions of life play in scientific research practices.

We focus on the possible contributions of definition to the research carried out in those domains in which are used most (Synthetic Biology, Origins of Life, Alife, and Astrobiology).

We use the term 'operational' in a wide sense, referring both to (1) the possibility to define something by means of operations (e.g., defining an entity by measuring or building it following a specific procedure), and (2) the idea that the contents of the definition (e.g., the conditions for life) can be operationalised for empirical research, that is, can be built, manipulated and tested in the laboratory.

Philosophical Challenges to the Attempts to Define Life

In light of the proliferation of definitions of life, Machery [111] argues that if defining life is a scientific project like defining gene, virus, cell etc., then the project is pointless because different scientific disciplines, or even research groups within the same discipline, do not reach a consensus.

The prospects of finding one unified definition that covers all and only the relevant life forms are challenged by the fact that it is "unclear whether living beings form a single natural kind since nature rarely yields a unique way of classifying the world" (Machery [111], p. 159).

Machery's and Cleland's criticisms seem to leave no other choice but to abandon the enterprise of defining life, since definitions cannot provide univocal necessary and sufficient conditions for life.

Defining and Redefining Life in Practice

Research based on the autopoietic definition of life traditionally focused on individual cells (represented as individual vesicles).

The case of Luisi's group shows that definitions are used and can be useful as operational tools to guide research in Origins of Life and Synthetic Biology.

Definitions play an important role in fields where research is guided, at least in part, by explicit or implicit assumptions on the difference between living and non-living systems, and where the goal is theoretical advancement through analysis, challenging, testing, and subsequently improvement of models of living systems.

In the development of fields working on problems regarding the characterisation, detection, design and synthesis of life, definitions of life play important roles as boundary concepts for specifying research approaches and specific operations.

Operational Definitions and Their Virtues

Operational definitions coherently combine, or integrate into a theoretical model, a set of mutually dependent necessary and satisficing criteria for life that imply observable operations, and that are considered pertinent and relevant for research.

Our aim is simply to show that (operational) definitions of life can be useful in scientific practice, by guiding the development of a research program, providing criteria for the evaluation of results, and in generating new lines of investigations.

A crucial aspect that distinguishes definitions from criteria for life in general is precisely that in definitions criteria are put together into a model or a set of mutually dependent necessary conditions.

While scattered criteria of life suggested by Cleland, Griesemer, and Bains avoid the criticism targeted towards strong ontological definitions, this approach misses one of the aspects that make living systems both interesting and difficult to study from the theoretical and practical points of view, namely their integrated character.

Concluding Remarks

Strong definitions aim at providing answers to questions regarding natural kinds, by specifying necessary and sufficient conditions.

As we demonstrated with examples, operational definitions better capture the use of definitions in Origins of Life and Synthetic Biology.

Strong definitions attempt to provide a complete set of necessary and sufficient conditions.

Operational definitions focus on an open–ended set of possible necessary and satisficing conditions for life according to the specific requirements and goals of different research programs.

We have questioned the assumption that the only (or even main) purpose of definitions of life is to establish a set of universal criteria that strongly demarcate natural kinds.

Even if the criticisms against strong definitions of life are accepted, there is still a role for definitions of life in science, which is played by operational definitions.

Acknowledgments

A machine generated summary based on the work of Bich, Leonardo; Green, Sara (2017 in Synthese).

What Are Definitions of Life Good for? Transdisciplinary and Other Definitions in Astrobiology

https://doi.org/10.1007/s10539-017-9600-4

Abstract-Summary

In a series of articles, Cleland, Chyba, and Machery claim that definitions of life seek to provide necessary and sufficient conditions for applying the concept of life— something that such definitions cannot, and should not do.

Cleland, Chyba, and Machery approach definitions of life as classifying devices, thereby neglecting their other epistemic roles.

We identify within the discussions of the nature and origin of life three other types of definitions: theoretical, transdisciplinary, and diagnostic definitions.

We focus on the definitions of life within the budding field of astrobiology, paying particular attention to transdisciplinary definitions, and diagnostic definitions in the search for biosignatures from other planets.

Extended

In a series of highly influential articles, Cleland and Chyba [112, 113] argue that defining life is "fundamentally misguided": "[…] the idea that one can answer the question 'What is life'?

Introduction

In so doing Cleland, Chyba, and Machery focus on the classificatory nature of definitions of life, yet this is but one of the uses they can be put to.

While we think that such operational definitions can be important in experimental contexts that require the combination of theoretical insights to experimental procedures, operational definitions can hardly cover the activity of defining life in all its heterogeneousness—and neither do they capture the kinds of definitions that Cleland, Chyba, and Machery are primarily addressing.

We suggest that the attempt of defining life—especially in the context of astrobiology that was the original target of Cleland and Chyba's criticisms—is better understood from three other interrelated perspectives, each of which highlights one particular type of definition of life: First, many definitions of life are theoretical definitions that embody and draw together theoretical ideas and experimental results by often employing some focal theoretical concept or idea.

The Philosophical Critique of Defining Life

Having thus established that definitions of life aim to capture life as a natural kind by singling out its necessary and sufficient conditions, Cleland notes the futility of this task: such definitions are not apt for capturing natural kinds.

As for the philosophical reasons, the notion of life cannot be given a definition because the idea of defining a concept in terms of necessary and sufficient conditions is based on an erroneous theory of how concepts such as life acquire their meaning.

From the scientific perspective, this activity is not successful, since it does not lead to a univocal definition of life, either because different scientific perspectives lead to different definitions (Machery), or those definitions do not succeed in classifying concepts according to natural kinds (Cleland).

Definitions of Life in Astrobiology

Even though the seven pillar and NASA definitions do not provide in their generality and interpretative openness theoretically and conceptually unambiguous definitions of life, they nevertheless function as valuable heuristic and communicative tools for the exploration of different aspects of life.

In making use of diagnostic definitions, scientists attempt to causally link the measured molecules in the atmosphere of planets in our solar system, as well as in exoplanets, to some metabolic processes of possible extraterrestrial life.

As our discussion of detecting possible life in exoplanets shows, diagnostic definitions of life form but a part of an interrelated fabric making use of sophisticated detection technologies and simulations, empirical data from the research of metabolic processes on Earth, and theoretical knowledge on general physical, chemical, geological and biochemical processes.

Some Notes on Classification

The basic function of definitions of life is classificatory, aiming to distinguish between living and nonliving entities.

In order to assist life scientists in such a classificatory task, definitions of life attempt to spell out the necessary and sufficient conditions for unambiguously applying the notion of life.

In response to these claims, we studied several different definitions of life in order to show, on the one hand, that they have many other uses than classificatory ones.

One might want to argue that instead of providing necessary and sufficient conditions, definitions of life only attempt to spell out some general characteristics of life in the form of necessary conditions of life.

Metabolic processes—independent of their specific form—as well as the fact that organisms take part in evolution, are pivotal parts of many, if not most, current definitions of life.

Acknowledgments
A machine generated summary based on the work of Knuuttila, Tarja; Loettgers, Andrea (2017 in Biology & Philosophy).

Life in Icy Habitats: New Insights Supporting Panspermia Theory

https://doi.org/10.1007/s12210-011-0136-2

Abstract-Summary
Panspermia theory holds that microbial life is present in space or on bodies like comets or asteroids, and it can be safely delivered to Earth and start life there.

For the theory of panspermia to have credence, it is necessary to demonstrate that life could exist in the harsh condition of space.

The discovery of microorganisms able to survive permanently in frozen environments, such as ancient permafrost and ice, makes in principle realistic for bacteria to be carried through space in ice-cold comets.

Panspermia Theory
Amongst the numerous theories, some of them based on religious doctrine, others relying on scientific assumptions, there is the panspermia theory, which proposes that life on Earth did not originate on our planet, but was transported here from else where in the universe.

The panspermia hypothesis, that is based on the ideas of Anaxagoras, the Greek philosopher of the fifth century bc, states that the "seeds" of life exist all over the universe and also can be propagated through space from one stellar body to another.

Basic panspermia holds that microbial life is present in space or on bodies like comets or asteroids, and it can be safely delivered to planets and start life there.

For the theory of panspermia to have credence, we must show that life could exist in the harsh condition of space.

Ice as a Microbial Habitat
Liquid water within ice can exist as the liquid vein network within crystalline ice (Mizuno and Kuroiwa [114]) and as thin films on sediment grains within ice (Anderson and Tice [115]).

In polycrystalline ice the solute ions concentrate at the triple junctions, leading to a vein structure.

The structure of polycrystalline ice in glaciers and ice sheets contains a concentrated water phase that forms an interconnected network of highly concentrated water-filled veins and films around the crystals (Baker et al. [116]; Barnes et al. [117]).

They calculate the expected concentrations of soluble impurities in the veins for typical bulk concentrations found in natural ice (3.5 M total ions at -10 °C).

Typical bacterial concentrations in clean ice (10^2–10^3 cells/mL) would result in concentrations of 10^6–10^8 cells/mL of vein fluid, but bacteria occupy only a minor fraction of the total available vein volume (0.2%).

Eventhough the veins and mineral surfaces offer habitats for numerous bacteria, observations suggest that these are not the only places for life existence in glacial ice.

Microbial Metabolism in Icy Habitats

As for growth, only a few laboratory studies of reproduction at low temperatures, relying on laboratory cultures of isolates and standard measurement techniques, have reported bacterial activity at temperature below -10 °C (Bakermans et al. [118]).

Microbial metabolism at low temperatures is often studied by investigating selected metabolic processes, such as incorporation of radiolabeled substrates into biomass, as proxies for growth and activity (Bakermans [119]).

Taking into account the results obtained he concluded that bacteria can synthesize macromolecules in ice at -15 °C at a rate that is very low but suitable to repair macromolecular damage and speculated that cells remain metabolically active in solute-enriched water films on the surfaces of mineral grains in ice or in liquid veins at triple junctions (Price [120]) in ice.

Measurements of [^{3}H]-thymidine incorporation demonstrate the physiological potential for DNA metabolism at -15 °C and propose that sufficient activity is possible to offset chromosomal damage incurred in near-subsurface terrestrial permafrost.

Issues Imposed by Cold to Living Organisms and Observed Solutions

Current knowledge on the microbial ecology of extreme cold and frozen environments has shown that the main feature of all cold adapted microorganisms is the successful overcoming of the negative effects of low temperatures by a range of structural and functional adaptations (Bej et al. [121]).

Although genomic data from individual microorganisms can provide a better understanding of biological adaptations relevant to particular low temperature ecosystems, the absence of common features shared by all studied psychrophilic genomes has suggested that cold adaptation superimposes on pre-existing cellular organization and that the adaptive strategies may be different between the various microorganisms (Bowman [122]).

Comparative genome-wide analysis indicated that charged residues are statistically less frequent in the proteome of cold adapted bacteria, with a major trend for a higher content in polar residues, which is consistent with increased protein flexibility at low temperatures.

Conclusions

The understanding that life (as we know it) can thrive or, at least, survive in very cold conditions casts a new light upon the panspermia theory, although many other evidences will be needed to validate it.

Recent literature is full of papers describing the discovery of remnants of alien forms of life trapped in meteorites.

Acknowledgments

A machine generated summary based on the work of Parrilli, Ermenegilda; Sannino, Filomena; Marino, Gennaro; Tutino, Maria Luisa (2011 in Rendiconti Lincei. Scienze Fisiche e Naturali).

Astrobiology's Cosmopolitics and the Search for an Origin Myth for the Anthropocene

https://doi.org/10.1007/s13752-017-0281-7

Abstract-Summary

This article traces the ways in which astrobiology employs scientific methodologies and engages with popular culture in ways that do four kinds of major work commonly found in origin myths: telling the origin story, demarcating the boundaries between self and the other, giving normative guidance, and declaring a shared societal purpose.

Extended

This article will argue that the astrobiological cosmopolitical project is not just one of seeking alien life and gathering data about the possible origins of microbial life on Earth, but rather that astrobiology is engaged in a far deeper endeavor, the construction of a new set of myths that help the Moderns make sense of their place in a future replete with an ever-unfolding climate crisis.

Introduction: Astrobiology as Cosmopolitics

By using scientific practice to describe the galactic history of Earth's life-world, and to materially contextualize humanity's place and role in the larger fabric of space and time, astrobiologists contribute to the reconfiguration of the Modern cultural mythos.

While the fishscale approach to building a reflexive astrobiology practice by philosophically contextualizing its potential empirical findings is a laudable goal, to position the social sciences and the humanities as helpful adjuncts to scientific inquiry—either by ruminating on the moral consequences of astrobiological findings or by determining how to best educate the public about astrobiological findings—reinforces the problematic Modern purification of the sciences.

The goal of Modern science to generate "pure" data may still be understood as possible, given appropriate laboratory conditions; and the results of scientific practice themselves, rather than being inherently political, may be understood as becoming politicized upon their entry into the human social world.

The Anthropocene: Myths for the "Quake in Being"

To say that astrobiology is engaged in the crafting of an Anthropocene origin myth is not to say that astrobiologists are authors of fiction, but rather to say that astrobiologists are authors of ontologies; of metaphysical structures that help the Moderns make sense of our embeddedness in the world.

Astrobiology's methods and influence extend well beyond what are traditionally considered the boundaries of science via the ways in which astrobiological inquiry crafts origin myths, and in particular, how astrobiology works to answer material-cultural questions about modern humanity's origins, identity, ethics, and future.

Normative claims in astrobiology "after the end of the world" then, contend with (1) the material-cultural histories out of which species and ecosystems emerged; (2) the needs, myths, desires, and material realities of the contemporary "local" as imagined by modernity; and (3) the radical potentialities of modernity's spatial, biological, and temporal frontiers.

Conclusion

Astrobiology was born out of this slow-burning crisis, and the field carries the marks of the Anthropocene upon its soul.

To their credit, however, practicing astrobiologists, distributed across the fields of biology, astronomy, political science, philosophy, anthropology, Hollywood, government, and the private sector, have embraced their role as the mythmakers of a new era, if perhaps unconsciously.

This article was written as an attempt to trace the ways in which these various participants in astrobiological practice across academic and nonacademic spaces are engaged in constructing a new origin myth, and thus a new metaphysics for after the end of the world.

Acknowledgments
A machine generated summary based on the work of Malazita, James W. (2017 in Biological Theory).

Preparing for Extraterrestrial Contact

https://doi.org/10.1057/rm.2014.4

Abstract-Summary
This article introduces the issue of extraterrestrial life to the study of disasters, their prevention and their management.

Major journals in the field of risk and disaster research have ignored the potential threats posed by the existence of extraterrestrial life.

With increasing scientific support for the existence of planets able to support life, and the rapid development of scientific disciplines such as astrobiology, the article argues that limiting the scope of disaster research to terrestrial matters is increasingly intellectually untenable.

Extended

As posing risks and threats, the possibility of ETI can be seen as a key intellectual challenge to us all—one that may lead to unexpected and positive reconfigurations in conceptions of ourselves, our place in the universe and the nature and relative importance of the risks we face.

Introduction

Of disaster management and prevention, the risks posed to humanity by extraterrestrial contact or visitation have largely been ignored, reflecting a wider marginalisation of academic discussion of these issues to narrow fields in the natural sciences, the social sciences and the humanities [123, 124].

Published social scientific and humanities research into issues related to extraterrestrial life have remained resolutely social constructionist in their analysis of such topics as marginal group leader dynamics [125], psychodynamic trauma [126] and media trends [127].

Such pronouncements by senior natural scientists, and the increasing evidence produced by astrobiology mean that issues concerning the risks and consequences of encounters with extraterrestrial life forms should be critically scrutinised and discussed in disaster management and prevention studies discourse.

In discourses of risks to humanity, then, the very academic field dedicated to understanding the nature of hazards, risks, vulnerability, resilience and preparedness—and communicating its analyses and findings to policymakers—has treated extraterrestrial life as a non-issue [128, 129].

Five Theoretical Scenarios and their Consequences

One implication of such a message, however, is perhaps more immediate and significant from a risk management point of view—with one message received, probability estimates of the likelihood of other intelligent civilisations existing out there would probably be revised upwards quite dramatically; as would our estimates of the likelihood of alien visitation or contact.

In terms of levels of risk, this scenario would resemble the previous one concerning remote alien contact—with some risks associated with socio-cultural and philosophical disjuncture, and a reassessment of the likelihood of other extraterrestrial civilisations.

Rather than ignoring such discourses, it is more prudent to accept that they are—at least in theory—possible; and, from a risk management perspective, this scenario of prior visitation should be critically considered as seriously as the perhaps more unlikely Scenario 1, that we are alone in the universe—the scenario that appears to inform government policies around the world.

Conclusion

Leading on from this article, it is important that each contact scenario is explored, criticised and refined, such that the risks associated with each—and their interconnections—are classified and brought into clearer analytical focus, through employing data and models from emerging astrobiological research.

In light of these analytical developments, we will be better able to evaluate the relative likelihood of each scenario, and commission research on methodologies, procedures and technologies to prepare for—and deal with—such eventualities.

3 By exploring risks from extraterrestrial life is there a chance that other fields of risk research will benefit?: As astrobiological research develops, it is vital that a related programme of research evaluates the risks associated with its findings, and helps to prepare humanity for any such encounters.

With astrobiological research revealing increased likelihoods of extraterrestrial life, our earth-centric models of ecology must be challenged, and eventually expanded, reconfiguring the concept of ecological systems.

Acknowledgments

A machine generated summary based on the work of Neal, Mark (2014 in Risk Management).

Book Reading List

Astrobiology

by Horneck, G. (Ed), Baumstark-Khan, C. (Ed) *(2002).*

How did life originate in the universe? How did it all start after the creation of matter and the formation of elements in the stars? What are the pathways from the first organic molecules in space to the evolution of complex life forms on Earth and perhaps elsewhere? And how will it all end? The Universe itself sets the stage for the very interdisciplinary field of astrobiology that attempts to answer such questions, the central one being: What is the (cosmic) recipe for life? Currently there are only very few known elements in this vast mosaic. This book bridges a gap in the literature by bringing together leading specialists from different backgrounds who lecture on their fields, with close relevance to astrobiology, providing tutorial accounts that lead all the way to the forefront of research.

Please see https://www.springer.com/gp/book/9783642639579 for original source.

Astrobiology

by Chela-Flores, J. (Ed), Lemarchand, G. A. (Ed), Oró, J. (Ed) *(2000).*

Origins from the Big-Bang to Civilisation Proceedings of the Iberoamerican School of Astrobiology Caracas, Venezuela, 28 November–8 December, 1999.

Please see https://www.springer.com/gp/book/9780792365877 for original source.

Astrobiology: Future Perspectives

by Ehrenfreund, P. (Ed), Despois, D. (Ed), Lazcano, A. (Ed), Robert, F. (Ed), Irvine, W. (Ed), Owen, T. (Ed), Becker, L. (Ed), Blank, J. (Ed), Brucato, J. (Ed), Colangeli, L. (Ed), Derenne, S. (Ed), Dutrey, A. (Ed) *(2005)*.

Astrobiology, a new exciting interdisciplinary research field, seeks to unravel the origin and evolution of life wherever it might exist in the Universe. The current view of the origin of life on Earth is that it is strongly connected to the origin and evolution of our planet and, indeed, of the Universe as a whole.

We are fortunate to be living in an era where centuries of speculation about the two ancient and fundamental problems: the origin of life and its prevalence in the Universe are being replaced by experimental science.

Please see https://www.springer.com/gp/book/9781402023040 for original source.

Astrobiology

by Yamagishi, A. (Ed), Kakegawa, T. (Ed), Usui, T. (Ed) *(2019)*.

This book provides concise and cutting-edge reviews in astrobiology, a young and still emerging multidisciplinary field of science that addresses the fundamental questions of how life originated and diversified on Earth, whether life exists beyond Earth, and what is the future for life on Earth. Readers will find coverage of the latest understanding of a wide range of fascinating topics, including, for example, solar system formation, the origins of life, the history of Earth as revealed by geology, the evolution of intelligence on Earth, the implications of genome data, insights from extremophile research, and the possible existence of life on other planets within and beyond the solar system.

Please see https://www.springer.com/gp/book/9789811336386 for original source.

Astrobiology, History, and Society

by Vakoch, D. A. (Ed) *(2013)*.

This book addresses important current and historical topics in astrobiology and the search for life beyond Earth, including the search for extraterrestrial intelligence (SETI). The first section covers the plurality of worlds debate from antiquity through the nineteenth century, while section two covers the extraterrestrial life debate from the twentieth century to the present. The final section examines the societal impact of discovering life beyond Earth, including both cultural and religious dimensions.

Please see https://www.springer.com/gp/book/9783642359828 for original source.

Astrochemistry and Astrobiology

by Smith, I. W. M. (Ed), Cockell, C. S. (Ed), Leach, S. (Ed) *(2013).*

Astrochemistry and Astrobiology is the debut volume in the new series *Physical Chemistry in Action.* Aimed at both the novice and experienced researcher, this volume outlines the physico-chemical principles which underpin our attempts to understand astrochemistry and predict astrobiology.

Please see https://www.springer.com/gp/book/9783642317293 for original source.

Astrobiology on the International Space Station

by de Vera, J. *(2020).*

This volume on astrobiology of the Springer Briefs in Life Sciences book series addresses the three fundamental questions on origin, evolution, distribution and future of life in the universe: how does life begin and evolve? Is there life beyond Earth and, if so, how can we detect it? What is the future of life on Earth and in the universe? The book provides insights into astrobiological experiments that are being performed on the International Space Station, ISS, and discusses their findings.

Please see https://www.springer.com/gp/book/9783030616908 for original source.

Adaption of Microbial Life to Environmental Extremes

by Stan-Lotter, H. (Ed), Fendrihan, S. (Ed) *(2017).*

This entirely updated second edition provides an overview on the biology, ecology and biodiversity of extremophiles. Unusual and less explored ecosystems inhabited by extremophiles such as marine hypersaline deeps, extreme cold, desert sands, and man-made clean rooms for spacecraft assembly are presented. An additional focus is put on the role of these highly specialized microorganism in applied research fields, ranging from biotechnology and nanotechnology to astrobiology.

Please see https://www.springer.com/gp/book/9783319483252 for original source.

Encyclopedia of Astrobiology

by Gargaud, M. (Ed), Viso, M. (Ed), Irvine, W. M. (Ed), Amils, R. (Ed), Cleaves II, H. J. (Ed), Pinti, D. (Ed), Cernicharo Quintanilla, J. (Ed), Rouan, D. (Ed), Spohn, T. (Ed), Tirard, S. (Ed) *(2015).*

The interdisciplinary field of Astrobiology constitutes a joint arena where provocative discoveries are coalescing concerning, e.g. the prevalence of exoplanets, the diversity and hardiness of life, and its increasingly likely chances for its emergence. Biologists, astrophysicists, biochemists, geoscientists and space scientists share this exciting mission of revealing the origin and commonality of life in the Universe.

Please see https://www.springer.com/gp/book/9783662441862 for original source.

Between Necessity and Probability: Searching for the Definition and Origin of Life

by Popa, R. *(2004)*.

This study investigates the major theories of the origins of life in light of modern research with the aim of distinguishing between the necessary and the optional and between deterministic and random influences in the emergence of what we call 'life.' Life is treated as a cosmic phenomenon whose emergence and driving force should be viewed independently from its Earth-bound natural history.

Please see https://www.springer.com/gp/book/9783540204909 for original source.

Biosignatures for Astrobiology

by Cavalazzi, B. (Ed), Westall, F. (Ed) *(2019)*.

This book aims at providing a brief but broad overview of biosignatures. The topics addressed range from prebiotic signatures in extraterrestrial materials to the signatures characterising extant life as well as fossilised life, biosignatures related to space, and space flight instrumentation to detect biosignatures either in situ or from orbit. The book ends with philosophical reflections on the implications of life elsewhere.

Please see https://www.springer.com/gp/book/9783319961743 for original source.

Chemical Evolution and the Origin of Life

by Rauchfuss, H. *(2008)*.

Up to now, we do not have a generally accepted theory about the origin of life and about the process of development of life, we only have a great number of—to some extent even contradictory—hypotheses. Meanwhile there came up some scientific findings beyond thought only a few years ago.

Horst Rauchfuss is comparing the different theories from the view of the latest results and is giving an exciting and easy understandable insight into the present state of research.

Please see https://www.springer.com/gp/book/9783540788225 for original source.

From Influence to Inhabitation

by Christie, J. E. *(2019)*.

This book describes how and why the early modern period witnessed the marginalisation of astrology in Western natural philosophy, and the re-adoption of the cosmological view of the existence of a plurality of worlds in the universe, allowing the possibility of extraterrestrial life.

Please see https://www.springer.com/gp/book/9783030221683 for original source.

Journey to Diverse Microbial Worlds

by Seckbach, J. (Ed) *(2000)*.

In this *Journey to Microbial Worlds* we present the diversity of microorganisms, from the state of fossil microbes in Archaean age rocks to the possibilities of extraterrestrial life. This volume discusses the extremophiles living in harsh environments (from our anthropocentric point) and describes them in considerable detail. Some chapters also review topics such as symbiosis, bacterial luminescence, methanogens, and petroleum-grown cells. The final chapters of this book shed new light on astrobiology and speculate on extremophiles as candidates for extraterrestrial life.

Please see https://www.springer.com/gp/book/9780792360209 for original source.

Lectures in Astrobiology

by Barbier, B. (Ed), Martin, H. (Ed), Reisse, J. (Ed) *(2006)*.

This is the second of a divided two-part softcover edition of the "Lectures in Astrobiology Volume I" containing the sections "General Introduction", "From Prebiotic Chemistry to the Origin of Life on Earth" and "Appendices" including an extensive glossary on Astrobiology.

"Lectures in Astrobiology" is the first comprehensive textbook at graduate level encompassing all aspects of the emerging field of astrobiology.

Please see https://www.springer.com/gp/book/9783540290049 for original source.

Life on Earth and other Planetary Bodies

by Hanslmeier, A. (Ed), Kempe, S. (Ed), Seckbach, J. (Ed) *(2012)*.

A trio of editors [Professors from Austria, Germany and Israel] present Life on Earth and other Planetary Bodies. The contributors are from twenty various countries and present their research on life here as well as the possibility for extraterrestrial life. This volume covers concepts such as life's origin, hypothesis of Panspermia and of life possibility in the Cosmos.

Please see https://www.springer.com/gp/book/9789400749658 for original source.

Life in the Universe

by Seckbach, J. (Ed), Chela-Flores, J. (Ed), Owen, T. (Ed), Raulin, F. (Ed) *(2004)*.
From the Miller Experiment to the Search for Life on other Worlds.

Please see https://www.springer.com/gp/book/9781402023712 for original source.

The Science of Astrobiology

by Chela-Flores, J. *(2011)*.

Since the publication of *The New Science of Astrobiology* in the year 2001—the first edition of the present book—two significant events have taken place raising the subject from the beginning of the century to its present maturity. Firstly, in 2001 the Galileo Mission still had two years to complete its task, which turned out to be an outstanding survey of the Jovian system, especially of its intriguing satellite Europa. Secondly, the Cassini Huygens Mission was on its way to Saturn.

Please see https://www.springer.com/gp/book/9789400716261 for original source.

The New Science of Astrobiology

by Chela-Flores, J. *(2001)*.

Astrobiology is a very broad interdisciplinary field covering the origin, evolution, distribution, and destiny of life in the universe, as well as the design and implementation of missions for solar system exploration. A review covering its complete spectrum has been missing at a level accessible even to the non-specialist.

The last section of the book consists of a supplement, including a glossary, notes, and tables, which represent highly condensed 'windows' into research ranging from basic sciences to earth and life sciences, as well as the humanities.

Please see https://www.springer.com/gp/book/9780792371250 for original source.

Travels with Curiosity

by Byrne, C. J. *(2020)*.

The Mars Curiosity Rover is the most sophisticated mobile laboratory ever deployed on a planet. For over seven years, scores of investigators have planned its daily route and activities, poring over the overwhelming images and data and revising our understanding of planetary surfaces, geology, and potential habitability.

This book takes readers right down to the surface of Mars, chronicling Curiosity's physical and scientific journey across the planet's Earth-like, yet strikingly alien vistas.

Please see https://www.springer.com/gp/book/9783030538040 for original source.

The Evolving Universe and the Origin of Life

by Teerikorpi, P., Valtonen, M., Lehto, K., Lehto, H., Byrd, G., Chernin, A. *(2019)*.

Regarding his discoveries, Sir Isaac Newton famously said, "If I have seen further it is by standing upon the shoulders of giants."

The Evolving Universe and the Origin of Life describes, complete with fascinating biographical details of the thinkers involved, a history of the universe as interpreted by the expanding body of knowledge of humankind. From subatomic particles to the protein chains that form life, and expanding in scale to the entire universe, this book covers the science that explains how we came to be.

Please see https://www.springer.com/gp/book/9783030179205 for original source.

References

1. Hubbart SG (2015) What is astrobiology? In: NASA. https://www.nasa.gov/feature/what-is-astrobiology. Accessed 5 Oct 2019
2. NASA (2018) NASA Astrobiology Institute. https://nai.nasa.gov/about/. Accessed 5 Oct 2019
3. Merino N et al (2019) Living at the extremes: extremophiles and the limits of life in a planetary context. Front Microbiol 10:780
4. Marlow JJ et al (2011) Organic host analogues and the search for life on Mars. Int J Astrobiol 10:31–44
5. Preston LJ, Dartnell LR (2014) Planetary habitability: lessons learned from terrestrial analogues. Int J Astrobiol 13:81–98
6. Carr MH, Bell JF (2014) Chapter 17—Mars: surface and interior. In: Spohn T et al (eds) Encyclopedia of the solar system, 3rd edn. Elsevier, Boston, pp 359–377
7. Golombek MP, McSween HY (2014) Chapter 19—Mars: landing site geology, mineralogy, and geochemistry. In: Spohn T et al (eds) Encyclopedia of the solar system, 3rd edn. Elsevier, Boston, pp 397–420
8. Domagal-Goldman SD, Segura A (2013) Exoplanet climates. In: Mackwell SJ et al (eds) Comparative climatology of terrestrial planets, pp 121–135
9. Luger R, Barnes R (2015) Extreme water loss and abiotic O_2 buildup on planets throughout the habitable zones of M dwarfs. Astrobiology 15:119–143
10. Barnes R et al. (2018) The habitability of Proxima Centauri b I: evolutionary scenarios. ArXiv: 160806919 [astro-ph]
11. Lozada-Chávez I et al (2009) Metanogenic diversity through mcrA gene in hypersaline conditions. In: Origins of life and evolution of biospheres, vol 39, pp 382–383
12. Montoya L et al (2011) The sulfate-rich and extreme saline sediment of the ephemeral Tirez lagoon: a biotope for Acetoclastic sulfate-reducing bacteria and hydrogenotrophic methanogenic archaea. Int J Microbiol 2011:1–22
13. Meadows VS (2017) Reflections on O_2 as a biosignature in exoplanetary atmospheres. Astrobiology 17:1022–1052
14. Catling DC et al (2018) Exoplanet biosignatures: a framework for their assessment. Astrobiology 18:709–738
15. Kiang NY et al (2018) Exoplanet biosignatures: at the Dawn of a new era of planetary observations. Astrobiology 18:619–629
16. Fujii Y et al (2018) Exoplanet biosignatures: observational prospects. Astrobiology 18:739–778
17. Walker SI et al (2018) Exoplanet biosignatures: future directions. Astrobiology 18:779–824
18. Cocconi G, Morrison P (1959) Searching for interstellar communications. Nature 184:844
19. Cervantes-de la Cruz KE et al (2015) Experimental chondrules by melting samples of olivine, clays and carbon with a CO_2 laser. Bol Soc Geol Mex 67:401–412
20. Reyes-Salas AM et al (2010) Petrography and mineral chemistry of Escalón meteorite, an H4 chondrite, México. Rev Mex Cienc Geol 27:148–161
21. Corona-Chávez P et al (2018) Petrology, phase equilibria modelling, noble gas chronology and thermal constraints of the El Pozo L5 meteorite. Geochemistry 78:248–253
22. Flores-Gutiérrez D et al (2010) Scanning electron microscopy characterization of iron, nickel and sulfur in chondrules from the Allende meteorite–further evidence for between-chondrules major compositional differences. Rev Mex Cienc Geol 27:338–346
23. Flores-Gutiérrez D et al (2010) Micromagnetic and microstructural analyses in chondrules of the Allende meteorite. Rev Mex Cienc Geol 27:162–174
24. Urrutia-Fucugauchi J et al (2014) Meteorite paleomagnetism—from magnetic domains to planetary fields and core dynamos. Geofis Int 53:343–363
25. Navarro-González R et al (2006) The limitations on organic detection in Mars-like soils by thermal volatilization–gas chromatography–MS and their implications for the Viking results. Proc Natl Acad Sci U S A 103(44):16089–16094

26. Wolszczan A, Frail DA (1992) A planetary system around the millisecond pulsar PSR1257 + 12. Nature 355:145–147
27. Wolszczan A (1994) Confirmation of Earth-mass planets orbiting the millisecond pulsar PSR B1257+12. Science 264:538–542
28. Demory B-O, Seager S (2011) Lack of inflated radii for Kepler giant planet candidates receiving modest stellar irradiation. Astrophys J Suppl S 197(1):12
29. Segura A et al (2003) Ozone concentrations and ultraviolet fluxes on Earth-like planets around other stars. Astrobiology 3(4):689–708
30. Scalo J et al (2007) M stars as targets for terrestrial exoplanet. Astrobiology 7(1):85–166
31. Tarter JC et al (2007) A reappraisal of the habitability of planets around M dwarf stars. Astrobiology 7(1):30–65
32. Billings L (2011) Astronomy: exoplanets on the cheap. Nature News 470(7332):27–29
33. Shields AL et al (2016) The habitability of planets orbiting M-dwarf stars. Phys Rep 663:1–38
34. Anglada-Escudé G et al (2016) A terrestrial planet candidate in a temperate orbit around Proxima Centauri. Nature 536:437–440
35. Brack A, Fitton B, Raulin F (1999) Exobiology in the solar system & the search for life on Mars. ESA Scientific Publication SP, vol 1231
36. Elkins-Tanton L (2013) Evolutionary dichotomy for rocky planets. Nature 497:570–572
37. Morbidelli A, Lunine JI, O'Brien DP, Raymond SN, Walsh KJ (2012) Ann Rev Earth Planetary Sci 40. Jeanloz R (ed) pp 251–275
38. Albarède F (2009) Volatile accretion history of the terrestrial planets and dynamic implications. Nature 461:1227–1233
39. Albarède F, Ballhaus C, Blichert-Toft J, Lee C-T, Marty B, Moynier F, Yin Q-Z (2013) Asteroidal impacts and the origin of terrestrial and lunar volatiles. Icarus 222:44–52
40. Hamano K, Abe Y, Genda H (2013) Emergence of two types of terrestrial planet on solidification of magma ocean. Nature 497:607–610
41. Wood B, Halliday A, Rehkämper M (2010) Volatile accretion history of the Earth. Nature 467:7
42. Kindermann M, Stahl I, Reimold M, Pankau WM, von Kiedrowski G (2005) Angew Chem Int Ed Engl 44:6750–6755
43. Ludlow RF, Otto S (2008) Systems chemistry. Chem Soc Rev 37:101–108
44. Hoehler TM, Amend JP, Shock EL (2007) A "follow the energy" approach for astrobiology. Astrobiology 7:819–823
45. Marion GM, Fritsen CH, Eicken H, Payne MC (2003) The search for life on Europa: limiting environmental factors, potential habitats, and Earth analogues. Astrobiology 3:785–811
46. Rothschild LJ, Mancinelli RL (2001) Life in extreme environments. Nature 409:1092–1101
47. Hecht MH, Kounaves SP, Quinn RC, West SJ, Young SMM, Ming DW, Catling DC, Clark BC, Boynton WV, Hoffman J, DeFlores LP, Gospodinova K, Kapit J, Smith PH (2009) Detection of Perchlorate and the Soluble Chemistry of Martian Soil at the Phoenix Lander Site. Science 325:64–67
48. Formisano V, Atreya S, Encrenaz T, Ignatiev N, Giuranna M (2004) Detection of methane in the atmosphere of Mars. Science 306:1758–1761
49. Mumma MJ, Villanueva GL, Novak RE, Hewagama T, Bonev BP, DiSanti MA, Mandell AM, Smith MD (2009) Science 323:1041–1045
50. Webster CR, Mahaffy PR, Atreya SK, Flesch GJ, Mischna MA, Meslin P-Y, Farley KA, Conrad PG, Christensen LE, Pavlov AA, Martín-Torres J, Zorzano M-P, McConnochie TH, Owen T, Eigenbrode JL, Glavin DP, Steele A, Malespin CA, Archer PD, Sutter B, Coll P, Freissinet C, McKay CP, Moores JE, Schwenzer SP, Bridges JC, Navarro-Gonzalez R, Gellert R, Lemmon MT, t.M.S. Team (2014) Science 347:415–417
51. Jakosky B, Nealson K, Bakermans C, Ley R, Mellon M (2003) Subfreezing activity of microorganisms and the potential habitability of Mars' polar regions. Astrobiology 3:343–350
52. Stoker CR, Zent A, Catling DC, Douglas S, Marshall JR, Archer D, Clark B, Kounaves SP, Lemmon MT, Quinn R, Renno N, Smith PH, Young SMM (2010) J Geophys ResPlanets 115:E00E20

53. Ulrich M, Wagner D, Hauber E, de Vera JP, Schirrmeister L (2012) Strong release of methane on Mars in northern summer 2003. Icarus 219:345–357
54. de la Vega UP, Rettberg P, Reitz G (2007) Simulation of the environmental climate conditions on martian surface and its effect on Deinococcus radiodurans. Adv Space Res 40:1672–1677
55. de Vera J-P, Moehlmann D, Butina F, Lorek A, Wernecke R, Ott S (2010) Survival potential and photosynthetic activity of lichens under Mars-like conditions: a laboratory study. Astrobiology 10:215–227
56. Nicholson WL, Krivushin K, Gilichinsky D, Schuerger AC (2013) Proc Natl Acad Sci USA 110:666–671
57. Schirmack J, Böhm M, Brauer C, Löhmannsröben H-G, de Vera J-P, Möhlmann D, Wagner D (2013) Growth of Carnobacterium spp. from permafrost under low pressure, temperature, and anoxic atmosphere has implications for Earth microbes on Mars. Planet Space Sci
58. Wassmann M, Moeller R, Rabbow E, Panitz C, Horneck G, Reitz G, Douki T, Cadet J, Stan-Lotter H, Cockell CS, Rettberg P (2012) Astrobiology 12:498–507
59. Bersini H, Reisse J (2007) Comment définir la vie ? Les réponses de la biologie, de l'intelligence artificielle et de la philosophie des sciences. Vuibert, Paris
60. Gayon J, Malaterre C, Morange M, Raulin-Cerceau F, Tirard S (2008) Proceeding of the conference: defining life, origin of life and evolution of the biosphere. Springer, Paris
61. Oparin AI (1953) The origin of life. Dover, New York. (Republication of the 1938 edition with the addition of a new Introduction by the translator)
62. Cottin H, Kotler JM, Bartik K, Cleaves HJ, Cockell CS, de Vera J-PP, Ehrenfreund P, Leuko S, ten Kate IL, Martins Z, Pascal R, Quinn R, Rettberg P, Westall F (2015) Astrobiology and the possibility of life on earth and elsewhere.... Space Sci Rev. https://doi.org/10.1007/s11 214-015-0196-1
63. Hansen, European Space Agency, ESA Communications SP-1299 (2007)
64. Westall F, Foucher F, Bost N, Bertrand M, Loizeau D, Vago JL, Kminek G, Gaboyer F, Campbell KA, Bréhéret J-B, Gautret P, Cockell CS (2015) Biosignatures on Mars: what, where and how? Implications for the search for Martian life. Astrobiology 15:998–1029
65. Westall F, Campbell KA, Bréhéret JG, Foucher F, Gautret P, Hubert A, Sorieul S, Grassineau N, Guido DM (2015) Archean (3.33 Ga) microbe-sediment systems were diverse and flourished in a hydrothermal context. Geology 43:615–618
66. Schlacht IL, Voute S, Irwin S, Mikolajczak M, Foing B, Westenberg A, Stoker C, Masali M, Rötting M (2010) (Crew 91 & Mission Support) Moon-Mars analogue mission at the MDRS. EuroMoonMars-1 Mission. In: IAC GLUC Global Lunar Conference 2010, Beijing
67. Direito SOL, Ehrenfreund P, Marees A, Staats M, Foing B, Röling W (2011) A wide variety of putative extremophiles and large beta-diversity at the Mars Desert Research Station (Utah). Int J Astrobiol 10:191–208
68. Ehrenfreund P et al (2011) Astrobiology and habitability studies in preparation for future Mars missions: trends from investigating minerals, organics and biota. Int J Astrobiol 10:239–254
69. Foing BH, Stoker C, Zavaleta J, Ehrenfreund P, Thiel C, Sarrazin P, Blake D, Page J, Pletser V, Hendrikse J, Direito S, Kotler JM, Martins Z, Orzechowska G, Gross C, Wendt L, Clarke J, Borst AM, Peters STM, Wilhelm MB, Davies GR (ILEWG EuroGeoMars 2009 Team) (2011) Field astrobiology research in Moon–Mars analogue environments: instruments and methods. Int J Astrobiol 10:141–160
70. Foing BH, Stoker C, Rodrigues L, Svendsen Å, Rammos I et al (2013) Astrobiology, geology and habitability field studies supporting Mars research. In: LPI, vol 44, p 3057
71. Foing BH, Orgel C, Stoker C, Ehrenfreund P et al (2014) Gale crater analogue geology studies at multiple scales. In: LPI, vol 45, p 2675
72. Kotler RC, Quinn Z, Martins BH, Foing P (2011) Ehrenfreund, analysis of mineral matrices of planetary soils analogs from the Utah Desert. Int J Astrobiol 10:221–230
73. Martins Z, Sephton MA, Foing BH, Ehrenfreund P (2011) Extraction of amino acids from soils close to the Mars Desert Research Station (MDRS) (Utah). Int J Astrobiol 10:231–238
74. Orzechowska G, Kidd RD, Foing BH, Kanik I, Stoker C, Ehrenfreund P (2011) Analysis of Mars analog soil samples using solid phase microextraction, organic solvent extraction and gas chromatography/mass spectrometry. Int. J. Astrobiol. 10:209–220

75. Thiel P, Ehrenfreund B, Foing V, Pletser O (2011) Ullrich, PCR-based analysis of microbial communities during the EuroGeoMars campaign at Mars Desert Research Station (Uath). Int J Astrobiol 10:177–190

76. Woese CRA (1979) Proposal concerning the origin of life on the planet Earth. J Mol Evol 13:95–101

77. Hazen RM, Boctor N, Brandes JA, Cody GD, Hemley RJ, Sharma A, Yoder HS Jr (2002) High pressure and the origin of life. J Phys Condens Matter 14:11489–11494

78. Kral TA, Altheide TS, Lueders AE, Schuerger AC (2011) Low pressure and desiccation effects on methanogens: implications for life on Mars. Planet Space Sci 59:264–270

79. Taubner R-S, Leitner J, Firneis M, Hitzenberger R (2016) Modelling the interior structure of Enceladus based on the 2014's Cassini gravity data. Orig Life Evol Biospheres 46:283–288

80. Taubner R-S, Pappenreiter P, Zwicker J, Smrzka D, Pruckner C, Kolar P, Bernacchi S, Seifert AH, Krajete A, Bach W, Peckmann J, Paulik C, Firneis MG, Schleper C, Rittmann SK-MR (2018) Biological methane production under putative Enceladus-like conditions. Nat Commun 9:748. https://doi.org/10.1038/s41467-018-02876-y

81. van Zuilen MA, Chaussidon M, Rollion-Bard C, Marty B (2007) Carbonaceous cherts of the Barberton Greenstone Belt, South Africa: isotopic, chemical and structural characteristics of individual microstructures. Geochim Cosmochim Acta 71:655–669

82. Marshall CP, Love GD, Snape CE, Hill AC, Allwood AC, Walter MR, Van Kranendonk MJ, Bowden SA, Sylva SP, Summons RE (2007) Structural characterization of kerogen in 3.4 Ga Archaean cherts from the Pilbara Craton, Western Australia. Precambrian Res 155:1–23

83. Takahashi E, Scarfe CM (1985) Nature 315:566

84. Arndt NT, Francis D, Hynes AJ (1979) The field characteristics and petrology of Archean-Proterozoic komatiites. Can Mineral 17:147

85. Huppert HE, Sparks SJ, Turner JS, Arndt NT (1984) Emplacement and cooling of komatiite lavas. Nature 309:19

86. Green DH (1972) Archaean greenstone belts may include terrestrial equivalents of lunar maria?. Earth Planet Sci Lett 15(3):263

87. Jones AP (2002) Komatiites: new information on the type locality (Barberton), and some new ideas. Geol Today 18(1):23

88. Jones AP, Price DG, DeCarli PS, Price N, Clegg R (2003) Impact markers in the stratigraphic record. Koeberl C, Martinez Ruiz F (eds) Springer, Berlin, pp 91

89. Arndt NT, Kerr AC, Tarney J (1997) Dynamic melting in plume heads: the formation of Gorgona komatiites and basalts. Earth Planet Sci Lett 146:289

90. Echeverria LM (1980) Tertiary or Mesozoic komatiites from Gorgona Island, Colombia: field relations and geochemistry. Contrib Mineral Petrol 73:253

91. Brandon AD, Walker RJ, Puchtel IS, Becker H, Humayun M, Revillon S (2003) 1860s–1870s systematics of Gorgona Island komatiites: implications for early growth of the inner core. Earth Planet Sci Lett 206:411

92. Inoue T, Rapp RP, Zhang J, Gasparik T, Weidner DJ, Irifune T (2000) Garnet fractionation in a hydrous magma ocean and the origin of Al-depleted komatiites: melting experiments of hydrous pyrolite with REEs at high pressure. Earth Planet Sci Lett 177:81

93. Campbell IH, Griffiths RW, Hill RI (1989) Melting in an Archaean mantle plume: heads it's basalts, tails it's komatiites. Nature 339:697

94. Kerrich R, Wyman D, Hollings P, Polat A (1999) Variability of Nb/U and Th/La in 3.0 to 2.7 Ga Superior Province ocean plateau basalts: implications for the timing of continental growth and lithosphere recycling. Earth Planet Sci Lett 168:101

95. Polat A, Kerrich R (2000) Archean greenstone belt magmatism and the continental growth–mantle evolution connection: constraints from Th–U–Nb–LREE systematics of the 2.7 Ga Wawa subprovince, Superior Province, Canada. Earth Planet Sci Lett 175:41

96. Polat A, Kerrich R, Wyman DA (1998) The late Archean Schreiber–Hemlo and White River–Dayohessarah greenstone belts, Superior Province: collages of oceanic plateaus, oceanic arcs, and subduction–accretion complexes. Techonophysics 294:295

97. Puchtel IS, Hofmann AW, Amelin YV, Garbe-Schönberg CD, Samsonov AV, Shchipansky AA (1999) Combined mantle plume-island arc model for the formation of the 2.9 Ga Sumozero-Kenozero greenstone belt SE Baltic Shield: isotope and trace element constraints. Geochim Cosmochim Acta 63:3579
98. Parman SW, Grove TL, Dann JC (2001) The production of Barberton komatiites in an Archean Subduction Zone. Geophys Res Lett 28:2513
99. Gangopadhyay A, Walker RJ, Sproule RA (2003) In: Major and trace element geochemistry and Os isotopic compositions of komatiites from Dundonald Beach, Abitibi Greenstone Belt, Canada, American Geophysical Union, Fall Meeting 2003, Abstract V42C-0374
100. Capdevila R, Arndt N, Letendre J, Sauvage JF (1999) Nature 399:456
101. Woodhead J, Kent AJ, Hergt J, Bolhar R, Rowe MC (2005) Volatile contents of komatiite magmas and the Archaean mantle: Insights from melt inclusions in komatiites. EOS Trans AGU 86(52), Fall Meet Suppl Abstract
102. Fiorentini ML, Beresford SW, Stone WE, Deloule E, Hanski E (2005) In: The role of volatiles in the genesis of Komatiite and Perrocrite-Hosted Ni–Cu-(PGE) systems. In: 10th international platinum symposium, "Platinum-Group Elements—from Genesis to Beneficiation and Environmental Impact", 7–11 Aug 2005, Oulu, Finland (Abstract)
103. Reyes DP, Christensen PR (1994) Geophys. Res. Lett. 21(10):887
104. Williams DA, Wilson AH, Greeley R (1999) In: Komatiites from the commondale greenstone belt. A Potential Analog to Ionian Ultramafics, South Africa. 30th annual lunar and planetary science conference, 15–29 Mar 1999, Houston, Texas, Abstract no. 1353 (1999b)
105. De Witt MJ, Ashwal LD (1997) Greenstone belts. Oxford monographs on geology and geophysics, vol 35. Oxford University Press, pp. 840
106. Nisbet EG, Fowler CMR (1996) In: Tectonic, magmatic, hydrothermal and biological segmentation of mid-ocean ridges the hydrothermal imprint on life: did heat-shock proteins, metalloproteins and photosynthesis begin around hydrothermal vents? MacLeod CJ, Tyler PA, Walker CL (eds) Geological Society Special Publication No. 118 (1996), pp 239–251
107. Powell R, Mariscal C (2015) Convergent evolution as natural experiment: the tape of life reconsidered. Interface Focus 5(6):20150040
108. Deardoff J, Haisch B, Puthoff HW (2005) Inflation-theory implications for extraterrestrial visitation. J Br Interplanet Soc 58:43–50
109. Luisi PL (2006) The emergence of life: From chemical origins to synthetic biology. Cambridge University Press, Cambridge
110. Ruiz-Mirazo K, Peretó J, Moreno A (2010) Defining life or bringing biology to life. Origins of Life and Evolution of Biospheres 40(2):203–213
111. Machery E (2012) Why I stopped worrying about the definition of life … and why you should as well. Synthese 185(1):145–164
112. Cleland C, Chyba C (2002) Defining 'life.' Orig Life Evol Biosph 32:387–393
113. Cleland C, Chyba C (2010) Does 'life' have definition? In: Bedau M, Cleland C (eds) The nature of life: classical and contemporary perspectives from philosophy and science. Cambridge University Press, New York, pp 326–339
114. Mizuno Y, Kuroiwa D (1970) Solute segregation in ice observed by autoradiography. J Glaciol 9(55):117–124
115. Anderson D, Tice A (1973) The unfrozen interfacial phase in frozen soil water systems. In: Hadas A, Swartzendruber D, Rijtema PE, Fuchs M, Yaron B (eds) Ecological studies. Analysis and synthesis, vol 4. Springer, New York, pp 107–124
116. Baker I, Cullen D, Iliescu D (2003) The microstructural location of impurities in ice. Can J Phys 81:1–9
117. Barnes PRF, Wolff EW, Mallard DC, Mader HM (2003) SEM studies of the morphology and chemistry of polar ice. Microsc Res Tech 62(1):62–69
118. Bakermans C, Tsapin AI, Souza-Egipsy V, Gilichinsky DA, Nealson KH (2003) Reproduction and metabolism at −10°C of bacteria isolated from Siberian permafrost. Environ Microbiol 5:321–326

119. Bakermans C (2008) Limits to microbial life at subzero temperatures. In: Margesin R, Schinner F, Marx JC, Gerday C (eds) Psychrophiles: from biodiversity to biotechnology. Springer, Berlin, pp 17–28
120. Price PB (2000) A habitat for psychrophiles in deep Antarctic ice. Proc Natl Acad Sci USA 97:1247–1251
121. Bej AK, Aislabie J, Atlas RM (2010) Polar microbiology. In: The ecology, biodiversity and bioremediation potential of microorganisms in extremely cold environments. CRC Press, Boca Raton
122. Bowman J (2008) Genomic analysis of psychrophilic prokaryotes. In: Margesin R, Schinner F, Marx JC, Gerday C (eds) Psychrophiles: from biodiversity to biotechnology. Springer, Berlin, pp 265–285
123. Tipler FJ (1981) A brief history of the extraterrestrial intelligence concept. Q J R Astron Soc 22:133–145
124. Jones M (2013) Mainstream media and social media reactions to the discovery of extraterrestrial life. In: Vakoch DA (ed) Astrobiology, history and society. Springer, Berlin and Heidelberg, pp 313–328
125. Walton M (2013) The dark side of transformational leadership: a critical perspective. Ind Commer Train 45(6):369–370
126. Karon BP (1996) On being abducted by aliens. Psychoanal Psychol 13(3):417
127. Bick IJ (1989) Aliens among us: a representation of children in science fiction. J Am Psychoanal Assoc 37(3):737–759
128. Alexander D (1997) The study of natural disasters, 1977–97: some reflections on a changing field of knowledge. Disasters 21(4):284–304
129. Bankoff G (2001) Rendering the world unsafe: 'Vulnerability' as Western discourse. Disasters 25(1):19–35

Chapter 8
Planets and Exoplanets, Habitability Sustainability and Time

Introduction by Guido Visconti

The first extrasolar planet (exoplanet) was discovered in 1995 by the 2019 Nobel prize winners Quiloz and Mayor and since then the scientific community counts something like 4300 confirmed exoplanets. The planet discovered by Quiloz and Mayor was orbiting the stars 51 Cygnus at about 50 light years from Earth. The planets come from every size and chemical or mineralogical composition and are grouped in five main categories that are Gas Giants, composed mainly of gases and similar or larger than Jupiter and Saturn. Neptunian, similar to Neptune or Uranus with a hydrogen–helium atmosphere, and a rocky core. Super Earth that is mainly rocky planets with masses from 2 to 10 times the Earth and finally Terrestrial planets that are rocky bodies similar to Venus, Earth and Mars with a possible atmosphere. The distribution of exoplanets with distance is peaked within about 100 pc (parsec, 1pc = 3.26 light years) and the rest shows a rather flat distribution up to about 500 pc. A possible explanation for the distribution is that the nearest planets are easier to detect with the various, sophisticated techniques used by the astronomers. Some planets are not bound to any star but "run free" in the galaxy and these are known as Rogue planets.

One can get all the possible information on exoplanets for example at the NASA site and there you also find some interesting clue why this scientific research is receiving so much attention. NASA (or ESA for that matter) receives government money and so it has to find ways to get support from the taxpayer and the decoys they agitated is the search for extraterrestrial life and consequently the so-called biosignatures. The first places where to look for an inhabited planet around a star is the habitable zone (HZ). For some kind of geocentric reasoning life is associated to the presence of water so that the HZ is calculated (once one knows the power emitted by the star) as that region where physical conditions are such to maintain liquid water. For our Sun the HZ is roughly located between the orbits of Venus and Mars so that Earth it just in the middle (for this reason HZ is also called the Goldilocks zone). However, there are many other mechanisms that determine the temperature of a planet like the atmospheric composition or gravitational tides so that this concept is continually evolving. At the same time a Galactic Habitable Zone (GHZ) has been proposed that is located in an annular region between 7 and 9 kiloparsec around the Milky Way center.

Once the HZ is calculated we may expect that if life exists there is some kind of signature from that planet not necessarily in the form of a TV broadcast. The idea is that similarly to what happens for the Earth, the presence of some gas in the atmosphere, may reveal some biological activity. That happened when methane was discovered on Mars and was attributed to some bacterial activity or the more recent discovery of phosphine on Venus. So for exoplanets one looks at some spectroscopic feature that may reveal biological activity. This approach has been strongly criticized in a commentary by David Stevenson where he compares the search for extraterrestrial life on exoplanets to the hunting of the Snark of Lewis Carroll Memory. In particular, he noticed that excessive emphasis on habitability is bad science policy. The same definition of HZ is misleading because for example in the solar system most of the water is outside the HZ as a matter of fact we may have satellites like Europe or Enceladus which are water planets. The presence of water outside the HZ is related to the formation mechanisms of the planetary system. The approach to find life on exoplanets is the same used by a drunkard that after losing the car keys looks mainly under the street lamps. The same geocentric reasoning was criticized at the time of the first attempt by the Viking landers to find life on Mars. The same experiments run on Earth would find no trace of life on our planet. The dimension of this enterprise is confirmed by the Seager equation. This formula is a variation of the famous Drake equation and has been applied to estimate how many exoplanets with biosignature could be observed with TESS or JWST space mission. The results are that over a base of 30,000 observable stars there are only 1–2 planets.

We have to recognize that similarly to other topics in Astrobiology these are fascinating matters that capture our emotions but do not bear directly on our everyday life. As for other scientific endeavors the most tangible results could be the development of technologies that could be applied to more practical matters. However, the game is worth the candle. As stated in the 2007 NRC report, *The Limits of Organic Life in Planetary Systems*, "nothing would be more tragic in the exploration of space than to encounter alien life and fail to recognize it." A reminder not to trust geocentric reasoning.

Machine-Generated Summaries

Keywords: *mass, sun, environment, stellar, evolution, atmospheric, formation, jupiter, question, moon, orbital, observe, distance, feature, planetary system.*

Extra-Solar Planetary Systems

https://doi.org/10.1007/978-1-4614-9090-6_7

Abstract-Summary
We begin with the historical searches for planets around other stars to the present day, then survey the methods used to detect and study extra-solar planets (aka exoplanets), and discuss several of the historically important as well as some of the more interesting cases.

We tackle the thorny question of how to define a planet (and a star, and of those intermediate objects, brown dwarfs).

Extended
We begin with a guess for M_2.

Historical Perspective
Planets are common among stars like the Sun, especially those richer in metals and they are not absent near other types of stars, such as sub-giants, giants, a certain type of pulsar, and around very cool, low-mass late K and M dwarfs.

The HD 168,443 system, with a main sequence star and a likely brown dwarf, also harbors a possible planet with $M_J \sin i = 7.7$ (Udry and Others [1]).

The deficit continues to be apparent as more and more planets, including objects suspected of being "free-floating planets," continue to be discovered, whereas brown dwarf numbers in binary systems are being discovered at much lower rates.

To the present, extra-solar planets have been detected primarily through low-amplitude radial velocity surveys and through transit detections, conducted in the visual region of the spectrum, whereas brown dwarfs have been uncovered through long-wavelength photometric surveys and through direct visual and infrared imaging.

Methods to Find "Small"-Mass Companions
There are more or less six direct and a few indirect methods to detect the presence of a brown dwarf or of a planet near a star: Radial velocity variations Transit eclipses Astrometric variations Direct imaging/spectroscopy Gravitational lensing Timings (of minima or maxima in periodic variable stars, variations in transit timings, or time delays for pulsar planets) Indirect effects (e.g., stability analyses for already known planetary systems) We now discuss each of these methods in turn.

Benedict and Others [2] have made astrometric measurements of a star perturbed by an orbiting planet previously detected from radial velocities of stellar reflective motions, Gliese 876, an M4 dwarf, also known as Ross 780.

Seifahrt and Others (2007) find a radius 3.5 $R_\odot$ and likely mass <36 MJ. The companion to the K2 V star AB Pic is another candidate with spectral type L0–L1 and a mass 13.5 ± 0.5 MJ, right at the boundary between planets and brown dwarfs, at a projected separation of 260 au from the star (Bonnefoy and Others 2010).

Definitions of Planets and Brown Dwarfs

Pluto is relegated to a class known as dwarf planets, defined, in the same IAU resolution, as celestial bodies that fulfill planetary criteria (a) and (b) but fail (c), and are not satellites.

Sub-stellar objects above this mass are assumed to be brown dwarfs.

As a practical matter, in at least some cases, one may be able to distinguish a planet from a brown dwarf, and a brown dwarf from a low-mass star, through observational means: brightness, color and/or spectral characteristics.

According to Chabrier and Baraffe [3], the characteristic spectral sequence features for ranges of temperature at the cool end of the spectral sequence (with examples of brown dwarfs added) can be summarized as follows: A planetary temperature is typically less than ~1200 K, but this depends on the surface temperature of the star, the proximity to it, and the planet's albedo.

Extra-Solar Planets Detected or Awaiting Confirmation

A larger size for such a planet (relative to our Jupiter) is expected because of its proximity to its star, which would increase the equilibrium temperature (1449 ± 12 K, as tabulated in Torres and others 2008, most likely computed for the 1:1 spin–orbit coupled "slow rotator" case) and therefore increase the pressure scale height of the atmosphere; but the radius of HD 209458b exceeds predictions of models that take this into account.

These authors investigated the varying fluxes of solar-type stars at different ages for the orbital radii of HD 209458b (0.046 au) and OGLE-TR-56b (0.023 au) and found that the planets in the mass range of 0.5–5 MJ have evaporation rates varying from 10^{-8} MJ/y at young ages, to ~10^{-12} for ages greater than 5 Gy.

Origins of Brown Dwarfs and Planets

It is quite likely that planets exist around hotter non-solar type stars, such as Vega; at least, the hypothesis has been put forward that clumpy features observed in the infrared are due to the trapping of disk material in mean motion resonances of a Neptune-mass planet in an orbit 65 au from the star (Marsh and Others [4]).

A birth within a disk around a proto-star appears very likely, but is this due to core-accretion, where accretion through collisions results in objects sufficiently massive to attract large amounts of gas, or is it due to the onset of gravitational instabilities (where the thermal gas pressure is unable to prevent the collapse of material into gravitationally viable clumps), as suggested by Boss [5]?

The latter mechanism may occur in the cooler, outer portions of the disk, and over short time scales, ~1000 years or less, while core-accretion may take ~8 My to form a Jupiter-mass planet.

Acknowledgement

A machine generated summary based on the work of Milone, Eugene F.; Wilson, William J. F. (2013 in Astronomy and Astrophysics Library).

The Worlds Out There

https://doi.org/10.1007/978-1-4419-1684-6_7

Abstract-Summary

A planet is one of the possible byproducts accompanying the formation of a star.

The four rocky planets show us a relatively wide range of physical conditions that may be present on a terrestrial-type planet.

Nature offers us a broad range of possibilities as well, and so we can describe potential worlds that may be discovered in future surveys when looking for Earth-like planets, some of which already have observational evidence.

Definition of a Planet

A classical planet is a celestial body that (a) is in orbit around the Sun, (b) has sufficient mass for its self-gravity to overcame rigid body forces so that it assumes a hydrostatic equilibrium (nearly round) shape, and (c) has cleared the neighbourhood around its orbit.

A dwarf planet is a celestial body that (a) is in orbit around the Sun, (b) has sufficient mass for its self-gravity to overcame rigid body forces so that it assumes a hydrostatic equilibrium (nearly round) shape, (c) has not cleared the neighbourhood around its orbit, and (d) is not a satellite.

All other objects orbiting the Sun shall be referred to collectively as 'Small Solar System bodies'.

The upper mass limit for a planet is often taken to be about 13 Jupiter masses (4100 Earth masses), above which deuterium fusion occurs and the body is called a brown dwarf.

Our Solar System

Most of the angular momentum of the solar system is concentrated in the planets and especially in the gas giants.

The presence on Earth and perhaps on the other planets of high abundances of light elements such as lithium, beryllium and boron that are unstable under the nuclear reactions inside the stars is consistent with the idea that planet-forming matter came from regions outside the star.

Refractory elements are condensed close to the star, where also less materials are available, forming the terrestrial or rocky planets with an upper mass limit estimated in 13 Earth masses.

Its water is delivered by collisions with embryos formed beyond the snow line, a_{snow}, a minimum distance from the star with luminosity L_{star} at which ice could have condensed out of the gas at $T_{snow} \sim 50$–200 K, corresponding approximately to a distance of 2.7 AU from a solar-like star.

Planetary Atmospheres

The existence of an atmosphere is a characteristic factor of an astronomical body of planetary mass.

Atmospheres can gain material through the following processes: Bombardment by other small-mass bodies of the planetary system.

Mechanisms we can mention the following: Bombardment by other small-mass bodies of the planetary system.

Statistical Properties of the Extrasolar Giant Planets

Quoted values of the frequency, f_p, of giant planets around solar-type stars range between 4 and 9% for planetary masses in the range $1 \, MJ \leq M_p < 10 \, MJ$ and orbital radii a ≤ 3 AU, and 0.

The details of the demographics of substellar-mass objects will be worked out in the next decade, providing basic constraints to our understanding of star and planet formation.

Any theory that attempts to explain the process of stellar formation in clusters should reproduce the initial census of stars, brown dwarfs and planets, the so-called initial mass function.

Other common characteristics of Hot Jupiters are the following: They have a much greater chance of transiting their star as seen from a further outlying point than planets of the same mass in larger orbits Because of high levels of insolation, they have a lower density than otherwise would be the case.

Types of Terrestrial Planets

Along its history, a planet like the Earth has undergone major changes in the chemical composition of its atmosphere, something that we can also expect in other Earth-like planets to be detected in the future.

Kuchner [6] figured that planets formed beyond the snow line with masses M_p < $10 M_E$ could retain their atmospheres, rich in volatiles like H_2O and NH_3, when migrating toward the inner parts of the planetary system.

These volatile-rich planets, also known as Hot-Earths, could be common in the habitable zones of young stars and around M-stars.

The different behaviours of plate tectonics between planets of similar mass, like Venus and Earth, indicate that parameters other than mass may be important for the dynamics of the lithosphere.

The Sun shows a C/O ratio of about 0.5 and therefore this oxidative characteristic has led to the formation of rocky planets with CO_2 atmospheres and surfaces composed mainly of silicates.

Characterization of Exoplanets

For a hypothetical homogeneous body this ratio is simple, but real planets have multiple layers of different chemical composition and therefore detailed theoretical models must be elaborated to be compared with the observations.

These results suggest an unforeseen variety of extrasolar worlds Baraffe and Others [7] have modelled the R_p/M_p ratio for planets in the range [1 MJ $-$ 10 M_E] at different ages and levels of heavy metals enrichment, providing grids of planetary evolution to be compared with future transit observations.

For a given mass, the planetary radius increases significantly from the iron-rich to the atmospheric-rich planet.

Charbonneau and Others [8] first observed sodium in its atmosphere when the planet transited across the stellar disk.

Tinetti and Others [9] developed a spatially and spectrally resolved model of the Earth, based on observations of our planet from different spacecrafts.

Terraformed Planets

Terraforming is the hypothetical process by which the climate of a planet would be changed into something similar to that if the Earth, with the goal of making it habitable.

Fogg (1995) and Zubrin and Mc Kay (1997) discusses the technological requirements for this work.

An increase in the current temperatures to levels comparable to the Earth's could be possible by adding powerful greenhouse gases to the atmosphere (Gerstell and Others [10]).

The solar flux in the Mars orbit is sufficient for photosynthesis, but probably the level of UV solar radiation would be lethal for the microorganisms growing after the first phase of terraforming.

Oxygen producing bacteria could mimic the process occurring on Earth.

This author deems that this may turn out to be a long-term solution to the energy crisis and problems of population growth that civilization on Earth is now experiencing.

Expect the Unexpected

The window for observing other Earths still needs to be opened.

When dealing with extrasolar planets, we will need to expect the unexpected and all kinds of situations are possible.

Our final tribute goes to the artists who imagined these worlds some decades in advance of the future astronomical observations (e.g. Carroll 2007).

From the beginning, the Earth's evolution was well constrained by certain fixed parameters: mass, distance to the Sun, chemical composition, etc. We can easily imagine that by changing some of these parameters, other types of 'Earth-like' planets could be obtained.

We are interested in describing this potentially rich variety of planets.

From a certain value, we have stars undergoing some kind of thermonuclear fusion and below this value we have planets.

Let us start by describing the physical parameters of what we can call a planet, and the two broad types, giants and terrestrial (rocky), that we can expect to find in other systems.

Acknowledgement

A machine generated summary based on the work of Vázquez, M.; Pallé, E.; Rodríguez, P. Montañés (2010 in Astronomy and Astrophysics Library).

Extrasolar Planetary Systems

https://doi.org/10.1007/978-1-4419-1684-6_8

Abstract-Summary

A unique process, gravitational collapse of a molecular cloud, can give rise to different planetary systems.

The detailed configuration of this cloud, such as its mass, rotation and metallicity, will give the genetic print of the newborn system, conditioning the destiny of the different planets.

The Origin of the Solar System: Early Attempts

This theory was severely criticized by J. C. Maxwell (1831–1879), who argued that progressive condensation of annular rings to form the planets could not be possible under self gravitation as differential rotation in the outer and inner parts of the rings would destroy any initial condensation.

Another major difficulty for the theory was based on the distribution of angular momentum in the Solar System as there was no known mechanism to explain why the planets would have most of the angular momentum while the Sun has most of the mass.

We should mention the Capture Theory (Woolfson [11], 1978) that considers the idea of planets formed from captured and fissioned filaments (blobs) by an almost condensed Sun from other less massive diffuse proto-stars in a primitive stellar cluster scenery.

Formation of Planetary Systems

The temperature profile across a typical disk is usually given by Planets are known to have formed around stars and brown dwarfs.

The mass of the disk around a classical T Tauri star is about 1–3% of the stellar mass, mostly made of gas with only 1% of the disk in the form of dust (Natta and Others [12]).

Far-IR observations at 70 μm suggest that 10–15% of solar-like stars possess cool outer dust disks that are massive analogs of the Solar Systems's Kuiper Belt (Bryden and Others [5]).

Boss (2006) have studied the mechanisms of formation of Super-Earths around M dwarfs, concluding that these stars should form significantly more Super-Earths than giant planets.

Raymond and Others [13] suggest that the fraction of systems with sufficient disk mass to form planets with $M > 0.3 \, M_E$ decreases for low-mass stars.

Planetary Orbits

Mean anomaly of the epoch, M, the fraction of the orbital period that has elapsed since the last passage at periapsis, expressed as an angle: $M - M_0 = n(t - t_0)$, where M_0 is the mean anomaly at time t_0 and n the mean motion, a measure of how fast a body progresses around its orbit.

If the barycentre is located within the more massive body, like a star, that body will appear to 'wobble' rather than following a discernible orbit.

J.L. Lagrange (1736–1813) found an analytical solution for the three-body problem when the mass of the third is negligible compared with the other two, and its movement is in the same orbital plane.

If the masses, eccentricities and inclinations of the planets are small enough, then many initial conditions lead to quasi-periodic orbits.

Mean motion resonances occur when two bodies have orbital periods that fulfill this condition.

The Dynamically Habitable Zone

We have defined the habitable zone (HZ) of a planetary system as the region where water can exist in liquid phase on the surface of a planet by virtue of a combination of stellar and planetary parameters.

Its dynamically habitable zone (DHZ) is defined as the region of the planetary system where a terrestrial planet can survive without suffering major perturbations in its orbit due to the influence of giant planets.

Stable orbits of terrestrial planets inside the DHZ exist only if the orbits of the giant planets are located sufficiently far away from either the inner or the outer edge of the habitable zone (Noble 2002).

Architecture of Planetary Systems

An initially formed planetary system can evolve via one of the following options: Ejection of one or more planets An increase in the orbital separation of the planets, toward a more stable configuration Collisions between planets Many extrasolar systems formed by two giant planets lie near instability, and then at least one additional planet must exist in the stable regions of well-separated extrasolar planetary systems to push these systems to the edge of stability (Barnes and Raymond [14]; Raymond and Barnes [15]).

Sándor and Others (2007) have compiled a stability catalogue for a dynamical model consisting of a star, a giant planet and an Earth-like planet, establishing also the stability properties of known planetary systems.

The planetary system around 47 UMa, a G0 dwarf star located 46 light-years away, is probably one of the best analogues of the Solar System observed to date with two giant planets similar to Jupiter/Saturn, with masses of 2.6 and 0.46 MJ.

Violence and Harmony

We began this chapter by presenting the current knowledge on the process of formation of planetary systems, starting with our own.

According to current knowledge, all planetary systems seem to have a common process for their formation.

After the formation of the first protoplanetary bodies, the gravitational interactions between them will leave only the winners in a process where collisions, migrations and different kinds of perturbations are dominant.

A great deal of observational work remains to be done in the coming decades to fill in the gaps in our knowledge about different types of planets and planetary systems.

Our Solar System seems to be just one member of a large community with some peculiarities determined by the position and masses of the different components (star, planets, satellites and minor bodies).

This applies perfectly to the type of planetary system to which our terrestrial planet belongs.

Acknowledgement

A machine generated summary based on the work of Vázquez, M.; Pallé, E.; Rodríguez, P. Montañés (2010 in Astronomy and Astrophysics Library).

Planetary Atmospheres and Chemical Markers for Extraterrestrial Life

https://doi.org/10.1007/978-3-642-31730-9_5

Abstract-Summary

Observation techniques have reached the sensitivity to explore the chemical composition of the atmospheres as well as physical structure of some detected exoplanets and to detect planets of less than 10 Earth masses (M_{Earth}) and 2 Earth radii, so called Super-Earths, among them some that may be habitable.

To characterize a planet's atmosphere and its potential habitability, we explore absorption features in the emergent and transmission spectra of the planet that indicate the presence of biology.

Extended

Future missions will allow us to probe planets similar to our own for atmospheric features indicating habitable conditions.

Introduction

Observations of transits, that provide a radius estimate for the planet, combined with RV information, that provide a mass estimate for the planet, have provided estimates of the density of a subset of these planets, ranging from giant planets to rocky planets like Corot 7b [16] and Kepler 10b [17].

The discovery of transiting planets with masses below 10 M_{Earth} and radii consistent with rocky planetary models answered the important question as to whether planets more massive than Earth could be rocky.

Recent discoveries by ground based observations, as well as the Corot and Kepler space-missions, found planets with masses below 10 M_{Earth} and densities akin to Neptune as well as Earth, suggesting that there is not one cut-off mass above which a planet is like Neptune and below which it is rocky like Earth or Venus.

Characterizing a Habitable Planet (Learning from Earth)

Transmission spectroscopy of an Earth-like planet will have to be co-added for the first generation of space missions like JWST to provide enough signal to detect atmospheric absorption features.

The spectrum of potentially rocky planets in the HZ of M stars, like Gl 581d (see e.g. [18] and references therein) are being used to design instruments and observation strategies that will allow us to explore the atmosphere of the first temperate rocky worlds in the near future.

Whether Earth-analogue planets around stars other than Sun-analogues exist is still an open question that will be one of the first questions we can explore with future space and ground-based missions that can characterize planetary atmospheres.

The Spectral Fingerprint of an Earth-Like Atmosphere

Visible combined with near-infrared wavelengths as well as infrared spectral regions contain the signature of atmospheric gases that can be observed with low resolution and can indicate habitable conditions and, possibly, the presence of a biosphere: CO_2, H_2O, O_3, CH_4, and N_2O in the thermal infrared, and H_2O, O_3, O_2, CH_4 and CO_2 in the visible to near-infrared (see e.g. [19, 20], and references therein for detailed reviews).

Sagan and others [21] analysed a spectrum of the Earth taken by the Galileo probe as a direct image, searching for signatures of life and concluded that the large amount of O_2 and the simultaneous presence of a reducing gas like CH_4 traces are strongly suggestive of biology for a planet around a Sun-like star.

The presence or absence of these spectral features (detected collectively) will indicate similarities or differences with the atmospheres of terrestrial planets, and their astrobiological potential.

Characterizing Planetary Environments

A first estimate of the planetary effective temperature is obtained by calculating the stellar energy of the star that is received at the measured planet orbital distance and depends on the planetary albedo.

The effective temperature of our planet was due exclusively from radiation to the surface instead of an atmosphere layer (the average equilibrium temperature of Earth is about 265 K instead of the 288 K surface temperature), would only introduce a few percent error on the derived Earth radius.

If the secondary eclipse of the transiting planet can also be observed (when the planet passes behind the star), then the visible reflected starlight and the thermal emission of the planet allow the values of the mean albedo and the mean brightness temperature to be retrieved, based on knowledge of the radius measurements from the primary transit.

The Concept of the Habitable Zone

According to these models, the HZ is an annulus around a star where a rocky planet with a $CO_2/H_2O/N_2$ atmosphere and sufficiently large water content (such as in Earth) can host liquid water permanently on a solid surface.

This definition of the HZ makes several underlying assumptions: it assumes that the planet is rocky, water is present, the main atmospheric composition is $CO_2/H_2O/N_2$, and the abundance of H_2O and CO_2 in the atmosphere is regulated by a geophysical cycle similar to that of Earth, resulting in an H_2O- and CO_2-dominated atmosphere on the inner and the outer edges of the HZ, respectively.

The HZ would reduce to a distance from the host star, where the stellar flux and the atmospheric composition of the planet could maintain liquid water on the planet's surface.

At the limits of the HZ, the Bond albedo of a habitable planet analogue to a geologically active Earth is fully determined by its atmospheric composition and also depends on the spectral distribution of the stellar irradiation (see e.g. [22]).

Influences on Planetary Spectra

Using a numerical code that simulates the photochemistry of a wide range of planetary atmospheres several groups (see e.g. [23, 24]) have simulated the atmospheric composition of a replica of our planet orbiting different types of star: F-type star (more massive and hotter than the Sun) and a K-type star (smaller and cooler than the Sun).

In the IR, the cloud layers generally emit at lower temperatures than the surface would, decreasing the overall planetary flux, but can increase the absorption feature depth, because of the dependence of atmospheric absorption features in the IR on the temperature contrast between the absorbing/emitting layer and the continuum layer.

A planet orbiting a K star has a thin O_3 layer, compared to Earth's one, but still exhibits a deep O_3 absorption: because the low UV flux is absorbed at lower altitudes than on Earth which results in a less efficient warming (because of the higher heat capacity of the dense atmospheric layers).

Evolution of Biosignatures over Geological Times on Earth

The geological atmosphere model ranges from a CO_2-rich atmosphere (3.9 Gyr ago, epoch 0) to a CO_2/CH_4-rich atmosphere (epoch 3) to a present-day atmosphere (epoch 5, present-day Earth).

In our input model for this epoch we use 10% CO_2, current amounts of CH_4, and no O_2, O_3, or N_2O in the atmosphere.

The atmosphere in epoch 3 consists mainly of N_2, constant CO_2, and about equal amounts of CH_4 and O_2 (see e.g. [25] for an overview).

The atmosphere in epoch 4 consists mainly of N_2, 1% of CO_2, 0.04% CH_4, 2% O_2, and further increases in the trace species O_3 and N_2O.

CO_2 has negligible visible features at present abundance, but in a high-CO_2 atmosphere of 10%, seen in the early evolution stage of epoch 0, the weak 1.06 μm band could be observed.

Surface and Red Edge Features

On Earth around 440 million years ago, an extensive land plant cover developed, generating the red chlorophyll edge in the reflection spectrum between 700 and 750 nm.

The high SNR produced by EPOXI, due to its proximity to Earth during the measurements, show that cloud and surface features can be distinguished with high enough SNR for Earth and in turn for Earth-like planets with second or third generation space missions that provide the collecting area needed for such high signal.

Earth's hemispherical integrated vegetation red-edge signature is very weak, but planets with different rotation rates, obliquities, land–ocean fraction, and continental arrangement may have lower cloud-cover and higher vegetated fraction.

For Earth [26, 27] such high SNR measurements show a correlation to Earth's surface features because the individual measurements are time resolved as well as having individual high SNR, making it a very interesting concept for future generations of missions.

Conclusions

Spectroscopy of the atmosphere of extrasolar planets allows us to remotely explore a planet's environment, to distinguish Mini-Neptunes from rocky Super-Earths, and to explore atmospheric compositions as well as searching for indications of habitability.

The combination of spectral information in the visible (starlight reflected off the planet) as well as in the mid-IR (planet's thermal emission) allows a confirmation of atmospheric species, a more detailed characterization of individual planets but also to explore a wide domain of planet diversity.

Future missions will allow us to probe planets similar to our own for atmospheric features indicating habitable conditions.

Acknowledgement

A machine generated summary based on the work of Kaltenegger, Lisa (2012 in Physical Chemistry in Action).

Stellar Activity and CMEs: Important Factors of Planetary Evolution

https://doi.org/10.1007/978-3-319-10416-4_18

Abstract-Summary

CME activity of the Sun is known to be an important impacting factor for the magnetospheres, atmospheres, and surfaces of solar system planets.

The main planetary impact factors of the stellar CMEs include the associated interplanetary shocks, plasma density and velocity disturbances, energetic particles accelerated in the shock regions, as well as distortions of the magnetic field direction and modulus.

The planetary impact of the stellar CME activity may vary depending on stellar age, stellar spectral type and the orbital distance of a planet.

We discuss an issue of the stellar CME activity in the context of several actual problems of modern exoplanetology, including planetary atmosphere mass loss, planet survival at close orbits, and definition of a criterion for habitability.

Introduction

Broad observational material, together with the growing body of information about the Sun, shaped up the currently accepted paradigm regarding the stellar magnetic dynamo and its connection with the stellar activity in cool stars (Parker [28, 29]).

Stellar activity in the form of flares has been observed since the 1940s on UV Ceti type stars.

An additional special interest to the stellar flaring and CME activity is caused by their strong planetary effectiveness.

The classical concept of HZ around a star primarily relies on the stellar radiation flux allowing liquid water to exist on the surface of an Earth-like planet with a suitable atmosphere (Kasting et al. [30], Kasting [31]).

We consider the issue of the stellar CME activity in the context of several actual problems of modern exoplanetology, including planetary atmosphere mass loss, planet survival at close orbits, and definition of a criterion for habitability.

Exoplanets

Although most of the exoplanets discovered so far are thought to be gas or ice giants, like Jupiter or Neptune, some potentially rocky planets have been identified around M stars (Bonfils et al. [32]).

The study of close-orbit exoplanets under extreme stellar radiation and plasma conditions also helps to understand how terrestrial planets and their atmospheres, including early Venus, Earth and Mars evolved during the active early evolution phase of their host stars.

One of the major questions of the recent and future exoplanet search missions (COROT, Kepler, CHEOPS, PLATO, etc.) is, which of the main sequence star-types (M, K, G, F) may be good or at least preferred candidates for hosting habitable terrestrial type planets?

The search for the Earth-like exoplanets should not be limited only to the Sun-like G-type stars.

Impact of Stellar Radiation and Plasma Flows on Planetary Atmospheres

Close location of the majority of known exoplanets to their host stars results in intensive heating, ionization, and chemical modification of their upper atmospheres by the stellar X-ray/EUV (XUV) radiation with the subsequent expansion of the ionized atmospheric material and its loss due to interaction with the stellar wind (Lammer et al. [33]; Khodachenko et al. [34, 35]).

Questions a prominent position belongs to the problem of stellar—planetary interactions, including consideration of influences of stellar radiation and plasma flows (e.g., stellar wind and CMEs) on planetary environments and related erosion of upper atmospheres of exoplanets and their mass loss.

Interaction of short-periodic exoplanets with the stellar wind plasma and high XUV flux at close-orbit distances plays a crucial role regarding the ionization and ion loss processes of atmospheric species.

Stellar Radiation and Plasma Environment

Audard et al. [36] found that the energy of flares correlates with the stellar activity, characterized by L_X/L_{bol}, where L_X and L_{bol} are X-ray and bolometric luminosities of a star, respectively.

It has been found (Audard et al. [36]; Ribas et al. [37]; Scalo et al. [38]) that early K-stars and early M-stars may have XUV emissions level, and therefore flaring rates, of $\sim (3\text{--}4)$ and $\sim (10\text{--}100)$, respectively, times higher than solar-type G-stars of the same age.

Schaefer et al. [39] reported, from observations in various wavebands, nine super-flares with total radiative energies of $10^{33} - 10^{38}$ erg on solar (F-G type) analog stars, none of which is exceptionally young or extremely active, or a member of a close binary system.

From observations of our Sun that flaring activity of a star is accompanied by eruptions of coronal mass (e.g. CMEs), occurring sporadically and propagating in the stellar wind as large-scale plasma-magnetic structures.

The Problem of Magnetospheric Protection of Close-Orbit Exoplanets

The expanding partially ionized plasma of a Hot Jupiter atmosphere interacts with the planetary intrinsic magnetic field and appears a strong driver in formation and shaping of the planetary magnetospheres (Adams [40]; Trammell et al. [41]; Khodachenko et al. [42]), which in turn influences the overall mass loss of a planet.

Reiners and Christensen [43], based on scaling properties of convection-driven dynamos (Christensen and Aubert [44]), calculated the evolution of average magnetic fields of Hot Jupiters and found that (a) extrasolar gas giants may start their evolution with rather high intrinsic magnetic fields, which then decrease during the planet life time, and (b) the planetary magnetic moment may be independent of planetary rotation.

The influence of tidal locking on the value of an expected planetary magnetic dipole was studied for different planets (giants and terrestrial-type) in Grießmeier et al. [45, 46].

The estimates of the non-thermal mass loss of the weakly magnetically protected Hot Jupiter, HD 209458b, due the stellar wind ion pick-up, lead to significant and sometimes unrealistic values—up to several tens of planetary masses M_p lost during a planet life time (Khodachenko et al. [35]).

Magnetodisk-Dominated Magnetosphere of a Hot Jupiter

As applied to the Hot Jupiters, PMM reveals that the electric currents induced in the plasma disk produce an essential effect on the overall magnetic field structure around the planet, resulting in the formation of a magnetodisk-dominated magnetosphere of a Hot Jupiter.

Given the fact that the complete or partial tidal locking of such close-orbit exoplanets may lead to relatively weak intrinsic planetary magnetic moments, the encountering stellar wind and CMEs will push the planetary magnetospheres down to the heights at which direct exposure of the planetary atmosphere to the stellar CMEs plasma flow takes place.

A more realistic structure of the magnetodisk-dominated magnetosphere of a Hot Jupiter predicted by the Paraboloid Magnetospheric Model and its significantly larger size, as compared to a dipole-type magnetosphere, have important consequences for the study of magnetospheric protection of close-orbit exoplanets.

Acknowledgement
A machine generated summary based on the work of Khodachenko, Maxim L. (2014 in Astrophysics and Space Science Library).

Observational Techniques with Transiting Exoplanetary Atmospheres

https://doi.org/10.1007/978-3-319-89701-1_1

Abstract-Summary
Transiting exoplanets provide detailed access to their atmospheres, as the planet's signal can be effectively separated from that of its host star.

For transiting exoplanets three fundamental atmospheric measurements are possible: transmission spectra—where atmospheric absorption features are detected across an exoplanets limb during transit, emission spectra—where the day-side average emission of the planet is detected during secondary eclipse events, and phase curves—where the spectral emission of the planet is mapped globally following the planet around its orbit.

A few of the major science related questions are also discussed, which range from broad questions about planet formation and migration, to detailed atmospheric physics questions about how a planet's atmosphere responds under extreme conditions.

As a transmission spectra derived from primary transit is a unique measurement outside of our solar system, I discuss its physical interpretation and the underlying degeneracies associated with the measurement.

Background and History of Exoplanet Atmosphere Observations
When a planet is observed over the course of a full orbit, the flux contribution from the exoplanet will modulate the total star-plus-planet flux as its orbital viewing geometry changes, with atmospheric information obtainable throughout the phase curve.

A year earlier the Spitzer Space Telescope was launched, and while it was not designed for exoplanet transit observations, the first secondary eclipse measurements were achieved shortly after in 2005 by Deming et al. [47] for HD 209458b, while Charbonneau et al. [48] targeted TrES-1b.

For an exoplanet transmission spectrum, a good indication of the expected signal can be estimated by calculating the contrast in area between the annular region of the atmosphere observed during transit and that of the star.

On the bright end, the discovery of a planet transiting around HD 97658b (Dragomir et al. [49] permitted atmospheric investigations of a super-Earth around a much more massive K1V star.

Exoplanet Atmosphere Science Topics

A hot stratosphere is caused by strong optical absorbers, which absorb stellar radiation at altitudes higher than they thermally radiate energy, which heats the upper atmosphere and causes a stratospheric layer (Hubeny et al. [50]; Burrows et al. [51]; Fortney et al. [52]).

In highly irradiated gas giant exoplanets that lack a strong optical absorber, the incident stellar irradiation is absorbed deep in the atmosphere, near pressures of 1 bar (see Burrows et al. [53]).

The presence or absence of clouds and hazes have strong implications on all aspects of a planet's atmosphere including the radiation transport, chemistry, total energy budget, and advection (Marley et al. [54]).

Most of the exoplanets characterized thus-far show some levels of clouds (Sing et al. [55]), though the strong diversity of cloud and haze covers found thus far indicates there will be a sizeable population hot Jupiters with largely clear atmospheres, especially in the infrared where the scattering opacity of hazes and clouds is likely to become greatly reduced.

Analysing Transmission Spectral Data

A widely adopted method for most all time-series transit/eclipse data is to keep the point-spread-function of the telescope on the same pixel (or sub-pixel if possible) during the entire course of the observations.

Given the extremely high count levels of the science frames that are obtained during a transit, very large numbers of well exposed flat-field images are needed such that the photon noise levels per-pixel in the combined flat-field is comparable to the integrated photon noise levels per pixel in the science frames.

Be sure to obtain as many flat-field and bias images as possible, as it is unlikely taking one or two flat-field images would prove useful when a time-series dataset is aiming for high photometric precisions measuring a transit depth over perhaps hundreds or thousands of images.

There are several important initial steps which need to be taken before fitting transit light curves: Calculate accurate time stamps for the time series images.

Interpreting a Transmission Spectrum

A transmission spectrum is not a typical astrophysical measurement, the derivation helps to illustrate how the spectrum depends upon several key parameters (such as the temperature and molecular abundances), what quantities can be derived from a transmission spectra, and the nature of modelling degeneracies.

Na and K are both doublets, so by including only four absorption lines and a scattering component, the majority of an hot Jupiter optical transmission spectrum can be modelled.

The analytic model reproduces the Na and K lines profiles well, and certainly better than the accuracy of any transmission spectral data to date.

It is typically difficult to measure absolute abundances in a transmission spectrum, which will be limited by the degeneracy between the abundance and baseline pressure.

The H itself can typically be well determined from a well-measured transmission spectra.

Acknowledgement

A machine generated summary based on the work of Sing, David K. (2018 in Astrophysics and Space Science Library).

The Earth in Time

https://doi.org/10.1007/978-1-4419-1684-6_2

Abstract-Summary

During this time, it has undergone multiple changes that have clearly affected its global properties as seen from space.

After an introduction on the present structure of the planet, we make a journey from the early days of the Earth to the present day, in its current state of global warming.

The activities of life, both in the past and in the present, have played an active role in the global evolution of the planet.

At the end of the chapter, we stress the importance of the current age, the Anthropocene, in which humans are starting to influence the global environment.

The Earth at the Present Time

The Earth's surface, aerosols in the atmosphere and clouds all reflect some of the incoming solar short-wavelength radiation back to space, preventing that energy from warming the planet.

The temperature of the Earth is determined by the balance between the received shortwave (visible) flux, F_{in}, and the emitted infrared (long-wave) radiation, F_{out}, so that Assuming that the Earth radiates as a black body $F_E = \sigma T^4$, we can define an equilibrium temperature as where σ is the Stefan–Boltzmann constant and T_{eq} ($\sim$255 K) is the equilibrium temperature of the Earth, a physical averaged long-wave emission temperature at about 5.5 km height in the atmosphere (depending on wavelength and cloud cover, altitudes from 0 to 30 km contribute to this emission).

Two pioneers must be mentioned in this context: Joseph Fourier (1768–1830) established the concept of planetary energy balance (Fourier [56]), and J. Tyndall (1820–1893) began to study the capacities of various gases to absorb or transmit the heat emitted by the Earth (infrared radiation).

The Precambrian Era (4500–4550 Ma BP)

The deep ocean may have become oxic at 1.8 Ga. Instead of a tectonic driver for this transition, Schwartzman and Others [57] favoured a biological mechanism based on a methane atmosphere producing the emergence of oxygenic cyanobacteria at about 2.8–3.0 Ga. At that time the mean temperature was $\sim$60 °C, the most adequate for cyanobacteria.

The variations in oxygen content can be explained in terms of a balance between the following factors (Claire and Others [58]): $[O_2]$ is the total reservoir of molecular oxygen in units of teramoles, F_B the flux of oxygen due to organic carbon burial, F_E the flux of oxygen to the Earth due to hydrogen escape, F_V and F_M represent flux of reducing gases (i.e. H_2, H_2S, CO, CH_4) from volcanic/hydrothermal and metamorphic/geothermal processes, respectively, and F_W the oxygen sink due to oxidative weathering of continental rocks.

The Phanerozoic Era

The extinction occurring at the end of the Ordovician period (443 Ma ago) was probably caused by a glaciation produced by the position of the supercontinent Gondwana, close to the South Pole (Sheehan [59]).

The end of the Permian period occurred 250 Ma before the present and was characterized by the extinction of 80% of all ocean-dwelling creatures and 70% of those on land (cf. Erwin 1994, 2006; Berner [60]; Benton [61]).

At the boundary between the Triassic and Jurassic periods, about 200 million years ago, a mass extinction, occurring in a very short interval of time, destroyed at least half of the species on Earth (Ward and Others 2001).

The Quaternary

The Quaternary Period is the geologic time period from the end of the Pliocene Epoch, roughly 1.8–1.6 million years ago to the present.

The climate fluctuations in this period are mainly driven by variations in the orbital parameters of our planet, namely the inclination of the rotation axis, the eccentricity and the climate precession determining the time of the year where perihelion takes place.

In 1941, M. Milankovitch (1879–1958) calculated the insolation associated with these changes and proposed that the triggering of a glaciation was associated to the increase of ice-sheets in the northern hemisphere, produced at times of reduced temperature contrast between summer and winter.

The 41,000 year periodicity, driven by obliquity changes, was dominant during the early Pleistocene.

The impacts of current human activities will continue over long periods and the climate may depart significantly from natural behaviour over the next thousands of years.

The Future of Earth

It seems that the time period in the Earth's history in which complex life exists is very short.

Of increasing temperatures, the Earth will enter a runaway greenhouse phase, similar to that suffered by Venus, probably 1.2 Ga from now; as a first consequence, the oceans will evaporate.

The strong mass loss during this period will expand the Earth's orbit, probably to 1.6 AU, saving it temporarily from being destroyed.

The minimum orbital radius for a planet to be able to survive is found to be about 1.15 AU.

They show that planets with masses smaller than one Jupiter mass do not survive the planetary nebula phase if located initially at orbital distances smaller than 3–5 AU.

Silvotti and Others [62] have detected at 1.7 AU a giant planet orbiting the red giant V391 Pegasi, which is already burning helium in its core, whereas Mullally and Others [63] suggest the presence of a giant planet (2.1 Jupiter masses) orbiting a white dwarf at 2.36 AU.

The Earth is now characterized by a set of physical parameters and features observed on its surface.

The Earth in its present state represents just one stage in an evolutionary process starting from the moment the planet formed to the present day.

In our future search for Earth-like planets, we may encounter a planet at any stage in its evolution, and therefore, we can feasibly use the history of the Earth as a guide to characterize these planets.

We study the different components of the Earth system over time, starting with an introduction to the Earth as it is at present.

Most of the observed changes in the surface of the Earth have come about as a consequence of the dissipation of the energy stored in its interior.

Acknowledgement

A machine generated summary based on the work of Vázquez, M.; Pallé, E.; Rodríguez, P. Montañés (2010 in Astronomy and Astrophysics Library).

Is Our Environment Special?

https://doi.org/10.1007/978-1-4419-1684-6_9

Abstract-Summary

We are using our planet as the reference point for interpreting future observations of terrestrial exoplanets.

To gain some insight on this fundamental problem, in this last chapter we will discuss three basic questions for which we do not yet have complete answers: Is our Sun special?

Is the Sun Anomalous?

The Sun is a yellow dwarf star, a common stellar type in our Galaxy.

These two factors (metallicity and Supernova explosions) have led to the formation of the hypothesis of galactic habitable zone (Gonzalez and Others [64]; Lineweaver and Others 2003), an annular region between 7 and 9 kparsec from the Galactic centre, where 75% of the stars are older than the Sun.

Edvardsson and Others [65] and Gustafsson and Others (1999) propose that the Sun has a low [C/O] ratio relative to solar-like stars at similar galactocentric radii.

Stars show all the phenomena associated with solar magnetic activity, the base of the solar–stellar connection, but under different conditions of luminosity, mass, age, chemical composition, etc. Magnetic activity decays with the stellar age (Skumanich [66]) and is proportional to the rotation period, P_{rot}, and the depth of the convection zone, usually expressed by convective turnover time (Noyes and Others [67]).

Lockwood and Others [68] have recently stated, contradicting previous claims, that the amplitude of solar variations related to the magnetic activity is typical, compared to those of sun-like stars of similar age.

Is the Solar System Unique?

Gas giant planets form within 1–10 Ma, during the time that their parent stars possess a gas disk (see Haisch and Others [69]).

They suggest that planetary systems like our own solar system can exist only around singleton stars and estimate the singleton frequency with masses similar to that of the Sun to be between 90 and 95% in the solar neighbourhood.

Solar-like planetary systems form from disks with initial masses of 0.02 solar masses and angular momenta $J = 3 \times 10^{52}$ g cm^{2}s^{-1}.

The solar system was probably dustier in the past, but interactions between the giant planets and encounters with nearby stars have configured the present situation, as we have described previously.

A danger zone can be established around these hot stars, where no planetary systems can be formed and where, therefore, no Earth-like planets are expected.

Is the Earth Something Special?

The low eccentricity of the Earth's orbit was also an essential feature of the stability of its climate and therefore its aptitude to be habitable during such a long period of time.

Franck and Others [70] have estimated that the number of stellar systems with Earth-like habitable planets reached a maximum at the time of the Earth's formation.

Following the pioneering work of James Croll (1821–1890), Milutin Milankovitch (1879–1958) calculated variations in solar insolation driven by changes in the orbital parameters of the Earth at the scale of the last millions of years (see Berger [71]; Berger and Others [72]).

The lunar and solar tidal torques on Earth's ellipticity give rise to the 26,000 years' precession with the gravitational pull of other planets (mainly Saturn and Jupiter) slowly perturbing the orientation of the ecliptic in space.

At the early Earth, the Moon was ~10 times closer to our planet and tides were ~ 1000 times larger.

The Ultimate Factor: Life

The formation of a star and its planetary system is a consequence of a process of collapse in a molecular cloud.

The LHB event was decisive for the final arrangement of planets and perhaps for the emergence of life in the third planet.

The Earth has a set of properties that as the genetic print of a living being makes it something special, and also similar to other members of its family: the rocky planets.

What makes our environment so special is the presence of life in the third planet rotating around the star Sol.

At the time this book went into print, several space missions and ground-based projects are starting to give data on Earth-like planets.

Only a detailed cartography of Earth-like planets in our Galaxy will give us the necessary empirical data to have an answer to this fundamental question.

To validate this approach we must discuss how special is our planet, and its nearby environment, compared with the rest of exoplanets detected and to be discovered.

In mind, we are now prepared to discuss three classical questions: 'Is the Sun something special?', '

and 'Is the solar system something rare or unique?'.

The answer is obvious: for the habitability of the planetary system and the existence of life.

Studying the planetary habitability, it is important to avoid a circular reasoning, where the conclusion is in the assumption, and the abuse of mere correlations without a physical background.

Acknowledgement

A machine generated summary based on the work of Vázquez, M.;Pallé, E.; Rodríguez, P. Montañés (2010 in Astronomy and Astrophysics Library).

Biosignatures and the Search for Life on Earth

https://doi.org/10.1007/978-1-4419-1684-6_5

Abstract-Summary

Earth is the only life-hosting planet in the Universe that we currently know of; however, it is just a matter of time until other Earth-like planets are detected.

By analysing their spectral signatures, we will likely be able to approach the question of how common life is in the universe.

There are certain features of living systems that, because of the effects they have on the environment, allow us to distinguish them from other aspects of planetary geology, that is their presence or absence requires life.

The Physical Concept of Life
Living systems seem to challenge this law The food chain of life on Earth is based on plants, which trap solar radiation from the Sun and turn it into highly structured sugars through the process of photosynthesis (Deisenhofer and Michel 1989).

Photosynthesis dissipates solar energy and in this way terrestrial plants seem to create order, apparently defying the second law It was Schrödinger (1944) who first explained that the only way for organisms (or living systems) to remain alive is 'by continually drawing from its environment negative entropy', that is Schrödinger proposed that the level of complexity (order) in a body can only increase provided that the amount of disorder in its surroundings increases by a greater amount.

If systems that generally would be deemed non-biological can exhibit self-organization, self-replication, and even disequilibria (Feinberg and Shapiro 1980), then the mere detection of a chemical disequilibrium in a planetary atmosphere is not an unequivocal biosignature.

Astrobiology: New Perspectives for an Old Question
Astrobiology, the scientific cross-disciplinary study of life as part of the cosmic evolution, is quickly becoming a major research area.

This new discipline considers questions such as: (1) how do inhabited (or habitable) worlds form and evolve? (2) how did living systems emerge?

Requirements for Life
Life structures, the genetic code and metabolism, all need complex organic molecules that are based on carbon, a chemical element with a strong tendency to form covalent bonds.

Hydrogen bonds are also formed between water and organic molecules if the latter contain $-OH$, $-NH$ or $-SH$ groups in addition to carbon and hydrogen.

The photosynthesis reaction for anaerobics is given by The overall reaction for aerobic photosynthesis is given by The anaerobic process of photosynthesis is followed up by fermentation, which breaks down the carbohydrates to supply the energy needed to build amino acids and proteins, as shown by For aerobic organisms we have the cellular respiration By means of photosynthesis, all green plants, blue-green algae (cyanophyta) and certain bacteria synthesize organic compounds, primarily sugars (Alberts and Others 2003), from inorganic sources, such as carbon dioxide, water or other electron donors, following the previous reactions.

Biosignatures on Present Earth
Kasting (1997) shows that the detection of large amounts of atmospheric oxygen in an Earth-like planet whose surface temperature is adequate to harbour liquid water is a strong indicator of continuous production through biological activity (photosynthesis).

While the remote sensing detection of atmospheric components will give an indirect indication of the presence of biological material, the detection of surface signals, like the red edge, will give a direct confirmation of the existence not only of exobiological organisms, but of an advanced degree of complexity and evolution.

Although this signal is a reliable chlorophyll indicator for spatially resolved areas, in the case of globally integrated planetary observations it remains, in general, an ambiguous biomarker, because other surface and atmospheric compounds have spectral features that contribute to enhance or to reduce the vegetation index.

Biosignatures on Early-Earth

Life has a strong influence on the Earth's atmosphere, and consequently early eras will have different life footprints in the atmosphere.

Sleep and Bird (2008) recently published a review on the evolution of ecology during this rise of O_2 in the Earth's atmosphere.

Kaltenegger and Others (2007) characterized the evolution of the Earth's atmosphere and surface in order to model the observable spectra of an Earth-like planet through its geological history.

Kaltenegger and Others (2007) chose six epochs that exhibit a wide range in atmospheric abundances, ranging from a CO_2-rich early atmosphere 4 Ga ago, to a CO_2/CH_4-rich atmosphere around 2 Ga ago, to a present-day atmosphere.

Life in the Universe

A planet is considered to be within the habitable zone around a star if there is sufficient temperature and pressure in its atmosphere to sustain stable liquid water at the planetary surface.

The time of permanence in this phase, T_{MS}, is related with the mass of the star, M_S, (Scalo and Miller 1979) For main sequence stars the luminosity of a star is related with its mass through a simple power law ratio of the type The relationship between the radius, R_S, and the mass of a main sequence star is given approximately by In summary, habitable planets are expected around stars in the main sequence.

According to their model, the extinction of life for an Earth-like planet around a star between 0.6 and 1.1 M_S is caused by planetary geodynamics, at the age of 6.5 Ga. On the basis of the terrestrial experience, we assume that a long time must elapse before multicellular life appears.

Signatures of Technological Civilizations

At radio wavelength, the signals of the ballistic missile early warnings systems (BMEWS), the Arecibo radiotelescope and TV and FM transmitters dominate the electromagnetic spectra of the Earth and are larger than the natural emissions from the Earth or the Sun (Sullivan and Others 1978; Billingham and Tarter 1992).

Radio emissions from the sunlit Earth should be detectable only at $f > f_p$ (at night f_p should be lower).

The radar of the US Naval Space Surveillance has a detectability range of leaking terrestrial signals to 60 light-years for an Arecibo-type antenna (Sullivan and Others 1978; Sullivan 1981).

Atmospheric nuclear explosions produce a unique signature: a short and intense flash lasting around 1 ms, followed by a much more prolonged and less intense emission of light lasting from a fraction of a second to several seconds.

The Test Ban Treaty of 1963 prohibited nuclear explosions in the atmosphere, outer space and under water.

These spectral signatures are the features that only living systems leave on the environment and are readily distinguishable from the planetary geology; their presence or absence requires life.

For these biosignatures to be the least biased by our particular understanding of life, we need to adopt a non-geocentric approach, and to find general physical features, common to all living systems, which are present in the planetary signal and detectable through remote sensing.

Acknowledgement

A machine generated summary based on the work of Vázquez, M.; Pallé, E.; Rodríguez, P. Montañés (2010 in Astronomy and Astrophysics Library).

Habitability and Cosmic Catastrophes

https://doi.org/10.1007/978-94-007-4966-5_10

Abstract-Summary

Catastrophes in our solar system have played a big role especially during its early evolution, e.g., the formation of the Earth's moon or the slow retrograde rotation of our sister planet Venus can be explained by catastrophic collisions of the early planets with other planet-sized objects.

Catastrophes may have happened also on later stages during the Earth's evolution.

We give a review about the possible catastrophe scenarios and address to the question whether they caused the known mass extinctions during the last 500 million years on Earth.

Introduction

During the solar system's history, several catastrophic events occurred.

These cosmic catastrophes can be further divided into events that are related to the special properties of the planetary system itself.

This chapter, more than 1000 extrasolar planets are known, and the study of cosmic catastrophes will be extended to some of these systems.

In the first section, we will discuss different catastrophes that can occur in a planetary system and give a comparison to what is known in the solar system.

Cosmic Catastrophes in the Early Solar System

Considering our solar system, there exist some facts that must be explained by any theory of its formation: the Sun contains most of the solar system's mass, but the largest part of the angular momentum is distributed in the motion of the planets about the Sun, the orbits of the planets is quite regular (they move in the same direction, coplanar nearly circular orbits), there exist two types of planets (terrestrial and gas planets), same sense of rotation and orbital motion of the planets (two exceptions: Venus and Uranus), belt of small asteroids, and the Earth has an unusual large satellite, the Moon.

Let us assume a smaller body (e.g., Earth) with mass mthat orbits a more massive body (e.g., Sun) of mass M with a semimajor axis and eccentricity e; then the radius of the Hill sphere for the smaller body is Habitability on an Earth-like planet depends on the perturbation of giant planets beyond the Hill sphere for close encounters.

Definitions of Habitability

There could exist habitable moons around giant planets that are located in the habitable zone around a central star, the planets themselves being not habitable.

Moons of more distant giant planets can become habitable because of tidal heating (this could be the case for the Jupiter's satellite Europa).

There exists possibly a correlation between large close-orbiting massive planets and high metallicity of the host star.

A probability $P_{GHZ}(r,t)$ can be written as (GHZ denotes habitable zone, tmeans time, r the location inside a galaxy): With too little metallicity, Earth-like planets are unable to form; with too high metallicity, giant planets destroy Earth-like planets because they preferably then form near their host stars.

The GHZ in a galaxy is restricted to a certain region: (1) too near to the center there occur too many supernovae, and (2) too far from the center the stars are metal poor.

Impacts and Supernovae

Large-body impacts may be the primary cause of mass extinctions of life on the Earth.

These events are strongly correlated with mass extinctions on Earth.

It is seen that a global catastrophic event is likely to occur every $10^5–10^6$ year; however, a mass extinction event like the KT impact occurs at intervals of about 10^8 year.

They investigated 86 extrasolar planetary systems and found that 60% of them could harbor habitable Earth-like planets on stable orbits.

As it has been discussed already, several shielding effects are provided by the Earth and the solar system environment.

The Earth lies within the heliosphere that is a result of the solar wind and the solar magnetic field and provides a shielding against cosmic radiation.

During such collisions, the density of stars and interstellar matter increases and therefore the perturbations in the Oort cloud, resulting in considerably large impact rates in the inner planetary system.

Solar and Stellar Activity

The major disturbances from the Sun are (1) solar flares, (2) coronal mass ejections (CMEs), (3) the solar wind, and (4) radiation.

Coronal mass ejections are also caused by magnetic reconnection, but this occurs higher in the outer solar atmosphere.

The solar wind is a stream of charged particles ejected from the upper atmosphere from the Sun, and it consists mainly of electrons and protons.

Solar flares and CMEs influence on the state of the ionosphere, trigger geomagnetic disturbances and even change the composition of the atmosphere above the troposphere by ozone depletion.

The influence of the solar wind and flares and CMEs on other planets becomes more severe when they have no magnetic field or atmosphere.

This causes large irregularities, and during eruptions of flares, the amount of radiation and particles will have extreme effects on planets in the habitable zones which lie at close distance around them.

Conclusions

For solar-like stars, the most probable cosmic catastrophe scenario is the impact by a large asteroid.

This has occurred on Earth several times, and as it was shown, smaller than planet-sized bodies may be quite common in extrasolar planetary systems.

Acknowledgement

A machine generated summary based on the work of Hanslmeier, Arnold (2012 in Cellular Origin, Life in Extreme Habitats and Astrobiology).

Life in the Saturnian Neighborhood

https://doi.org/10.1007/978-94-007-4966-5_27

Abstract-Summary

Titan, for example, is the only other body in the solar system besides Earth to possess a dense atmosphere composed essentially of nitrogen (97%) and in which the combination with methane (2%) gives rise to a host of organic compounds.

The presence of seasonal effects, unique geomorphological features, and a probable internal liquid water ocean make Titan one of the most astrobiologically interesting bodies.

Another Cassini discovery of tremendous relevance to astrobiology is the large organic-ladden plumes ejected from Enceladus' south pole cracks, mainly made of water vapor but suggestive of a complex organic chemistry occurring in the interior of this small satellite, in the presence of liquid water.

Introduction and Context

With the discovery of planets beyond our solar system and the search for current living organisms or for favorable conditions for past and future life in exotic places such as Mars, Europa, Titan, and Enceladus, the notion of habitability takes a new dimension.

With an environment very rich in organics, Titan, along with comets, is thus often considered as one of the best targets to search for prebiotic chemistry at a full planetary scale and a possible habitat for extraterrestrial life in all probability different from the terrestrial one.

Our understanding of the origin of life on Earth could greatly benefit from studying Titan, where the low solar influx, the composition of the atmosphere, and the possible presence of an internal water ocean give us the opportunity to study the conditions prevailing on the primitive Earth.

Habitability Issues for Outer Planetary Satellites

Liquid water is established as the necessary solvent in which life emerges and evolves.

In a deposit of pure water, life will probably never spontaneously originate and evolve, since while there are many organisms living in water, none we know of is capable of living on water alone, because life requires other essential elements such as nitrogen and phosphorus in addition to hydrogen and oxygen.

The requirements for extraterrestrial life do not have to be the same, suggesting that life could exist outside the habitable zone (Cohen and Stewart 2002), in particular if liquid water exists underneath the surface.

If it is long enough, the liquid water underground ocean may host life.

The icy satellites of the outer planets of the solar system, as well as the recently discovered exoplanets, host unique conditions which may inhibit the emergence of life precursors in isolating environments that can prevent the concentration of the ingredients necessary for life or the proper chemical inventory for the relevant biochemical reactions.

Titan: Organic Factory and Habitat

A subsurface ocean in the interior of the satellite of a gas giant (as on Europe, Enceladus, Titan, etc.) may be habitable for some kind of life form—even though not necessarily an Earth analogue-but also information on the terrestrial-like atmospheric and surface conditions on any planetary body can provide valuable information.

Trainer and Others [73] looked at the processes that formed haze on Titan and on early Earth and found many similarities for what could have served as a primary source of organic material to the surface.

Another possible indicator for life on Titan is the lack of acetylene on the surface since there is no clear evidence of this compound in the received data to date, while it is expected to have been deposited through the atmosphere.

An underground liquid ocean, several hundred kilometers deep at the surface of Titan, is suggested to be the source of cryomagma, hence outgassing methane into the atmosphere and thus replenishing the destroyed amounts.

Enceladus: Liquid Water Far Away from the Sun

The heat source for Enceladus is still an open question, as is the possibility for life to exist on this small satellite if an underground liquid water ocean or liquid water subsurface pockets exist to explain the plumes (Kieffer and Jakosky [74]).

Since the prerequisites for life to emerge are the simultaneous existence of energy, organic compounds, and liquid water that have been found on Enceladus, it seems that it possesses all the necessary components to support life.

The south-polar region of Enceladus presents extremely high temperatures while the exsoluted vapor has also been shown to contain simple organic compounds.

The standards for life that Enceladus' possible ocean does not fall into are the sunlight, the oxygen compounds, and the organics produced on a surficial-crust environment.

Specific questions relevant to the goal of understanding habitability on Enceladus include: Is liquid water present on Enceladus, either in a subsurface ocean, in the plume vent regions, or elsewhere?

Discussion: What Can Titan and Enceladus Tell Us About Life?
The surface of Titan appears, like the surface of Mars or Europa, as an unlikely location for extant life, at least for terrestrial-type life.

Even though Titan presents terrestrial-type geology with complex structures formed mostly from dynamic processes, the absence of water on the surface makes it unlikely to support terrestrial-type life.

Fortes [75] noted that Titan's internal water ocean might support terrestrial-type life that had been introduced there previously or formed when liquid water was in contact with silicates early in Titan's history.

According to McKay and Smith [76], photochemically derived sources of free energy on Titan's surface could maintain an exotic type of life, using liquid hydrocarbons as solvents.

Another possible location to look for life on Titan would be in a possible subsurface liquid water ocean, and thus, it seems astrobiologically essential to confirm its presence.

Future Exploration of the Kronian Satellites
The cycle is active but different from the Earth because Titan lacks a surface methane ocean.

Some drawbacks of the mission, such as insufficient global coverage which has inhibited a full mapping of the atmospheric structure, composition, and temporal variations as well as the surface features of Titan and Enceladus, point to the need for further studies.

The primary science goals of TSSM are to understand Titan's and Enceladus' atmospheres, surfaces, and interiors; to determine the pre- and proto-biotic chemistry that may be occurring on both objects; and to derive constraints on the satellites' origin and evolution, both individually and in the context of the complex Saturnian system as a whole.

Besides TSSM, other concepts for future missions to return to Titan have been proposed Such as a simple orbiter to perform close-up investigations of the surface and the atmosphere of Titan (JET, C. Sotin, PI).

Acknowledgement

A machine generated summary based on the work of Coustenis, Athena; Raulin, Francois; Bampasidis, Georgios; Solomonidou, Anezina (2012 in Cellular Origin, Life in Extreme Habitats and Astrobiology).

Plurality Enters the Scientific Era

https://doi.org/10.1007/978-3-030-41448-1_13

Abstract-Summary

Plurality in the Last Three Centuries

Its similarity to the other globes of the solar system with regard to its solidity, its atmosphere, and its diversified surface; the rotation upon its axis, and the fall of heavy bodies, leads us on to suppose that it is most probably also inhabited, like the rest of the planets, by beings whose organs are adapted to the peculiar circumstances of that vast globe… I thing myself authorized, upon astronomical principles, to propose the sun as an inhabitable world… Then he goes on to suppose that the Moon, the planets and their satellites are also inhabited "with a high probability."

In 1877, Giovanni Schiaparelli, director of the Brera observatory near Milan, announces that he sees many dark, straight lines on the surface of Mars, the so-called canali (the plural of canale).

Lowell believed that what he actually saw on Mars was not directly canals, but vegetation on their shores, and his collaboratoer Vesto Slipher tried without success to detect chlorophyll by spectroscopy.

The Time Coordinate of Plurality

As to past history, If all the dynamical and geological facts prove that the Earth globe originally had a very high temperature, as that of melting iron… one must conclude that a very long sequence of centuries must have passed before reaching the present state … from the Greek School of Alexandria to our own time the loss of central heat cannot have given a decrease as much as 1/300 degree (Fourier [77]).

The essence of the demonstration was still the Buffon-Fourier postulate: Earth-cooling from a melted original state down to the present temperature gives a maximum possible age; but Kelvin could rely on vastly advanced physics in all its branches.

Taking this spontaneous heat output and atomic disintegration as facts of Nature, physicists were to renovate the Earth-age problem entirely and at long last reach the present-day estimates.

How Life Appeared on Earth

Here, we will see that the long time span since the formation of the Earth was actually needed to produce the life as we see it at present.

The other possible hypothesis is that life appeared on Earth following a long process that gradually passed from inert matter to micro-organisms and more evolved ones, through mechanisms to be specified.

4100 million years (MY) before present (BP), hence 500 million years after the formation of the planet, there was liquid water on its surface and it is possible that the first unicellular beings appeared.

In the other planets of the Solar system and their satellites, the situation is much less favorable; indeed, space exploration has shown that life can only exist at present under the form of primitive, unicellular beings, and only in some locations (see Encrenaz and Others [78]).

Life on Exoplanets?

After the discovery in 1995 of the first planet around a solar-type star, an exoplanet, a new universe has opened.

The method has been used successfully since the early nineteenth century in the case of binary star systems, to detect the stellar companion; the principle here is the same, with the difference that the orbit described by the star is much smaller than that described by the planet, by the ratio of the masses planet/star.

The planet should not be too close to its star: such planets would always present the same side to the star, like the Moon to the Earth, and this does not seem to be very favorable for life.

SETI and Interstellar Communication

Nothing followed, and we have to wait for 1974 to see a message issued from the Arecibo 300-m diameter radio telescope towards M13, a globular cluster located at a distance of 20,000 light-years, which contains about 300,000 stars.

Physicists Giuseppe Cocconi and Philip Morrison investigated the best ways to search for signals from possible extraterrestrial civilizations, and concluded that the best solution would be to use radio telescopes at the wavelength of 21 cm, that of the line of atomic hydrogen discovered eight years before.

The longest observations made to detect extraterrestrial life were conducted with the Ohio State University radio telescope: a signal was received with this radio telescope on August 15, 1977, but it was not repeated and it is probable that it was terrestrial emissions reflected by an artificial satellite or a space debris.

They analysed radio signals between 10 and 22 cm wavelength received from the direction of 850 solar-type stars located at less than 150 light years.

Today's Plurality Myth

Considering that (1) the number of stars in the Galaxy is about 10^{11}, and perhaps about 10^{21} in the visible Universe at the latest count, and most of them have planets of their own; (2) among these planets, many are bound to be at a suitable distance from their star and fall in the so-called habitable zone, which roughly means: able to preserve some liquid water; (3) while the mechanism of Life's origin on Earth is unknown, the phenomenon itself is roughly datable, and appears to have been remarkably rapid once a decent enough temperature had been reached; thus (in a large logical leap) the generation of life must be deemed a probable happening on the surface of suitable planets; (4) the progression of life to intelligence has been a very slow process, but the major steps are well known, and the mechanism is broadly understood as evolution controlled by natural selection, thus it also appears a highly-probable phenomenon; (5) the last step, from intelligence to "advanced" technology has been so rapid that it must be simply unavoidable.

Appendix: The Drake Equation

It is when we consider recent discoveries in their relation to the existence of other worlds, when we attempt to form a conception of the immense varieties of the forms of life corresponding to the innumerable varieties of cosmical structure disclosed by modern researches, that we recognise the full significance of those discoveries.

After Huygens, the history of Plurality becomes relatively uninteresting for almost two centuries, for lack of new data.

At the end of the last century, came the events expected for so long: the discovery of planets about other stars than the Sun, and the realization that the conditions in many of these planets might be favorable for life.

While the exploration of the Solar system has shown that Internal Plurality can only be limited to primitive forms of life, External Plurality might flourish: what has long been a dream might become reality, perhaps even during the present century.

Acknowledgement

A machine generated summary based on the work of Connes, Pierre; Lequeux, James (2020 in Historical & Cultural Astronomy).

Are We from Outer Space?

https://doi.org/10.1007/978-94-007-2941-4_30

Abstract-Summary

The biological record suggests that life on Earth arose as soon as conditions were favorable, which indicates that life either originated quickly, or arrived from elsewhere to seed Earth.

While it is not unreasonable to consider the possibility that Earth's life originated elsewhere and potentially much earlier, we conclude that the current literature offers no definitive evidence to support this hypothesis.

Chladni's view, that they fall from the skies, pronounced in 1795, was ridiculed by the learned men of the times. (Rachel 1881) Evidence of life on Mars, even if only in the distant past, would finally answer the age-old question of whether living beings on Earth are alone in the universe.

Introduction

Aspects of "astrobiological" research, such as searching for exoplanets, characterizing extremophilic microbes, and searching for biosignatures of ancient life undoubtedly offer important insights into the origin and evolution of life on Earth.

This claim required researchers to critically assess fossil and geochemical evidence with a new level of scrutiny, and it has been suggested that despite the consensus that this claim has been refuted, the academic debate has better prepared scientists to assess future evidence for ET (extraterrestrial) life (Dick and Strick [79]).

Taken in this context, the hypothesis known as panspermia that posits life on Earth was introduced from elsewhere may indeed spur innovative research through healthy debate, but it is equally clear that inconclusive evidence is currently cited by some authors as strong evidence of the panspermia hypothesis (Wickramasinghe and Others [80]).

Can panspermia influence evolutionary progress or affect the origin of life debate?

Question 1: Is There Any Compelling Evidence for Life Outside Planet Earth?

Evidence, recent information obtained from Martian landers suggests an arid, acidic, and oxidizing early Mars (Squyres and Knoll [81]), and the recent unexpected discovery of perchlorate salts in the Martian soil (Hecht and Others [82]), gives evidence for the presence of an oxidant that may yet convincingly explain the results of the LR experiments as abiotic.

Certain proponents of panspermia take this ever-increasing inventory of organic molecules as evidence of the ubiquity of life, disputing the previously mentioned processes: In view of the impossibility of producing cosmically, by non-biological means, organic material from inorganics like water, carbon monoxide and nitrogen gas, it is clear that the immense amount of interstellar organics must be biological in origin. (Hoyle and Wickramasinghe [83]) Recent claims that biological breakdown products are responsible for the signal of organic carbon in the interstellar medium echo this opinion (Rauf and Wickramasinghe [84]).

While the presence of methane in large gas giants is not regarded as evidence of life in our own solar system or elsewhere, the recent discovery of several rocky planets dubbed "super Earths" (Charbonneau and Others [85]) gives hope that we will soon discover a rocky planet with an atmosphere indicative of active biology.

Question 2: Is It Conceivable that Life as We Know It Could Be Viably Transferred to Other Habitats?

Large particles may travel to another planet by random chance (lithopanspermia), and smaller particles may be accelerated outwards from their parent star by radiation pressure or carried by interstellar particle clouds (radiopanspermia) (Horneck [86]).

We can imagine a small number of large meteorites and a larger number of smaller particles exiting a planet's surface, all with the potential to carry resistant forms of life.

Radiation shielding is plausible with lithopanspermia events, but the small particle sizes necessary for radiopanspermia decrease radiation shielding greatly, although it has been suggested that a thin film of carbon may be sufficient to shelter organisms from inactivating UV radiation (Wickramasinghe and Others [87]).

Even if life was not viably transferred, these events may have an impact—DNA or other prebiotic chemicals could conceivably offer a "kick-start" for the origin of life on sterile planets (Secker and Others [88]).

Question 3: How Likely Is It That This Transferred Life Would Become Established and Go on to Seed Neighboring Systems, Thus Amplifying the Process?

Arguments that life is likely to be based on organic carbon polymers and existing in liquid water (the so-called carbaquist perspective (Fry [89]) constrain the type of environments where life may exist.

As evidence begins to accumulate for exoplanets that resemble Earth, Pace [90] notes that an equally important question becomes the metabolic strategies organisms can potentially exploit to survive in extraterrestrial environments—all organisms, no matter the diversity of their metabolic strategies, require an influx of energy and a means of removing waste products.

Speculation about how extremophile organisms may thrive in extraterrestrial habitats cannot be taken as proof of life's ability to colonize the universe—there may be unknown factors that prevent life from easily adapting to new environments that we are only just beginning to discover.

Question 4: Could Panspermia Influence Evolutionary Progress or Affect the Origin of Life Debate?

Evidence discussed in the previous sections suggests that Earth is unlikely to be uniquely capable of supporting life and therefore may be common throughout the cosmos (Lineweaver and Davis [91])—both panspermia and many independent origins of life could produce this inferred pattern.

It is also conceivable that the origin of life is much more probable than Hoyle and Wickramasinghe consider—this would be supported if we were able to discover biochemically distinct forms of life on Earth that are the product of a unique origin.

Any such discovery would, in the opinion of some authors, be more significant than evidence for a successful panspermia event since it would inform the probability of life originating not just on Earth, but throughout the universe (Shapiro and Schulze-Makuch [92]).

Conclusions and Future Research Directions

The study of astrobiology therefore is often significantly influenced by opinion—Sir Francis Younghusband's musings on this subject remain relevant even today to discussions of panspermia: The onus of proving that nowhere else throughout the whole mighty universe than on this tiny speck of a planet does life appear rests upon those who deny it. (p. 47 in (Younghusband [93])) This argument appears central in the writings of ardent supporters of panspermia but cannot in itself prove anything satisfactorily.

Harold C. Urey in: (Laurence [94]) Current research on the diversity and limits of microbial life, the discovery and investigation of extrasolar planets, and new scientific missions to bodies in our solar system are most likely to offer clear and uncontroversial evidence for astrobiology.

Whether life on Earth began solely here or was transferred from elsewhere, both panspermia and origin of life research will likely have complementary parts to play in probing these mysteries.

Acknowledgement
A machine generated summary based on the work of Mcnichol, Jesse C.; Gordon, Richard (2012 in Cellular Origin, Life in Extreme Habitats and Astrobiology).

The History and Philosophy of Biosignatures

https://doi.org/10.1007/978-3-319-96175-0_15

Abstract-Summary
It deals with four epistemological issues in the search for signs of life in outer space: (1) conceptualization, how we form concepts of life in astrobiology, how we define and categorize things, and the relation between our concepts and our knowledge of the world; (2) analogy, how we see similarities between things, and with inductive, analogical reasoning go from what we know to what we do not know, from the only example of life here on Earth, to possible extraterrestrial life; (3) perception, how we interpret what our senses convey in our search for biosignatures, how the information we get from the surrounding world is processed in our minds; and (4) the semiotics of biosignatures, how we, as interpreters, establish connections between things, between the expression (the biosignature) and the content (the living organism) in various forms of semiosis, as icons, indices, and symbols of life.

Introduction
Our endeavour depends on human cognition and our ability to understand and interpret what we observe in our surrounding world.

I will discuss four epistemological issues in the search for biosignatures: (1) conceptualization, how we form concepts of life in science, how we define and categorize things, and the relation between our concepts and our knowledge of the world; (2) analogy, how we see similarities between things, and with inductive, analogical reasoning go from what we know to what we do not know; (3) perception, how we interpret what our senses convey, how the information we get from the surrounding world is processed in our minds; and (4) the, how we, as interpreters, establish connections between things, between the expression and the content in various forms of semiosis, as icons, indices, and symbols of life.

The Concept of Life

The difficulty in arriving at an acceptable definition has to do with, among other things: (1) what we mean by "definition" and what such a definition should do for us in our search for life or what it should explain (cf. Persson [95]); (2) how we categorize things and how concepts are formed and used in our minds; and (3) that we know only one living planet in Universe, our own planet, and we still do not know how life emerged here on Earth.

Before we go into the epistemic and semiotic questions involved in the search for biosignatures, which are the main target of this chapter, we need to delve into the very concept that is in the focus for our search, life, and how it is connected to our understanding of definitions, categorization, and our own ignorance.

The Analogy of the Earth-Twins

There are a number of problems with such an analogical argument: first of all, do we know all the necessary and sufficient conditions for life?

Astrobiology as a great analogy is an inductive argument that cannot logically prove anything, just propose that life in Universe is a theoretical probability, but not an inductive probability.

Galileo never stretched his analogical reasoning so far as he could by clearly claiming the existence of life on other planets.

The analogy argument started from the general supposition that there were no actual differences between Earth and Venus.

He concluded that if Venus had land and sea, mountains and valleys, clearings and condensations occurred in its atmosphere, and it had a companion moon, then it was entirely similar to our Earth and consequently also habitable.

Epistemic Perception

Astronomer William Herschel's lunar observations from 28 May 1776 highlight the difficulty to see and arrive at a conclusive interpretation of the seen.

Beside the fact that the optical quality of the telescopes was not always reliable, and that weather conditions could considerably influence the quality of the observations, there is obviously also an epistemic perception that changes the interpretations of the seen.

The ring was re-observed on 3 June 1769, and its causes were still being debated even then, but it was unanimously taken as proof of the existence of an atmosphere on Venus.

The great observational astronomers Herschel and Schröter engaged in a heated argument as to the presence or absence of mountains on Venus.

Herschel re-observed Venus in 1793, and he questioned Schröter's findings.

The observers of Venus saw things that needed explanations and interpretations.

Semiotics of Biosignatures

The interpreter needs to identify the physical phenomenon as containing semiotic meaning, something that can be a sign of life, a biosignature, that has a particular meaning by referring to its content "life."

Phenomena that we call biosignatures become meaningful phenomena, when we interpret them as containing a meaning by making a connection between the expression and the object, in other words, between the "biosignature" and "the living organism."

The meaning of the relation between expression and content, that the interpreter experiences, is based on either similarity (iconicity), proximity (indexicality), or habits, rules, or conventions (symbolicity).

The clearest examples of bioindices are atmospheric, chemical biosignatures that refer to biological processes, such as the metabolism of living organisms.

This semiosis or meaning-making endeavour is however triadic, includes something more than expression and object, the biosignature and the living organism.

Conclusion

We have no conclusive evidence of the existence of extraterrestrial life.

There is no method or way of proving that life does not exist and cannot exist elsewhere in Universe.

If one exoplanet does not show any signs of life, we go on to the next, and so on.

We will never know empirically that life does not exist on other planets.

Even though life might not exist out there, it is we human beings with our brains, bodies, and cultures who are searching for it.

Certain cognitive processes are at work when astrobiologists encounter unknown things, when interpreting potential signs of life, when they gather and classify information, and make conclusions from the observational data.

The quest for signs of life rests on the cognitive and socio-cultural capabilities of that human species that makes this lonely planet alive and thinking.

Acknowledgement

A machine generated summary based on the work of Dunér, David (2018 in Advances in Astrobiology and Biogeophysics).

The Extraterrestrial Life Debate from Antiquity to 1900

https://doi.org/10.1007/978-3-642-35983-5_1

Abstract-Summary

This chapter provides an overview of the Western historical debate regarding extraterrestrial life from antiquity to the beginning of the twentieth century.

Popular and scientific writings, such as those by Fontenelle and Huygens, led to a reversal of fortunes for extraterrestrials, who by the end of the century were gaining recognition.

Physical evidence for the existence of extraterrestrials was minimal, and was always indirect, such as the sighting of polar caps on Mars, suggesting similarities between Earth and other places in the universe.

In the latter half of the century, increasing research in stellar astronomy would be carried out, heavily overlapping with an interest in extraterrestrial life.

By the end of the eighteenth century, belief in intelligent beings on solar system planets was nearly universal and certainly more common than it would be by 1900, or even today.

Natural theology led to most religious thinkers being comfortable with extraterrestrials, at least until 1793 when Thomas Paine vigorously argued that although belief in extraterrestrial intelligence was compatible with belief in God, it was irreconcilable with belief in God becoming incarnate and redeeming Earth's sinful inhabitants.

Criticism would become more prevalent throughout the nineteenth century, and especially after 1860, following such events as the "Moon Hoax" and Whewell's critique of belief in extraterrestrials.

Skepticism about reliance on arguments from analogy and on such broad metaphysical principles as the Principle of Plenitude also led scientists to be cautious about claims for higher forms of life elsewhere in the universe.

At the start of the twentieth century, the controversy over the canals of Mars further dampened enthusiasm for extraterrestrials.

Introduction

It was in that year that Copernicus published his De revolutionibus, the treatise in which he posited a universe centered upon the Sun, about which the Earth and the other planets revolved.

It is based in large part on two long and fully referenced books by Michael Crowe published in 1986 and in 2008 and on a course the two of us have co-taught over the last three years titled "The Extraterrestrial Life Debate: A Historical Perspective."

Before the Eighteenth Century

By the end of the seventeenth century, however, extraterrestrials were understood to populate an extended universe, a universe teeming with life in much the same way as the Earth was filled with a wide variety of living creatures.

Convinced of the Copernican claim that the Earth orbits the Sun, and decidedly anti-Aristotelian, Bruno posited that the universe is vastly extended in space and that it is filled with systems of stars and planets.

By the end of the seventeenth century, then, the size and nature of the universe had changed drastically from what it had been: from a bounded, geocentric universe in which life was restricted to the terrestrial region to a vastly extended universe of multiple solar systems that, despite a lack of observational evidence, were assumed to teem with planets and life, similar to ours in some ways but perhaps significantly different in the details.

The Extraterrestrial Life Debate in the Eighteenth Century

Pope's poem sets out this point in the following lines, which had such appeal to Thomas Wright and Immanuel Kant that they quoted them when in the 1750s they published their theories of the universe: He, who thro' vast immensity can pierce, See worlds on worlds compose one universe, Observe how system into system runs,

What other planets circle other suns, What vary'd Being peoples ev'ry star, May tell why Heav'n has made us as we are (Crowe [30, 96]).

Carl Becker did not discuss pluralism when in his Heavenly City of the Eighteenth-Century Philosophers he maintained that the Enlightenment "philosophes demolished the Heavenly City of St. Augustine only to rebuild it with more up-to-date materials" (Crowe [96]), but he might have found in such authors as Wright, Kant, Lambert, and Herschel evidence to support a parallel claim for the heavens of Enlightenment astronomy.

The Extraterrestrial Life Debate in the First Half of the Nineteenth Century
One example of an author who used these considerations against extraterrestrials was Thomas Young, who, commenting in 1807 on William Herschel's arguments for life on the Sun, suggested that the Sun's great mass would make human-sized solarians weigh over two tons (Crowe [96]).

Shortly thereafter, he offers the following example: "The planets, which appear only as stars somewhat brighter than the rest, are to (the astronomer) spacious, elaborate, and habitable worlds; several of them vastly greater and far more curiously furnished than the earth he inhabits … and the stars themselves properly so called … are to him suns of various and transcendent glory—effulgent centres of life and light to myriads of unseen worlds" (Crowe [97]).

The Extraterrestrial Life Debate in the Second Half of the Nineteenth Century
Of the nineteenth century, the theory that the thousands of nebular patches observed by William Herschel are in fact universes comparable to our own Milky Way system had a measure of support, which was important for advocates of extraterrestrials because it was assumed that these universes would, like ours, be well stocked with life.

It is especially striking that the co-discoverer of the theory of evolution by natural selection, Alfred Russel Wallace (1823–1913), emerged near the end of the nineteenth century as not only a critic of pluralism but also to publish in 1904 an appendix to his Man's Place in the Universe (1903), an appendix that lays out for the first time the type of argument based on evolutionary theory against extraterrestrials (Crowe [97], 427–437) that has been taken up by many evolutionary theorists in the last thirty years, for example, by Ernst Mayr, Simon Conway Morris, William Burger, and Jared Diamond.

Overview of Part I
We have provided a historical overview of the extraterrestrial life debate from antiquity through the start of the twentieth century.

Galileo's telescopic observations further reinforced the analogy between the Earth and other environs of the universe, leading some authors to consider the existence of extraterrestrials on other planets in our solar system.

He concludes the chapter with discussion of seventeenth-century examples of "reflexive telescopics," or the notion of considering the Earth from a point of view outside of the Earth itself, as an extraterrestrial would see it.

The focus of the chapter by Joseph T. Ross is the philosophy of the prominent German philosopher Georg Friedrich Wilhelm Hegel (1770–1831), especially Hegel's views on the trustworthiness of analogical arguments and on the credibility of claims for extraterrestrial life.

He was especially skeptical of claims for life on the Sun and on moons, believing that only planets are in any significant degree analogous to the Earth.

An Addendum by Michael J. Crowe: Updating My Extraterrestrial Life Debate (1986)

In my 1986 volume, where I analyzed about eight of Kant's writings discussing extraterrestrials, I attempted to show that Kant was not only referring to the three thousand or so stars visible on a clear night, but that he was thinking of those stars as typically encircled by inhabited planets, as worlds filled with life, indeed intelligent life.

That question is whether the numerous stars that Kant saw and we see on a dark night are typically seats of life.

Two factors are above all important in determining whether a star is visible to us: (1) its distance and (2) the luminosity of the star—how much light it produces.

Among the stars visible to Kant on a clear night nearly all must be barren of higher forms of life.

Acknowledgement

A machine generated summary based on the work of Crowe, Michael J.; Dowd, Matthew F. (2013 in Advances in Astrobiology and Biogeophysics).

Cultural Implications of the Search and Eventual Discovery of a Second Genesis

https://doi.org/10.1007/978-94-007-2941-4_44

Abstract-Summary

The question we wish to address in this chapter concerns one of the most rewarding scientific and cultural activities that are possible at present.

Extended

We need especially to understand the influence of philosophical doctrines that will tend to encourage any future constructive dialogue between the new physical sciences and the humanities—especially philosophy—pointing toward unified two-culture knowledge, aiming at consilience (with the Edward Wilson fortunate choice of word).

Introduction: Implications of a Second Genesis Elsewhere in the Universe

Two questions emerge out of the pioneering work in astrobiology: Firstly, we consider the implications of the search for a second Genesis: Since its search is already a fact, its implications are already observable in different aspects of our culture, such as in science, philosophy, and technology.

In this chapter, in a second stage, we shall later discuss the implications of an eventual discovery of a second Genesis.

If the search for a second Genesis is a fact, its eventual discovery is only a possibility, but a scientifically sound possibility, which places the study of its implications in the domain of speculation of high heurist interest.

Regarding the second stage of our discussion, we recall that the theory of Darwinian evolution is not a predictive theory.

Having assumed that Darwinian evolution is a universal process, we may argue in favor of the inevitability of the origin and evolution of life, including intelligence.

Cultural Implications of the Scientific Search of a Second Genesis

Astrobiology, besides transforming the question of the plurality of inhabited words from philosophy to science, is now contributing to a new way of doing science, transdisciplinarity.

Just as our perception from time immemorial of the Solar System was radically transformed due to Copernicus and Galileo, also astrobiology is making it possible to conceive the reality of living organisms from a new point of view, which earlier we have called sub specie Universorum: Under this category, life appears as a phenomenon intimately connected with the universe, as a natural tendency of its matter-energy substratum.

The sociocultural context in which astrobiology was born encouraged a constantly fluid and mutually beneficial interaction between this discipline and science fiction, but at the same time, it contributed to many misunderstandings as to the true nature and possibilities of such research.

Cultural Implications of the Discovery of a Second Genesis

Extrapolating elsewhere in the universe our present understanding for the emergence of life does provide us with an object for astrobiology, since it allows us: (a) To set some theoretical bases for the possible existence of the main object of our research (a second Genesis) (b) To suggest the elaboration of a coherent and realistic strategy, namely with possibilities of success, of the search and obtaining reliable data of extant, of extinct life or of simply the identification of reliable biomarkers Consequently, astrobiology can be understood as the scientific exploration of the universe searching for a second Genesis (Aretxaga [98]).

From the assumed universality of biology, it seems inevitable that intelligent behavior will emerge in the cosmos, provided certain conditions favorable to the presence of continuous life on a given planet (or satellite) are maintained.

Discussion: Some of the Larger Philosophical Issues Arising from a Second Genesis

The intelligibility of the universe raises questions that lie on the frontier between science and the humanities; and we need an open all-embracing approach.

The human being considered himself as a privileged manifestation of the natural universe in which the structure and ordering principle of everything in existence becomes intelligible.

When the open question of the intelligibility of the universe is posed in a wider cultural context, including the earth sciences and astrobiology, reductionism's restricted view becomes more evident.

The eventual success of astrobiology in its search for life elsewhere in the universe would force upon us a reconsideration of the question of intelligibility and comprehension of nature, a subject that is the very nucleus of Western thought.

The contrast between different views on the intelligibility of the universe, as discussed in this chapter, illustrates the need for philosophical understanding of meaning with fresh approaches in philosophy at the frontier of astrobiology.

Chart of the Implications of a Second Genesis Elsewhere

Acknowledgement
A machine generated summary based on the work of Aretxaga-Burgos, Roberto; Chela-Flores, Julian (2012 in Cellular Origin, Life in Extreme Habitats and Astrobiology).

In Search of a Living Planet

https://doi.org/10.1007/978-3-642-25550-2_22

Abstract-Summary
The only planet known to harbour life is our own Earth.

No definite signs that would prove life in other celestial bodies have been found from our solar system or from other planets orbiting distant stars.

Life: Its Characteristics, Requirements and Limits
Life does not function this way, and it tries to keep up the order by all means.

In Tellurian life, the only type of life we know of, the solvent is water and the space limiter is a membrane surrounding the cell.

One of the properties of life is the tendency to maintain various unstable states such as complex molecules and cellular structures.

Of the unstable condition is the high level of oxygen in the atmosphere, which is a consequence of photosynthetic life.

Would life dwindle from the Earth, oxygen would disappear in 10,000 years.

A large number of molecules take part in the complex molecular pathways of life.

Life has adapted to almost every conceivable environment on Earth (Stan-Lotter [99]).

Because of adaptation to very different environments, extraterrestrial life may show a different set of limits than what we are accustomed to here on Earth.

Searching for Life in Our Solar System
The temperature differences are not reduced the way they are at Earth, as Mercury has practically no atmosphere.

Mercury is possibly the most hostile place for life among planets and Moons within our solar system.

Venus, the sister planet of Earth, has an atmosphere, but the conditions can hardly be considered balmy.

The remaining bodies, planets, moons, asteroids and comets can be divided into three groups when considering the possibility of life.

The measurements by the Galileo mission revealed a narrow zone in the Jovian atmosphere where the pressure is a few bars, and the temperature is suitable for liquid water.

The atmospheres of other giant planets have not been measured to large enough depths, but according to atmospheric models water clouds may be present at pressure levels of 10–100 bars.

This cover can be an atmosphere (Earth, Mars or Titan) or it could be ice (Europa, Enceladus, and possibly comets).

Mars
The conditions on the Martian surface are deleterious to Earth-based life.

At Mars, our known life could survive on the surface for about 20 min, but just a few centimetres below the surface it would be protected from the harsh radiation.

Mars and the Earth may have exchanged early forms of life contained in ejecta from asteroid impacting the planets.

If such life is similar to life on Earth, then it would indicate that life was transferred between Mars and the Earth or that life can form only using similar chemistry and strategies.

Due to the differences in the atmospheric conditions, natural conditions on Earth cannot fully emulate Martian conditions, but the tolerance of life to dry and cold conditions can still be investigated in deserts such as the Atacama desert in Chile or in the dry valleys of Antarctica, where the temperatures can fall close to levels found on Mars.

Europa
The photographs and other data transmitted from Pioneer 10 and 11 and Voyager 1 and 2 missions showed that Europa is an ice-covered body.

The surface shows features that resemble ice cracks that have been formed and refrozen at different epochs.

Superficially the surface of Europa resembles the satellite images of the frozen Arctic Ocean and other Arctic seas and large lakes.

This suggested that Europa has a layer of conductive matter under the ice cover.

Under the ice cover, the environment may be much more suitable for forming and sustaining life.

The patterns reminiscent of frozen cracks on Europa's surface suggest that the ice cover breaks every now and then.

The lake resides below a 4 km thick ice cover and is 400,000 years old.

Technical problems and ethical questions relevant to a Europa life expedition have turned up during the drilling of the ice.

Titan

In July 2008 spectroscopic observations showed that these lakes are formed of light hydrocarbons, methane and ethane, as well as liquid nitrogen (Brown and Others [100]).

In Titan, the light hydrocarbons have atmospheric and liquid cycles that resemble the water cycle on Earth (e.g. Turtle and Others [101]).

It is not at all insignificant that liquid water, necessary for life as we know it, is found on Titan.

The main differences between Trinidadian and Titanian lakes are that in Trinidad, the temperature is above the freezing point of water, the hydrocarbons are long complex ones with high viscosity, although gaseous ethane and methane are present here too.

In Titan, the temperature is very low, thus the small hydrocarbons, ethane and methane, are in liquid phase.

The conditions only a few cm below the surface in the Pitch lake are anoxic, so it appears that the lake is indeed a good analogue for Titan's hydrocarbon lakes.

Enceladus

This is a direct indication of a water reservoir that is hidden under the ice cover.

It is not thought to be as extensive and deep as that of Europa, but since tidal forces exerted by Titan appear to provide energy, the necessary conditions for life, liquid water and energy, are fulfilled.

Comets

Many comets have significant amounts of water, carbon dioxide and carbon monoxide ice.

As the comet enters the Earth's orbital distance from the Sun, their volatile ices start to evaporate creating the tail.

Inside a comet the temperature can grow so high that water ice starts to melt.

Other Planetary Systems

If the mass of the star can be calculated from its spectral type, observed brightness and distance, then this indirect method can be used to determine the properties of planets and their orbits.

In planetary systems where the giant planet's orbits are rather circular, an Earth-like planet could survive for a long time.

For a planet to have the possibility for Earth-like life, it should be orbiting the star in the "habitable zone", at a distance, where water can remain liquid.

About one tenth of the planets orbit their star in the habitable zone.

We should keep in mind that the observational methods cause a strong bias towards detecting large planets on small orbits.

The former is particularly interesting because its mass is too small for a gaseous planet and because it orbits the central star at a distance that places it in the habitable zone.

Kepler
Kepler is looking at individual stars in the direction of the constellations of Cygnus and Lyra.

In February 2011 the Kepler mission released data of over 150,000 stars (Borucki and Others [102, 103]).

Within the habitable zone of the star 54 planets are found, six of these are less than two Earth masses.

The Keppler team in collaboration with other astronomers reported in December 2011 the discovery of a two Earth diameter planet, Keppler 20b, orbiting in the habitable zone of the star.

SuperWASP, MOST, COROT, GAIA, JWST
Some of the most noteworthy are SuperWASP, COROT and MOST.

The SuperWASP measures the brightness of millions of stars and searches for eclipses caused by planets.

The COROT satellite was launched in Earth orbit in 2007.

It searches for planets in the vicinity of nearby stars.

The ground based SuperWASP, and the space missions COROT and MOST have all produced exoplanet detections.

SETI
Radio frequencies are monitored in millions of narrow band frequencies, and amongst these one is searching for radio signals that would reveal artificial radio emission, such as radio transmitters or radar-like instruments.

This method is heavily based on the assumption that an advanced civilization "leaks" radio emission into outer space.

This assumption could also be its Achilles' heal, as it is quite possible that radio emission is not used for communication in highly advanced civilizations.

The second strategy for finding civilizations is based on searches in optical wavelengths.

Such short pulses are not known to occur naturally and their presence would indicate a civilization that is using nanosecond pulses for communication.

The idea is to intercept the communication e.g. between two civilizations or between host planets and e.g. an interstellar space mission.

The Consequences of Finding Extraterrestrial Life
The news that life has been found outside Earth is going to cause a state of general confusion in the general public, and after that, depending on the person, either a full ignorance or even a certain level of denial or alternatively great interest.

Understanding of another kind of life could result in new applications and discoveries, which could help us to save the Earth from the man made eco-catastrophy.

Life on another Solar system body would be the only possibility for us to study and see directly a different kind of life.

When a living planet is found outside our solar system, for example, on a planet orbiting within the habitable zone of a central star, and having an atmosphere containing water and oxygen or other gases that appear to be out of equilibrium, then the problem will have a different nature.

Obtaining further information after the first signs of life on a planet will be very challenging.

Acknowledgement

A machine generated summary based on the work of Lehto, Harry J. (2012 in Lecture Notes in Earth System Sciences).

Book Reading List

Dynamical Chaos in Planetary Systems

by Shevchenko, I. I. (2020).

This is the first monograph dedicated entirely to problems of stability and chaotic behaviour in planetary systems and its subsystems. The author explores the three rapidly developing interplaying fields of resonant and chaotic dynamics of Hamiltonian systems, the dynamics of Solar system bodies, and the dynamics of exoplanetary systems. The necessary concepts, methods and tools used to study dynamical chaos (such as symplectic maps, Lyapunov exponents and timescales, chaotic diffusion rates, stability diagrams and charts) are described and then used to show in detail how the observed dynamical architectures arise in the Solar system (and its subsystems) and in exoplanetary systems.

Please see https://www.springer.com/gp/book/9783030521431 for original source.

Dynamics of Small Solar System Bodies and Exoplanets

by Souchay, J. J. (Ed), Dvorak, R. (Ed) (2010).

This book offers an up-to-date overview of current research on *Dynamics of Small Solar System Bodies and Exoplanets*. In course-tested extensive chapters the authors cover topics of theoretical celestial mechanics, physics and dynamics of asteroids, comets, stability of exoplanets and numerical integration codes applied in dynamical astronomy.

Please see https://www.springer.com/gp/book/9783642044571 for original source.

Earth, Moon, and Planets

by Gilmour, J. (Ed).

Earth, Moon, and Planets: An International Journal of Solar System Science, publishes original contributions relevant to understanding our solar system including: The formation of the solar system as a planetary system, bodies in our solar system: Mercury, Venus, Earth, Mars, Jupiter, Saturn, Uranus and Neptune, their satellites, and minor bodies including asteroids, comets and Kuiper Belt objects, the formation and evolution of planetary bodies, etc.

Please see https://www.springer.com/journal/11038/aims-and-scope for original source.

Exoplanets

by Kitchin, C. R. (2012).

Exoplanets: Finding, Exploring, and Understanding Alien Worlds probes the basis for possible answers to the fundamentals questions asked about these planets orbiting stars other than our Sun. This book examines what such planets might be like, where they are, and how we find them.

Please see https://www.springer.com/gp/book/9781461406433 for original source.

Exoplanets

by Mason, J. (Ed) (2008).

The first collection of review articles in one volume covering the very latest developments in exoplanet research.

Please see https://www.springer.com/gp/book/9783540740070 for original source.

Extrasolar Planets

by Queloz, D. (Ed), Cassen, P., Udry, S. (Ed), Guillot, T., Quirrenbach, A., Mayor, M. (Ed), Benz, W. (Ed) (2006).

Research on extrasolar planets is one of the most exciting fields of activity in astrophysics. In a decade only, a huge step forward has been made from the early speculations on the existence of planets orbiting "other stars" to the first discoveries and to the characterization of extrasolar planets. This breakthrough is the result of a growing interest of a large community of researchers as well as the development of a wide range of new observational techniques and facilities.

Please see https://www.springer.com/gp/book/9783540292166 for original source.

Extrasolar Planets and Their Host Stars

by von Braun, K., Boyajian, T. (2017).

This book explores the relations between physical parameters of extrasolar planets and their respective parent stars. Planetary parameters are often directly dependent upon their stellar counterparts. In addition, the star is almost always the only visible component of the system and contains most of the system mass. Consequently, the parent star heavily influences every aspect of planetary physics and astrophysics.

Please see https://www.springer.com/gp/book/9783319611969 for original source.

Giant Planets of Our Solar System

by Irwin, P. (2009).

This book reviews the current state of knowledge of the atmospheres of the giant gaseous planets: Jupiter, Saturn, Uranus, and Neptune. The current theories of their formation are reviewed and their recently observed temperature, composition and cloud structures are contrasted and compared with simple thermodynamic, radiative transfer and dynamical models. The instruments and techniques that have been used to remotely measure their atmospheric properties are also reviewed, and the likely development of outer planet observations over the next two decades is outlined.

Please see https://www.springer.com/gp/book/9783540851578 for original source.

Habitability of Other Planets and Satellites

by de Vera, J. (Ed), Seckbach, J. (Ed) (2013).

Is the Earth the right model and the only universal key to understand habitability, the origin and maintenance of life? Are we able to detect life elsewhere in the universe by the existing techniques and by the upcoming space missions? This book tries to give answers by focusing on environmental properties, which are playing a major role in influencing planetary surfaces or the interior of planets and satellites. The book gives insights into the nature of planets or satellites and their potential to harbor life.

Please see https://www.springer.com/gp/book/9789400765450 for original source.

Handbook of Exoplanets

by Deeg, H. J. (Ed), Belmonte, J. A. (Ed) (2018).

The Handbook of Exoplanets contains 15 sections dealing with all aspects of exoplanets and exobiology research, including historic aspects, the Solar System as a template, objects at the planet-to-star transition, exoplanet detection and characterization with related instrumentation, technology and software tools, planet and planet-system statistics with recent and planned surveys, their atmosphere and formation and evolution processes, habitability and exobiology implications, and outlooks for future exploration and science development.

Please see https://www.springer.com/gp/book/9783319553320 for original source.

Methods of Detecting Exoplanets

by Bozza, V. (Ed), Mancini, L. (Ed), Sozzetti, A. (Ed) (2016).

In this book, renowned scientists describe the various techniques used to detect and characterize extrasolar planets, or exoplanets, with a view to unveiling the "tricks of the trade" of planet detection to a wider community. The radial velocity method, transit method, microlensing method, and direct imaging method are all clearly explained, drawing attention to their advantages and limitations and highlighting the complementary roles that they can play in improving the characterization of exoplanets' physical and orbital properties.

Please see https://www.springer.com/gp/book/9783319274560 for original source.

Planets in Binary Star Systems

by Haghighipour, N. (Ed) (2010).

The discovery of extrasolar planets over the past decade has had major impacts on our understanding of the formation and dynamical evolution of planetary systems. There are features and characteristics unseen in our solar system and unexplainable by the current theories of planet formation and dynamics. Among these new surprises is the discovery of planets in binary and multiple-star systems.

Please see https://www.springer.com/gp/book/9789048186860 for original source.

The Transits of Extrasolar Planets with Moons

by Kipping, D. M. (2011).

Can we detect the moons of extrasolar planets? For two decades, astronomers have made enormous progress in the detection and characterisation of exoplanetary systems but the identification of an "exomoon" is notably absent.

In this thesis, David Kipping shows how transiting planets may be used to infer the presence of exomoons through deviations in the time and duration of the planetary eclipses.

Please see https://www.springer.com/gp/book/9783642222689 for original source.

Three Body Dynamics and Its Applications to Exoplanets

by Musielak, Z., Quarles, B. (2017).

This brief book provides an overview of the gravitational orbital evolution of few-body systems, in particular those consisting of three bodies. The authors present the historical context that begins with the origin of the problem as defined by Newton, which was followed up by Euler, Lagrange, Laplace, and many others. Additionally, they consider the modern works from the twentieth and twenty-first centuries that describe the development of powerful analytical methods by Poincare and others.

Please see https://www.springer.com/gp/book/9783319582252 for original source.

References

1. Udry S, Mayor M, Naef D, Queloz D, Santos NC, Burenet M (2002) The CORALIE survey for southern extra-solar planets. VIII. The very low-mass companions of HD 141937, HD 168443 and HD 202206: brown dwarfs or "superplanets?" Astron Astrophys 390:267–279

2. Benedict GF, McArthur BE, Forveille T, Delfosse X, Nelan E, Butler RP, Spiesman W, Marcy G, Goldman B, Perrier C, Jeffreys WH, Mayor M (2002) A mass for the extrasolar planet Gliese 876b determined from Hubble space telescope fine guidance sensor 3 astrometry and high-precision radial velocities. Astrophys J 581:L115–L118

3. Chabrier G, Baraffe I (2000) Theory of low-mass stars and substellar objects. In: Burbidge G, Sandage A, Shu FH (eds) Annual review of astronomy and astrophysics, vol 38. Annual Review Inc., Palo Alto, CA, pp 337–377

4. Marsh KA, Dowell CD, Velusamy T, Grogan K, Beichman CA (2006) Images of the Vega dust ring at 350 and 450 μm: new clues to the trapping of multiple-sized dust particles in planetary resonances. Astrophys J 646:L77–L80

5. Boss AP (1997) Giant planet formation by gravitational instability. Science 276:1836–1839

6. Ehrenreich D, Cassan A (2007) Are extrasolar oceans common throughout the galaxy? Astron Nachr 328:789–792

7. Arnold L, Bouchy F, Moutou C (eds) (2006) Tenth anniversary of 51 Peg-b: status of and prospects for hot Jupiter studies

8. Charbonneau D, Brown TM, Noyes RW, Gilliland RL (2002) Detection of an extrasolar planet atmosphere. Astrophys J 568:377–384

9. Nieto MM (1972) The Titius-bode law of planetary distances: its history and theory. Pergamon Press

10. Elkins-Tanton L, Seager S (2008) Coreless terrestrial exoplanets. Astrophys J 808

11. Russell HN (1935) The solar system and its origin. The Macmillan Company, New York

12. Kowal CT, Drake S (1980) Galileo's observations of Neptune. Nature 287:311–313

13. Moulton FR (1905) On the evolution of the solar system. Astrophys J 22:165–181

14. Barnes R, Quinn T (2004) The (in)stability of planetary systems. Astrophys J 611:494–516

15. Militzer B, Hubbard WB, Vorberger J, Tamblyn I, Bonev SA (2008) A massive core in Jupiter predicted from first-principles simulations. Astrophys J 688:L45–L48

16. Léger A, Rouan D, Schneider J, Barge P, Fridlund M, Samuel B, Ollivier M, Guenther E et al (2009) Transiting exoplanets from the CoRoT space mission. VIII. CoRoT-7b: the first super-Earth with measured radius. Astron Astrophys 506:287

17. Batalha NM, Borucki WJ, Bryson ST, Buchhave LA, Caldwell DA, Christensen-Dalsgaard J, Ciardi D et al (2011) Kepler's first rocky planet: Kepler-10b. Astrophys J 729:27

18. Kaltenegger L, Segura A, Mohanty S (2011) Model spectra of the first potentially habitable super-Earth—Gl581d. Astrophys J 733:35

19. Des Marais DJ, Harwit MO, Jucks KW, Kasting JF, Lin DNC, Lunine JI, Schneider J, Seager S, Traub WA, Woolf NJ (2002) Remote sensing of planetary properties and biosignatures on extrasolar terrestrial planets. Astrobiology 2:153–181

20. Meadows V, Seager S (2010) Terrestrial planet atmospheres and biosignatures. In: Seager S (ed) Exoplanets. University of Arizona Press, Tucson, pp 441–470, 526 pp. ISBN 978-0-8165-2945-2

21. Sagan C, Thompson WR, Carlson R, Gurnett D, Hord C (1993) A search for life on Earth from the Galileo spacecraft. Nature 365:715–721

22. Selsis F, Kasting JF, Levrard B, Paillet J, Ribas I, Delfosse X (2007) Habitable planets around the star Gliese 581? Astron Astrophys 476(3):1373

23. Selsis F (2000) Review: physics of planets I: Darwin and the atmospheres of terrestrial planets. In: Darwin and astronomy—the infrared space interferometer', Stockholm, 17–19 Nov 1999. ESA, Noordwijk, SP 451, pp 133–142

24. Grenfell JL, Stracke B, von Paris P, Patzer B, Titz R, Segura A, Rauer H (2007) The response of atmospheric chemistry on earthlike planets around F, G and K Stars to small variations in orbital distance. Planet Space Sci 55:661–671

25. Kasting JF, Catling D (2003) Evolution of a habitable planet. Ann Rev Astron Astrophys 41:429
26. Cowan NB, Agol E, Meadows VS, Robinson T, Livengood TA, Deming D, Lisse CM, A'Hearn MF et al (2009) Alien maps of an ocean-bearing world. Astrophys J 700:915
27. Livengood TA, Deming LD, A'Hearn MF, Charbonneau D, Hewagama T, Lisse CM, McFadden LA et al (2011) Properties of an Earth-like planet orbiting a sun-like Star: Earth observed by the EPOXI mission. Astrobiology 11:907
28. Parker EN (1955) Hydromagnetic dynamo models. Astrophys J 122:293
29. Parker EN (1993) A solar dynamo surface wave at the interface between convection and nonuniform rotation. Astrophys J 408:707
30. Kasting JF, Whitmire DP, Reynolds H (1993) Habitable zones around main sequence Stars. Icarus 101:108–119
31. Kasting JF (1997) Habitable Zones around low mass stars and the search for extraterrestrial life. Origins Life Evol Biosphere 27(1/3):291
32. Bonfils X, Delfosse X, Udry S et al (2013). The HARPS search for southern extra-solar planets. XXXI. The M-dwarf sample. Astron Astrophys, 549, A109
33. Lammer H, Odert P, Leitzinger M et al (2009) Determining the mass loss limit for close-in exoplanets: what can we learn from transit observations? Astron Astrophys 506:399
34. Khodachenko ML, Ribas I, Lammer H et al (2007) Coronal Mass Ejection (CME) activity of low mass M stars as an important factor for the habitability of terrestrial exoplanets, Part I: CME impact on expected magnetospheres of Earth-like exoplanets in close-in habitable zones. Astrobiology 7:167
35. Khodachenko ML, Lammer H, Lichtenegger HIM et al (2007b) Mass loss of "Hot Jupiters"— implications for CoRoT discoveries. Part I: the importance of magnetospheric protection of a planet against ion loss caused by coronal mass ejections. Planet Space Sci 55:631
36. Audard M, Güdel M, Drake JJ et al (2000) Extreme-ultraviolet flare activity in late-type stars. Astrophys J 541:396
37. Ribas I, Guinan EF, Güdel M et al (2005) Evolution of the solar activity over time and effects on planetary atmospheres. I. High-energy irradiances (1–1700 Å). Astrophys J 622:680
38. Scalo J, Kaltenegger L, Segura AG, Fridlund M, Ribas I, Kulikov YN, Grenfell JL, Rauer H, Odert P, Leitzinger M, Selsis F, Khodachenko ML, Eiroa C, Kasting J, Lammer H (2007) M Stars as targets for terrestrial exoplanet searches and biosignature detection. Astrobiology 7:85
39. Schaefer BE, King JR, Deliyannis CP (2000) Superflares on ordinary solar-type stars. Astrophys J 529:1026
40. Adams FC (2011) Magnetically controlled outflows from hot Jupiters. Astrophys J 730, article id. 27
41. Trammell GB, Arras P, Li Z-Y (2011) Hot Jupiter magnetospheres. Astrophys J 728:152
42. Khodachenko ML, Alexeev II, Belenkaya E et al (2012) Magnetospheres of 'Hot Jupiters': the importance of magnetodisks for shaping of magnetospheric obstacle. Astrophys J 744, article id. 70
43. Reiners A, Christensen UR (2010) A magnetic field evolution scenario for brown dwarfs and giant planets. Astron Astrophys 522:A13
44. Christensen UR, Aubert J (2006) Scaling properties of convection-driven dynamos in rotating spherical shells and application to planetary magnetic fields. Geophys J Int 166:97
45. Grießmeier J-M, Motschmann U, Stadelmann A et al (2004) The effect of tidal locking on the magnetospheric and atmospheric evolution of "Hot Jupiters." Astron Astrophys 425:753
46. Grießmeier J-M, Preusse S, Khodachenko ML et al (2007) Exoplanetary radio emission under different stellar wind conditions. Planet Space Sci 55:618
47. Deming D, Seager S, Richardson LJ, Harrington J (2005) Nature 434:740
48. Charbonneau D, Allen LE, Megeath ST et al (2005) ApJ 626:523
49. Dragomir D, Matthews JM, Eastman JD et al (2013) ApJ 772:L2
50. Hubeny I, Burrows A, Sudarsky D (2003) ApJ 594:1011
51. Burrows A, Hubeny I, Budaj J, Knutson HA, Charbonneau D (2007) ApJ 668:L171

52. Fortney JJ, Lodders K, Marley MS, Freedman RS (2008) ApJ 678:1419
53. Burrows A, Budaj J, Hubeny I (2008) ApJ 678:1436
54. Marley MS, Ackerman AS, Cuzzi JN, Kitzmann D (2013) In: Mackwell SJ, Simon-Miller AA, Harder JW, Bullock MA (eds) Clouds and hazes in exoplanet atmospheres, p 367
55. Sing DK, Fortney JJ, Nikolov N et al (2016) Nature 529:59
56. Condie KC (2001) Mantle plumes and their record in earth history. Cambridge University Press, London
57. Schwartzman D, Caldeira K, Pavlov A (2008) Cyanobacterial emergence at 2.8 Gya and greenhouse feedbacks. Astrobiology 8:187–203
58. Canuto VM, Levine JS, Augustsson TR, Imhoff CL (1982) UV radiation from the young sun and oxygen and ozone levels in the prebiological palaeoatmosphere. Nature 296:816–820
59. Sheehan PM (2001) The late ordovician mass extinction. Annu Rev Earth Planet Sci 29:331–364
60. Berner RA (2002) Examination of hyphotheses for the permo-triassic boundary extinction by carbon cycle modeling. Proc Natl Acad Sci 99:4172–4173
61. Benton MJ (2003) When life nearly died. Thames and Hudson, London
62. Silvotti R, Schuh S, Janulis R, Solheim JE, Bernabei S, Østensen R, Oswalt TD, Bruni I, Gualandi R, Bonanno A, Vauclair G, Reed M, Chen CW, Leibowitz E, Paparo M, Baran A, Charpinet S, Dolez N, Kawaler S, Kurtz D, Moskalik P, Riddle R, Zola S (2007) A giant planet orbiting the extreme horizontal branch star V391 Pegasi. Nature 449:189–191
63. Mullally F, Winget DE, Degennaro S, Jeffery E, Thompson SE, Chandler D, Kepler SO (2008) Limits on planets around pulsating white dwarf stars. Astrophys J 676:573–583
64. Bounama C, Von Bloh W, Franck S (2007) How rare is complex life in the milky way? Astrobiology 7:745–756
65. Edvardsson B, Andersen J, Gustafsson B, Lambert DL, Nissen PE, Tomkin J (1993) The chemical evolution of the galactic disk—Part one—analysis and results. Astron Astrophys 275:101–152
66. Meléndez J, Dodds-Eden K, Robles JA (2006) HD 98618: A star closely resembling our sun. Astrophys J 641:L133–L136
67. Livio M (1999) How rare are extraterrestrial civilizations, and when did they emerge? Astrophys J 511:429–431
68. Gaudi BS et al (2008) Discovery of a Jupiter/Saturn analog with gravitational microlensing. Science 319:927–930
69. Fischer DA, Valenti J (2005) The planet-metallicity correlation. Astrophys J 622:1102–1117
70. Franck S, von Bloh W, Bounama C (2007) Maximum number of habitable planets at the time of Earth's origin: new hints for panspermia and the mediocrity principle. Int J Astrobiol 6:153–157
71. Berger A (1980) The Milankovitch astronomical theory of paleoclimates: a modern review. Vistas Astron 24:103–122
72. Berger A, Imbrie J, Hays J, Kukla G, Saltzman B (eds) Milankovitch and climate: understanding the response to astronomical forcing. Dordrecht, D. Reidel
73. Trainer MG, Pavlov AA, DeWitt HL, Jimenez JL, McKay CP, Toon OB, Tolbert MA (2006) Organic haze on Titan and the early Earth. Proc Natl Acad Sci USA 103:18035–18042
74. Kieffer SW, Jakosky BM (2008) Enceladus—oasis or ice ball? Science 320:1432–1433
75. Fortes AD (2000) Exobiological implications of a possible ammonia-water ocean inside Titan. Icarus 146:444–452
76. McKay CP, Smith HD (2005) Possibilities for methanogenic life in liquid methane on the surface of Titan. Icarus 178:274–276
77. Fourier J-B (1824) Remarques généraes sur les températures du Globe terrestre et des espaces planétaires. Annales De Chimie Et De Physique 27:136–167
78. Encrenaz T, Lequeux J, Casoli F (2019) Les planètes et la vie. EDP Sciences, Les Ulis
79. Dick SJ, Strick JE (2004) The living universe: NASA and the development of astrobiology. Rutgers University Press, New Brunswick

80. Wickramasinghe NC, Wainwright M, Narlikar JV, Rajaratnam P, Harris MJ, Lloyd D (2003) Progress towards the vindication of panspermia. Astrophys Space Sci 283(3):403–413

81. Squyres SW, Knoll AH (2005) Sedimentary rocks at Meridiani planum: origin, diagetiesis, and implications for life on Mars. Earth Planet Sci Lett 240(1):1–10

82. Hecht MH, Kounaves SP, Quinn RC, West SJ, Young SMM, Ming DW, Catling DC, Clark BC, Boynton WV, Hoffman J, DeFlores LP, Gospodinova K, Kapit J, Smith PH (2009) Detection of perchlorate and the soluble chemistry of Martian soil at the Phoenix Lander site. Science 325(5936):64–67

83. Hoyle F, Wickramasinghe NC (1996) Our place in the Cosmos: the unfinished revolution. Phoenix, London

84. Rauf K, Wickramasinghe C (2010) Evidence for biodegradation products in the interstellar medium. Int J Astrobiol 9(1):29–34

85. Charbonneau D, Berta ZK, Irwin J, Burke CJ, Nutzman P, Buchhave LA, Lovis C, Bonfils X, Latham DW, Udry S, Murray-Clay RA, Holman MJ, Falco EE, Winn JN, Queloz D, Pepe F, Mayor M, Delfosse X, Forveille T (2009) A super-Earth transiting a nearby low-mass star. Nature 462(7275):891–894

86. Horneck G (1995) Exobiology, the study of the origin, evolution and distribution of life within the context of cosmic evolution: a review. Planet Space Sci 43(1–2):189–217

87. Wickramasinghe JT, Wickramasinghe NC, Wallis MK (2009) Liquid water and organics in comets: implications for exobiology. Int J Astrobiol 8(4):281–290

88. Secker J, Wesson PS, Lepock JR (1996) Astrophysical and biological constraints on radiopanspermia. J R Astron Soc Can 90(4):184–192

89. Fry I (2000) The emergence of life on Earth: a historical and scientific overview. Rutgers University Press, New Brunswide

90. Pace NR (2001) The universal nature of biochemistry. Proc Natl Acad Sci USA 98(3):805–808

91. Lineweaver CH, Davis TM (2002) Does the rapid appearance of life on Earth suggest that life is common in the universe? Astrobiology 2(3):293–304

92. Shapiro R, Schulze-Makuch D (2009) The search for alien life in our solar system: strategies and priorities. Astrobiology 9(4):335–343

93. Younghusband F (1927) Life in the stars: an exposition of the view that on some planets of some stars exist beings higher than ourselves, and on one a world-leader, the supreme embodiment of the eternal spirit which animates the whole. John Murray, London

94. Laurence WL (1961) Science; life in space? Analysts of French meteorite say it contains organic matter. New York Times (March 19): E9

95. Persson E (2013) Philosophical aspects of astrobiology. In: Dunér D, Parthemore J, Persson E et al (eds) The history and philosophy of astrobiology: perspectives on extraterrestrial life and the human mind. Cambridge Scholars Publishing, Newcastle-upon-Tyne, pp 29–48

96. Crowe, Michael J (1986) The extraterrestrial life debate 1750–1900: the idea of a plurality of worlds from Kant to Lowell. Cambridge: Cambridge Univ. Press. Repr. 1999, as The extraterrestrial life debate 1750–1900. Dover, Mineola

97. Crowe MJ (ed) (2008) The extraterrestrial life debate, antiquity to 1900: a source book. University of Notre Dame Press, Notre Dame

98. Aretxaga R (2008) Astrobiología: entre la ciencia y la exploración. Letras De Deusto 38(118):13–27

99. Stan-Lotter H (2007) Extremophiles, the physiochemical limits of life (growth and survival). In: Horneck G, Rettberg P (eds) Complete course in astrobiology. Wiley-VCH, Weinheim, p 121

100. Brown RH, Soderblom LA, Soderblom JM, Clark RN, Jaumann R, Barnes JW, Sotin C, Burratti B, Baines KH, Nicholson PD (2008) The identification of liquid ethane in Titan's Ontario Lacus. Nature 454:607–610

101. Turtle EP, Del Genio AD, Barbara JM, Perry JE, Schaller EL, McEwen AS, West RA, Ray TL (2011) Seasonal changes in Titan's meteorology. Geophys Res Lett 38:L03203. https://doi.org/10.1029/2010GL046266

102. Borucki WJ et al (2011) Characteristics of Kepler planetary candidates based on the first data set. Astrophys J 728:117. https://doi.org/10.1088/0004-637X/728/2/117
103. Borucki WJ et al (2011b) Characteristics of planetary candidates observed by Kepler, II: analysis of the first four months of data. Astrophys J arXiv:1102.0541

Chapter 9
Geobiology

Introduction by Guido Visconti

Following the accepted definition, Geobiology is a scientific discipline in which the principles and tools of biology are applied to study of the Earth. In that geobiology is parallel to the use of physics and chemistry methods that transformed geology in geophysics and geochemistry. The origin of geobiology can be traced to the 1970s when the Gaia hypothesis was formulated. Actually the so-called "strong Gaia" claimed that life influenced the Earth system for its own benefit. On the other hand, the geobiology community does not accept this view while supporting the idea that life has a quite strong influence on the environment. The importance of the biosphere was however pointed out already in the late twenties by the Russian scientist Vladimir Vernadsky. We see already the strong connections that geobiology has with Earth's science and we have seen also the importance in relation to astrobiology. The American Geophysical Union (AGU) has a Biogeoscience section that "emphasizes linkages between biological sciences and geophysical sciences fundamental to study of the Earth and other planets. The research areas encompassed within the section include biogeochemistry, biogeophysics, astrobiology, and planetary scale ecosystem science".

The most obvious chapters of Geobiology is the study of the biogeochemical cycles that is to understand how the different basic elements (Carbon, Nitrogen, Sulfur, Oxygen, etc.) are cycled between the reservoirs present in the Earth system (atmosphere, soil, hydrosphere). All of them show that biological processes play a fundamental role. In the case of carbon beside the inorganic forms (CO_2) present in the atmosphere and in the ocean (CO_2, HCO_3-, CO_{2-}) photosynthesis converts CO_2 in organic compounds subtracting carbon dioxide to the atmosphere and producing oxygen while respiration does the opposite. The slight imbalance between these two processes could not justify the global oxygen balance because of the limited amount of carbon in the biosphere. The obvious candidate is a different cycle where the pyrite mineral (FeS_2) and organic carbon (C) are brought to the surface crust by tectonic processes and volcanism. These are oxidized to ematite (Fe_2O_3) sulfuric

G. Visconti (ed.), *Climate, Planetary and Evolutionary Sciences*,
https://doi.org/10.1007/978-3-030-74713-8_9

acid (H_2SO_4) and CO_2. When brought down to the oceanic sediments and under bacterial action (biology) these compounds are converted in organic material and oxygen. Roughly 2 molecules of ematite and 4 molecules of sulfuric acid produce 15 molecules of O_2 and 4 molecules of pyrite. The lifetime for these processes is several million years and the tectonic cycle takes about 100 million years of turnover time. Plate tectonics is then a fundamental process for the presence of life on Earth. Recently a link has been proposed between the build up of continents and the rise of oxygen.

The rather recent discovery of the snowball Earth and the following period of global deglaciation introduced a new division in the geological chronology. The period going from 720 to 635 Ma was named cryogenian and the following period up to the beginning of Cambrian (541 Ma) takes the name of Ediacaran from Ediacara Hills in South Australia. The biota in this period suggests that during this time the global ocean switched from being in an anoxic state to an oxygenated state that was similar to the levels we see today. This episode follows the great oxidation events of 2 billion years ago. Data also show in this period an increase in the total carbon and sulfur throughput in the Earth system that increased the rate of organic carbon and pyrite sulfur burial and hence atmospheric O_2. It is presumed that oxygen increased by roughly 50% during the Ediacaran Period reaching ~0.25 of the present atmospheric level (PAL). This increase is consistent with the estimated requirement of oxygen levels necessary for the large, mobile and predatory animals appearing during the Cambrian explosion. One of the major hurdles of geobiology is that the evidence of past changes is not preserved because the most recent changes may cancel the previous records.

It is quite evident the importance of geobiology in connection with the origin of life on our planet. Traces of early life are found in strata and must be interpreted in terms of the environmental conditions present at that time. For example, doubts remain on the nature (animals or plants) of some samples of the Ediacaran period. The study of the persistence of life in extreme conditions (like the so called 'extremophiles') is of great importance in the attempts to find life on other planets like the oceans of Europe (the Jovian satellite) or the dunes of Mars.

Machine-Generated Summaries

Keywords: *earth, process, environment, organic, geobiofacie, tectonic, difference, life, study, system, metal, clearly, surface, term, source*

The Isotopic Imprint of Life on an Evolving Planet

https://doi.org/10.1007/s11214-020-00730-6

Abstract-Summary

Stable isotope compositions of biologically cycled elements encode information about the interaction between life and environment.

Extracting biological information from stable isotopic compositions requires untangling the interconnected nature of the Earth's biogeochemical system, and must be viewed through the lens of evolving metabolisms on an evolving planet.

We discuss the isotope biogeochemistry of the biologically essential elements carbon, nitrogen and sulfur, and we summarize their distribution on the modern Earth as an interconnected network of isotopically fractionated reservoirs with contrasting residence times.

Extended

We discuss the isotopic imprint of life on materials found on Earth.

Introduction

Enzymatically catalyzed chemical reactions are often associated with (relatively) large isotope effects, whereby a reaction product is enriched or depleted in the rare isotope of a given element compared with the substrate by tens of permil.

In chemical reactions, a kinetic isotope effect reflects the difference in energy required to break a bond when it involves a heavy vs. a light isotope of the same element.

Kinetic isotope effects are usually negative in both the forward and reverse directions for a given chemical reaction, though there are apparent exceptions (e.g., Casciotti [1]).

The characteristic timescale of a feature may be imposed by an external 'forcing' on the biogeochemical cycle (such as temperature), by an independent process (such as biological evolution) driving a secular change in the factors controlling fluxes or isotopic fractionations between these pools, or by dynamics inherited from a pool with the appropriate residence time.

Carbon

The amount of carbon stored as carbonate rocks, terrestrial sediments, and as fossilised organic material in the continental and oceanic crust exceeds that of the deep ocean by 3 orders of magnitude (Ciais et al. [2]).

The deep ocean carbonate carbon isotope record, generated by targeting benthic species of foraminifera, is rich in environmental information and low in noise due to the relative homogeneity and low energy of the environment (Zachos et al. [3]).

These changes are reflected in the carbon isotopic composition of gaseous CO_2, possibly due to a smaller glacial terrestrial biosphere and a polar ocean with depressed productivity (Marino et al. [4]).

Atmospheric pCO_2 can be estimated from the carbon isotope compositions of alkenones, a set of lipids that are specific to a single known family of coccolithophorid algae, in ocean sediments (Popp et al. [5]; Pagani et al. [6, 7]).

Nitrogen

On Earth's surface, after the atmospheric reservoir of N_2 (0.7809 mol fraction) and its dissolved counterpart in the global ocean, the most abundant form of N is organic N buried in ocean sediments and sedimentary rocks (Bebout et al. [8]).

As with carbon, the extent to which nitrogen isotope effects associated with biological processes are expressed in the environment depends on the relative fraction of the N pool that is transformed or consumed.

Abiotic sources of N to the early biosphere (e.g., from lightning or hydrothermal sources, as described above) would have been limited (Navarro-González et al. [9]), requiring biological N_2 fixation to have evolved very early in Earth history in order to keep pace with the N requirements of a rapidly evolving biosphere.

Following the GOE, a hypothesized oxygen overshoot associated with the Lomagundi-Jatuli carbon isotope excursions could have increased the availability of nitrate, leading to enhanced denitrification rates and widespread loss of fixed N (Kump et al. [10]).

Sulfur

A diverse set of enzymatically-catalyzed metabolic processes interconvert sulfur compounds among redox states (e.g., see review in Jørgensen et al. [11]).

Since oxidation of H_2S is not associated with a large fractionation, repeated cycles of the production of H_2S via disproportionation, its subsequent oxidation to [34]S-depleted sulfate, and ultimate reduction of this sulfate to even more [34]S-depleted H2S have been invoked to explain large sulfate-sulfide sulfur isotope fractionations that have sometimes been observed in nature (Canfield and Teske [12]).

The sulfate–water oxygen isotope fractionations associated with a subset of sulfur metabolisms, including DSR (e.g., Böttcher et al. [13]; Brunner and Bernasconi [14]; Turchyn et al. [15]; Wortmann et al. [16]), elemental sulfur disproportionation (e.g., Böttcher et al. [17, 18]), and pyrite oxidation (Balci et al. [19]) have been studied in laboratory cultures and natural environments.

Experiments, observations from natural environments and models all suggest that isotopic fractionation during DSR may be large even when sulfate concentrations are low (Canfield et al. [20]; Crowe et al. [21]; Wing and Halevy [22]), casting doubt on the inference of sulfate-limited microbial fractionation of sulfur isotopes prior to the oxygenation of the atmosphere.

Synthesis: Carbon, Nitrogen, and Sulfur Cycles in the Absence of Life

Like the nitrogen cycle, the sulfur cycle is buffered against sub-Myr changes by a large reservoir with a long residence time (i.e., the atmospheric N_2 and marine sulfate pools).

Unlike both C and N cycles, however, in the modern (oxidized) world, the principal metabolic reaction of the S cycle that introduces reduced sulfur to the sedimentary record—dissimilatory sulfate reduction—largely occurs in depositional environments that are in inconsistent communication with the global substrate pool.

Under an anoxic atmosphere, which is expected to produce S-MIF, the absence of this isotopic anomaly in sedimentary sulfur repositories (e.g., sulfate evaporites, sedimentary pyrite) may be indicative of intense microbial sulfur cycling.

Synthesizing the above discussion, in the absence of life, across the C, N and S cycles, we might expect co-localized pools to be dominantly either reduced or oxidized because their simultaneous existence is unstable.

Acknowledgement
A machine generated summary based on the work of Lloyd, M. K.; McClelland, H. L. O.; Antler, G.; Bradley, A. S.; Halevy, I.; Junium, C. K.; Wankel, S. D.; Zerkle, A. L. (2020 in Space Science Reviews).

Geobiology of a Microbial Endolithic Community in the Yellowstone Geothermal Environment

https://doi.org/10.1038/nature03447

Abstract-Summary
In extreme environments, these 'endolithic' communities are often the main form of life.

Extreme is certainly the word for the pore space in the extremely acid (pH 1) rocks in the Yellowstone National Park geothermal environment, but a microbial community has now been discovered there, made up mainly of photosynthetic algae and large numbers of previously unknown Mycobacterium species.

An interesting aspect of this discovery is the possibility that such communities can deposit biosignatures in the geologic record, and provide clues about ancient life associated with geothermal environments.

Main
Although endolithic growth of red algae is known in other volcanic areas [23], this is the first comprehensive molecular analysis of the microbial diversity and composition of these unique communities and their potential mineralization and fossilization.

Similar fabrics are also observed in mineralized and recently fossilized microbial communities in other thermal spring environments [24–27].

The abundance of mycobacterial clones ($\sim$37%) detected in the community leads us to postulate that the ubiquitous curvilinear filamentous casts observed formed around a biofabric of primarily mycobacteria, although other organisms detected are probably also involved.

Mineralization of endolithic communities associated with geothermal environments might lead to the preservation of identifiable biosignatures.

We expect that the abundance of Yellowstone endolithic communities in environments favourable to mineralization leads to their preservation in the geological record.

Methods

Rock pH was measured as described [28].

Molecular phylogenetic community rRNA analysis was performed with DNA extracted from crushed rock samples as described [29].

BSE–SEM was performed as described [30, 31] with an FEI Quanta 600 SEM, BSE detector and Princeton Gamma Tech EDS system.

Photosynthetic pigments extracted from rock samples were analysed as described [32].

Acknowledgement

A machine generated summary based on the work of Walker, Jeffrey J.; Spear, John R.; Pace, Norman R. (2005 in Nature).

Breathing Metals as a Way of Life: Geobiology in Action

https://doi.org/10.1023/A:1020518818647

Abstract-Summary

Many microbes have the ability to reduce transition metals, coupling this reduction to the oxidation of energy sources in a dissimilatory fashion.

Many of the dissimilatory metal reducing bacteria (DMRB) also reduce other metals, including toxic metals like Cr(VI), and radioactive contaminants like U(VI), raising the expectations that these processes can be used for bioremediation.

The processes involved in metal reduction remain mysterious, and often progress is slow, as nearly all iron and manganese oxides are solids, which offer particular challenges with regard to imaging and chemical measurements.

As these studies progress, it should be possible to separate several processes involved with metal reduction, including surface recognition, attachment, metal destabilization and reduction, and secondary mineral formation.

Acknowledgement

A machine generated summary based on the work of Nealson, Kenneth H.; Belz, Andrea; McKee, Brent (2002 in Antonie van Leeuwenhoek).

Precambrian Geobiology

https://doi.org/10.1134/S0031030106100030

Abstract-Summary

The appearance of Bacteria sensu lato, Eukaryota, Metaphyta, Metazoa, etc., along with the oxygenization of the atmosphere, are shown to have occurred much earlier than was previously assumed.

Acknowledgement

A machine generated summary based on the work of Rozanov, A. Yu. (2006 in Paleontological Journal).

Geomicrobiology of Blood Falls: An Iron-Rich Saline Discharge at the Terminus of the Taylor Glacier, Antarctica

https://doi.org/10.1007/s10498-004-2259-x

Abstract-Summary

Blood Falls, a saline subglacial discharge from the Taylor Glacier, Antarctica provides an example of the diverse physical and chemical niches available for life in the polar desert of the McMurdo Dry Valleys.

Blood Falls influences the geochemistry of Lake Bonney, and provides organic carbon and viable microbes to the lake system.

The novel geological evolution of this subglacial environment makes Blood Falls an important site for the study of metabolic strategies in subglacial environments and the impact of subglacial efflux on associated lake ecosystems.

Acknowledgement

A machine generated summary based on the work of Mikucki, Jill A.; Foreman, Christine M.; Sattler, Birgit; Berry Lyons, W.; Priscu, John C. (2004 in Aquatic Geochemistry).

On the Geobiological Evaluation of Hydrocarbon Source Rocks

https://doi.org/10.1007/s11707-007-0041-2

Abstract-Summary
These variations throw constraints on the application of the conventional inversion evaluation of hydrocarbon potential by assessing the residual organic matter left in source rocks.

Geobiology, probing the interaction between the life system and the earth system, provides new principles in deciphering the whole dynamic processes related to the organic evolution history from living biomass to organic burial.

Geobiofacies, newly proposed herein, is terminologized to define the geobiological dynamic processes through the combination of biofacies with organic facies and sedimentary facies, and expressed by the biohabitat types, paleoproductivity, depositional and preserved organics.

Acknowledgement
A machine generated summary based on the work of Shucheng, Xie; Yin, Hongfu; Xie, Xinong; Qin, Jianzhong; Hu, Chaoyong; Yan, Jiaxin; Huang, Junhua; Zhou, Lian; Yang, Xianghua; Wang, Yongbiao; Xu, Sihuang (2007 in Frontiers of Earth Science).

Discussion On Geobiology, Biogeology and Geobiofacies

https://doi.org/10.1007/s11430-008-0120-6

Abstract-Summary
We first discuss the definition of and the difference between geobiology and biogeology following a brief introduction of recent geobiology research in China.

Geobiology is defined as an interdisciplinary study of life sciences and earth sciences, and biogeology as an interdisciplinary study of biology and geology.

We then propose the term geobiofacies, defined as the facies of a geologic body embodying the whole process of interaction between organisms and environments.

Differences among geobiofacies, biofacies, and organic facies are discussed.

Acknowledgement
A machine generated summary based on the work of Yin, HongFu; Xie, ShuCheng; Qing, JianZhong; Yan, JiaXin; Luo, GenMing (2008 in Science China Earth Sciences).

Tectonomicrobiology: A New Paradigm for Geobiological Research

https://doi.org/10.1007/s11430-017-9159-y

Abstract-Summary
There is a need to explicitly emphasize the biogeochemical processes performed by microorganisms associated with Earth's tectonic activities, especially under the framework of the modern theory of plate tectonics.

This explicit synergy should also foster better communications between solid Earth scientists and life scientists in terms of holistic Earth system dynamics at both tectonic and micro-scales.

Acknowledgement
A machine generated summary based on the work of Zhang, Chuanlun; Lin, Jian; Li, Sanzhong; Dong, Hailiang; Wang, Fengping; Xie, Shucheng (2018 in Science China Earth Sciences).

Progress and Perspective on Frontiers of Geobiology

https://doi.org/10.1007/s11430-013-4731-1

Abstract-Summary
Geobiology is a new discipline on the crossing interface between earth science and life science, and aims to understand the interaction and co-evolution between organisms and environments.

On the basis of the latest international achievements, the new data presented in the Beijing geobiology forum sponsored by Chinese Academy of Sciences in 2013, and the papers in this special issue, here we present an overview of the progress and perspectives on three important frontiers, including geobiology of the critical periods in Earth history, geomicrobes and their responses and feedbacks to global environmental changes, and geobiology in extreme environments.

Knowledge is greatly improved about the close relationship of some significant biotic events such as origin, radiation, extinction, and recovery of organisms with the deep Earth processes and the resultant environmental processes among oceans, land, and atmosphere in the critical periods, although the specific dynamics of the co-evolution between ancient life and paleoenvironments is still largely unknown.

Microbes of potential geobiology significance were found and isolated from some extreme environments with their biological properties partly understood, but little is known about their geobiological functions to change Earth environments.

The biotic processes to alter or modify the environments are thus proposed to be the very issue geobiology aims to decipher in the future.

Acknowledgement
A machine generated summary based on the work of Xie, ShuCheng; Yin, HongFu (2013 in Science China Earth Sciences).

Book Reading List

Advances in Stromatolite Geobiology

by Reitner, J., Quéric, N., Arp, G. (*2011*).

Stromatolites are the most intriguing geobiological structures of the entire earth history since the beginning of the fossil record in the Archaean. Stromatolites and microbialites are interpreted as biosedimentological remains of biofilms and microbial mats. These structures are important environmental and evolutionary archives which give us information about ancient habitats, biodiversity, and evolution of complex benthic ecosystems.

Please see https://www.springer.com/gp/book/9783642104145 for original source.

Encyclopedia of Geobiology

by Reitner, J. (Ed), Thiel, V. (Ed) (2011)

The interplay between Geology and Biology has shaped the Earth from the early Precambrian, 4 billion years ago. Moving beyond the borders of the classical core disciplines, Geobiology strives to identify cause-and-effect chains and synergisms between the geo- and the biospheres that have been driving evolution of life in modern and ancient environments. Combining modern methods, geobiological information can be extracted not only from visible remains of organisms, but also from organic molecules, rock fabrics, minerals, isotopes and other tracers.

Please see https://www.springer.com/gp/book/9781402092138 for original source.

Geobiology

by Noffke, N. (2010)

A murmur is heard from the depths of time. Life and Earth are engaged in a dialog that has lasted for four billion years. Sometimes it's a whisper, sometimes a roar. One part sometimes gets the upper hand, dominates the discussion and sets the agenda. But mostly the two have some kind of mutual understanding, and the murmur goes on. Most of us don't listen. Nora does. She listens, and she tries to understand. Nora Noffke has focused her scientific career on the interaction between the living and the non-living.

Please see https://www.springer.com/gp/book/9783642127717 for original source.

Magnetoreception and Magnetosomes in Bacteria

by Schüler, D. (Ed) (2007)

Magnetoreception or magnetotaxis in bacteria was discovered only some 30 years ago. All magnetotactic bacteria, which occur in many environments and display a remarkable diversity, synthesize magnetosomes, complex intracellular organelles that contain magnetic iron crystals.

Recent developments in the research on magnetotactic bacteria are presented in this volume.

Please see https://www.springer.com/gp/book/9783540374671 for original source.

Neoproterozoic Geobiology and Paleobiology

by Xiao, S. (Ed), Kaufman, A. J. (Ed) (2006)

The Neoproterozoic Era (1000–542 million years ago) is a geological period of dramatic climatic change and important evolutionary innovations. Repeated glaciations of unusual magnitude occurred throughout this tumultuous interval, and various eukaryotic clades independently achieved multicellularity, becoming more complex, abundant, and diverse at its termination. Animals made their first debut in the Neoproterozoic too. The intricate interaction among these geological and biological events is a centrepiece of Earth system history, and has been the focus of geobiological investigations in recent decades.

Please see https://www.springer.com/gp/book/9781402052019 for original source.

New Prospects in Environmental Geosciences and Hydrogeosciences

by Chenchouni, H. (Ed), Chaminé, H. I. (Ed), Khan, M. F. (Ed), Merkel, B. J. (Ed), Zhang, Z. (Ed), Li, P. (Ed), Kallel, A. (Ed), Khélifi, N. (Ed) (2021).

This edited book gives a general overview on current research, focusing on geoenvironmental issues and challenges in hydrogeosciences in model regions in Asia, Europe, and America, with a focus on the Middle East and Mediterranean region and surrounding areas. This proceedings book is based on the accepted papers for oral/poster presentations at the 2nd Springer Conference of the Arabian Journal of Geosciences (CAJG-2), Tunisia 2019. It offers a broad range of recent studies that discuss the latest advances in geoenvironmental and hydrogeosciences from diverse backgrounds including climate change, geoecology, biogeochemistry, water resources management, and environmental monitoring and assessment.

Please see https://www.springer.com/gp/book/9783030725426 for original source.

Topics in Geobiology

by Landman, N. H. (Ed), Harries, P. J. (Ed).

The Topics in Geobiology series covers the broad discipline of geobiology that is devoted to documenting life history of the Earth. A critical theme inherent in addressing this issue and one that is at the heart of the series is the interplay between the history of life and the changing environment. The series aims for high quality, scholarly volumes of original research as well as broad reviews.

Please see https://www.springer.com/series/6623 for original source.

The Geobiology and Ecology of Metasequoia

by LePage, B. A. (Ed), Williams, C. J. (Ed), Yang, H. (Ed) (2005).

The plant fossil record provides evidence that the genus Metasequoia was widely distributed and experienced a wide range of climatic and environmental conditions throughout the Northern Hemisphere from the early Late Cretaceous to the Plio-Pleistocene. Today the genus is limited to one species with approximately 5,000 mature individuals growing in the Xiahoe Valley in southeastern China. This book is a distillation of the collective efforts and results of the world's Metasequoia specialists and enthusiasts.

Please see https://www.springer.com/gp/book/9781402026317 for original source.

References

1. Casciotti KL (2009) Inverse kinetic isotope fractionation during bacterial nitrite oxidation. Geochim Cosmochim Acta 73(7):2061–2076
2. Ciais P, Sabine C, Bala G, Bopp L, Brovkin V, Canadell J, Chhabra A, DeFries R, Galloway J, Heimann M, Jones C, Quéré CL, Myneni R, Piao S, Thornton P (2013), Carbon and other biogeochemical cycles. In: Climate change 2013 the physical science basis: working group I contribution to the fifth assessment report of the intergovernmental panel on climate change, vol 9781107057, pp 465–570
3. Zachos J, Pagani H, Sloan L, Thomas E, Billups K (2001) Trends, rhythms, and aberrations in global climate 65 Ma to present. Science 292(5517):686–693
4. Marino BD, McElroy MB, Salawitch RJ, Spaulding WG (1992) Glacial-to-interglacial variations in the carbon isotopic composition of atmospheric CO_2. Nature 357(6378):461–466
5. Popp BN, Laws EA, Bidigare RR, Dore JE, Hanson KL, Wakeham SG (1998) Effect of phytoplankton cell geometry on carbon isotopic fractionation. Geochim Cosmochim Acta 62(1):69–77
6. Pagani M, Zachos JC, Freeman KH, Tipple B, Bohaty S (2005) Atmospheric science: marked decline in atmospheric carbon dioxide concentrations during the Paleogene. Science 309(5734):600–603
7. Pagani M, Huber M, Liu Z, Bohaty SM, Henderiks J, Sijp W, Krishnan S, DeConto RM (2011) The role of carbon dioxide during the onset of Antarctic glaciation. Science 334(6060):1261–1264
8. Bebout GE, Fogel ML, Cartigny P (2013) Nitrogen: highly volatile yet surprisingly compatible. Elements 9(5):333–338
9. Navarro-González R, McKay CP, Mvondo DN (2001) A possible nitrogen crisis for archaean life due to reduced nitrogen fixation by lightning. Nature 412(6842):61–64
10. Kump LR, Junium C, Arthur MA, Brasier A, Fallick A, Melezhik V, Lepland A, Črne AE, Luo G (2011) Isotopic evidence for massive oxidation of organic matter following the great oxidation event. Science 334(6063):1694–1696
11. Jørgensen BB, Findlay AJ, Pellerin A (2019) The biogeochemical sulfur cycle of marine sediments. Front Microbiol 10:849
12. Canfield DE, Teske A (1996) Late proterozoic rise in atmospheric oxygen concentration inferred from phylogenetic and sulphur-isotope studies. Nature 382(6587):127–132
13. Böttcher ME, Oelschläger B, Höpner T, Brumsack H-J, Rullkötter J (1998) Sulfate reduction related to the early diagenetic degradation of organic matter and "black spot" formation in tidal sandflats of the German Wadden Sea (southern North Sea): stable isotope (13C, 34S, 18O) and other geochemical results. Org Geochem 29(5–7):1517–1530

14. Brunner B, Bernasconi SM (2005) A revised isotope fractionation model for dissimilatory sulfate reduction in sulfate reducing bacteria. Geochim Cosmochim Acta 69(20):4759–4771
15. Turchyn AV, Sivan O, Schrag D (2006) Oxygen isotopic composition of sulfate in deep sea pore fluid: evidence for rapid sulfur cycling. Geochim Cosmochim Acta 70(18):A660
16. Wortmann UG, Chernyavsky B, Bernasconi SM, Brunner B, Böttcher ME, Swart PK (2007) Oxygen isotope biogeochemistry of pore water sulfate in the deep biosphere: Dominance of isotope exchange reactions with ambient water during microbial sulfate reduction (ODP Site 1130). Geochim Cosmochim Acta 71(17):4221–4232
17. Böttcher ME, Thamdrup B, Vennemann T (2001) Oxygen and sulfur isotope fractionation during anaerobic bacterial disproportionation of elemental sulfur. Geochim Cosmochim Acta 65(10):1601–1609
18. Böttcher ME, Thamdrup B, Gehre M, Theune A (2005) 34S/32S and 18O/16O fractionation during sulfur disproportionation by desulfobulbus propionicus. Geomicrobiol J 22(5):219–226
19. Balci N, Shanks WC, Mayer B, Mandernack KW (2007) Oxygen and sulfur isotope systematics of sulfate produced by bacterial and abiotic oxidation of pyrite. Geochim Cosmochim Acta 71(15):3796–3811
20. Canfield DE, Farquhar J, Zerkle AL (2010) High isotope fractionations during sulfate reduction in a low-sulfate euxinic ocean analog. Geology 38(5):415–418
21. Crowe SA, Paris G, Katsev S, Jones C, Kim S-T, Zerkle AL, Nomosatryo S, Fowle DA, Adkins JF, Sessions AL, Farquhar J, Canfield DE (2014) Sulfate was a trace constituent of Archean seawater. Science 346(6210):735–739
22. Wing BA, Halevy I (2014) Intracellular metabolite levels shape sulfur isotope fractionation during microbial sulfate respiration. Proc Natl Acad Sci USA 111(51):18116–18125
23. Gross W, Kuver J, Tischendorf G, Bouchaala N, Busch W (1998) Cryptoendolithic growth of the red alga *Galdieria sulphuraria* in volcanic areas. Eur J Phycol 33:25–31
24. Hofmann BA, Farmer JD (2000) Filamentous fabrics in low-temperature mineral assemblages: are they fossil biomarkers? Implications for the search for a subsurface fossil record on the early Earth and Mars. Planet Space Sci 48:1077–1086
25. Cady SL, Farmer JD (1996) In: Bock GR, Goode JA (eds) Evolution of hydrothermal ecosystems on earth (and Mars?). Wiley, New York, pp 150–173
26. Blank CE, Cady SL, Pace NR (2002) Microbial composition of near-boiling silica-depositing thermal springs throughout Yellowstone National Park. Appl Environ Microbiol 68:5123–5135
27. Zierenberg RA, Schiffman P (1990) Microbial control of silver mineralization at a sea-floor hydrothermal site on the northern Gorda Ridge. Nature 348:155–157
28. Doemel WN, Brock TD (1971) pH of very acid soils. Nature 229:574
29. Spear JR, Walker JJ, McCullem TM, Pace NR (2005) Hydrogen and bioenergetics in the Yellowstone geothermal ecosystem. Proc Natl Acad Sci USA 102:2555–2560
30. Ascaso C, Wierzchos J (2003) The search for biomarkers and microbial fossils in Antarctic rock microhabitats. Geomicrobiol J 20:439–450
31. Wierzchos J, Ascaso C (1994) Application of backscattered electron imaging to the study of the lichen rock interface. J Microsc 175:54–59
32. Frigaard NU, Larsen KL, Cox RP (1996) Spectrochromatography of photosynthetic pigments as a fingerprinting technique for microbial phototrophs. FEMS Microbiol Ecol 20:69–77

Chapter 10
The Fermi Paradox

Introduction by Guido Visconti

In a 1989 comic strip Calvin (of Calvin and Hobbes) makes the following consideration: "*Sometimes I think the surest sign that intelligent life exists elsewhere in the universe is that none of it has tried to contact us*". In a sense this is another way to solve the so-called Fermi Paradox. In the summer of 1950 during lunch Fermi was having a conversation about aliens (remember that the fifties were the period of the UFO fever in the US) and at some point he asked why, if they existed, there were no visitors from outer space. This very loosely question was since then considered as the *Fermi paradox*. In principle there are plenty of reasons to ask such a question. The dimensions of our galaxy are around 10^5 light years and if we assume an average velocity of c/1000 (with c velocity of light) it would take 10^8 years to transverse the Galaxy, which is a short time with respect to the age of our Galaxy (10^{10} years). Such a very simple calculation, typical of the Fermi approach, indicates that our solar system should have been visited (or contacted) several times during its existence by extraterrestrial intelligence. Another important point is that our solar system has an age of roughly 4.5 billion years so that is relatively young with respect to the age of the universe.

Again this is quite a typical academic question ("where are they?") and has generated countless papers and discussions (including the Calvin hypothesis). Answers to the question typically fall into three groups: (a) *we are alone*, because intelligence is very rare to develop and may be short-lived, (b) *aliens exist but not here*, due to limitations of our contact technology, very unlikely temporal synchronization and so on; (c) *they are here, but are no interacting* by any means. Calvin's idea falls in the latter category but seriously it refers to the so-called zoo hypothesis formulated in 1973 by John Ball. It assumes that intelligent life exists in the universe but intentionally avoids communicating with us to avoid cosmic contamination. Fermi paradox sometimes is also referred as "The great silence" because the projects to have contact through electromagnetic eavesdropping like SETI (Search for Extra Terrestrial Intelligence) have all failed.

Of particular interest, one of the explanations is based on the concept of phase transitions. Following this idea it is assumed that life is regulated by the Gamma-ray burst (GRB). These episodes, which have been observed in distant galaxies, can release immense quantities of energy (of the order of the conversion of a stellar mass) and destroy any form of life in the galaxy. GRB tends to decrease in frequency with time to a point when their period becomes of the same order of the time it takes for the evolution of intelligence. This time, following what happened on Earth, is of the order of 10^8 years. This model suggests that a galaxy undergoes a phase transition between an equilibrium state devoid of intelligent life to a different equilibrium state where it is full of intelligent life. It is obvious that contacts are possible only in coincidence with the latter state.

A more stringent version, of the Fermi Paradox, has been formulated by Ken Olum in that he refers not to visitors from our galaxy but rather to technological civilizations that could spread among galaxies. He claims that considering the inflation hypothesis our civilization could be part of this very large colony. However, the same conclusion is in contradiction with the anthropic principle. This view seems to be too optimistic considering the difficulties of intergalactic travel (Andromeda the nearest galaxy is 2.5 million light years away) and as someone has noted he could be a premature hypothesis.

The most popular explanation of the Fermi paradox is the so-called Great Filter (GF). This GF could be something happening during the lifetime of an intelligent life that precludes any possibility of further developments. For planets in our solar system (like Mars) the transition from bacterial life to more complex forms could have been prevented by a change in climate or for Earth an ecological catastrophe or a nuclear war could annihilate civilization. This implies that it is of primary importance to know if the GF has already happened or it is still to come. If we would discover traces of life on other planets (like Mars or some exoplanet) this would imply that life is not such an improbable eventuality and so we must expect that the GF lies ahead of us. On the other hand, if we continue to find no evidence of the presence of life this would mean that its origin is a very improbable event and we could be really alone in the universe.

The Fermi paradox is more an excuse to discuss its scientific and philosophical implications and surprisingly is based on very sparse and incomplete observations: it is academic stuff at its best!

Machine-Generated Summaries

Keywords: *astrobiology, civilization, universe, system, planet, infinity, solar, study, quantum, sun, apparent, planetary, olum, consequence, principle.*

Astrobiological Phase Transition: Towards Resolution of Fermi's Paradox

https://doi.org/10.1007/s11084-008-9149-y

Abstract-Summary

Based on the idea of James Annis, we develop a model of an astrobiological phase transition of the Milky Way, based on the concept of the global regulation mechanism(s).

Secular evolution of regulation mechanisms leads to the brief epoch of phase transition: from an essentially dead place, with pockets of low-complexity life restricted to planetary surfaces, it will, on a short (Fermi–Hart) timescale, become filled with high-complexity life.

An observation selection effect explains why we are not, in spite of the very small prior probability, to be surprised at being located in that brief phase of disequilibrium.

We show that, although the phase-transition model may explain the "Great Silence", it is not supportive of the "contact pessimist" position.

The phase-transition model offers a rational motivation for continuation and extension of our present-day Search for ExtraTerrestrial Intelligence (SETI) endeavours.

Some of the unequivocal and testable predictions of our model include the decrease of extinction risk in the history of terrestrial life, the absence of any traces of Galactic societies significantly older than human society, complete lack of any extragalactic intelligent signals or phenomena, and the presence of ubiquitous low-complexity life in the Milky Way.

Extended

The phase-transition model proposed here has the potential to address (1), but it is unclear to what extent it help solving more general problem (2); that said, an attempt in this direction has been made by one of the present authors (Ćirković [1]).

Introduction: Fermi's Paradox

Relevant is the result of Lineweaver [2], see also Lineweaver et al. [3] that the difference between the median age of Earth-like planets in the Milky Way and the age of Earth is: Such a huge difference (and this is only the median age difference; in fact, to assess the validity of Fermi's paradox we ought to consider the oldest habitable planets where, presumably, the oldest technological civilizations emerged first) makes Fermi's question significantly more puzzling.

The most significant contribution, in this respect, has been the phase transition idea of James Annis [4], which is a prototype disequilibrium hypothesis: there is no Fermi's paradox, since the relevant timescale is the time elapsed since the last "reset" of astrobiological clocks and this can be substantially smaller than the age of the Milky Way.

We attempt to (a) generalize Annis' model to a general neocatastrophic astrobiological regulation, and (b) to present results of a simple numerical model of GHZ in this manner, and to show that they have the capacity to resolve Fermi's paradox.

Catastrophes and Phase Transition

Global catastrophic events affecting large parts of GHZ will tend to reset many local astrobiological clocks nearly simultaneously, thus significantly decreasing the probability of existence of extremely old civilizations, conforming to Annis' scenario.

It is taken that the chain of events leading to life and intelligence can be reduced by a catastrophic event at any planet in our toy-model Galaxy with probability Q, and its astrobiological clock is then reset.

The frequency of resetting events decreased due to the astrophysical evolution of the Galaxy (the key point of Annis' model), and at some time which may lie in our past (as drawn in the sketch) or it may conceivably still be in our future, the balance will shift toward high probability of complex intelligent observers emerging and creating large interstellar civilizations.

Predictions

Some of the specific predictions of the present models are: We shall not find any traces or remnants of intelligent societies much older than ours anywhere in the Galaxy; no "interstellar archaeology" (cf. Freeman and Lampton [5]) will ever become a meaningful discipline.

This is necessary to qualify with the realistic possibility of biological exchange between Earth and our Solar System's other planets; stronger prediction would be finding life biochemically unrelated to the existing terrestrial example.

Improved geological and paleontological techniques will find coincidences between most of the extinction events, with catastrophes of global, Galactic origin; that is, investigations of less eroded environments in our Solar System will give ample evidence of high-energy γ- and cosmic-ray bombardment episodes approximately coincident with some of the major known Earth-biosphere extinction events.

Further development of the APT model, in particular assembling a realistic total risk function for Earth-like planets, will enable further specific predictions.

Discussion

We have outlined a quantitative model of astrobiological evolution of the Milky Way which can avoid Fermi's paradox, based on the qualitative phase-transition hypothesis of Annis [4].

V. The phase transition model strongly suggests that technological development and interplanetary/interstellar space colonization should be the foremost priorities in humanity's global policy-making.

The phase-transition model proposed here has the potential to address (1), but it is unclear to what extent it help solving more general problem (2); that said, an attempt in this direction has been made by one of the present authors (Ćirković [1]).

We have shown that 1-D models, like the toy model of GHZ presented here, can in principle offer support to Annis' phase transition hypothesis for explanation of Fermi's paradox.

The toy 1-D model can serve to undermine Carter's argument (as shown in Ćirković et al. [6]), but a stronger class of models is necessary to eliminate Fermi's paradox.

Acknowledgments
A machine generated summary based on the work of Ćirković, Milan M.; Vukotić, Branislav (2008 in Origins of Life and Evolution of Biospheres).

Natural Intelligence and Anthropic Reasoning

https://doi.org/10.1007/s12304-020-09388-7

Abstract-Summary
To justify these arguments, the neural-type intelligence represented by the form of reasoning known as anthropic reasoning will be compared and contrasted with types of intelligence explicated by four disciplines of biology—relational biology, evolutionary epistemology, biosemiotics and the systems view of life—not biased towards neural intelligence.

To answer the questions I will rely on a range of established concepts including SETI (search for extraterrestrial intelligence), Fermi's paradox, bacterial cognition, versions of the panspermia theory, as well as some newly introduced concepts including biocivilisations, cognitive/semiotic universes, and the cognitive/semiotic multiverse.

The key point emerging from the answers is that the process of cognition/semiosis—the essence of natural intelligence—is a biological universal.

Introduction
The study will compare arguments from four disciplines of biology not biased towards neural intelligence—relational biology, evolutionary epistemology, biosemiotics and the systems view of life—with arguments that are rooted in and biased towards the neural-type intelligence, such as those used in the Anthropic Principle (AP) reasoning or anthropic reasoning (Carter [7]; Barrow and Tipler [8]).

I will use anthropic reasoning to arrive at the position, broadly consistent with life principle, but also with biosemiotics, that natural intelligence is a biological universal—organisms from bacteria to animals are cognitive or semiotic agents—and that the human-type intelligence is only a fraction in the wide spectrum of natural intelligence which may be equivalent to semiotic scaffolding: "a tightly wound web of checks and balances gradually establishing itself through myriads of semiotic interactions between organisms" (Hoffmeyer [9], p 154).

A Brief Overview of AP and Associated Concepts
All forms of AP assume that the minimum requirement for the true observing capacity is the human-type intelligence (neural intelligence), which eventually peaks in science and technology.

If we accept this assumption, three types of observers are possible in the universe (i) humanity bound to Earth (and on the verge of the cosmic adventure) possessing techno-science at the present state of development as a form of knowledge acquisition, (ii) humanity-like civilisations living somewhere else in the universe, and (iii) civilisations with the observing capacities superior to the capacities of human or human-like civilisations, living somewhere else in the universe.

It also follows that AP assumes two distinct territories of life, as we know it on the planet Earth: (i) intelligent life represented by Homo sapiens and its technology in the form of techno-science and (ii) all other forms of life (microorganisms, plants, and animals) considered either non-intelligent or not intelligent enough from the perspective of observing capacities.

Arguing Against the Key AP Assumption

The AP assumption can be shortened into the following statement: There are no true observers in the universe below the human-type observers.

Substituting the word anthropic with the word cognition (AP becomes CP) provides a shortcut towards defining the concepts of observation and true observers.

From the perspective of AP, humanity self-selects evidence from the existing pool of knowledge, best suited to describe our position in the world/universe, in line with the prevailing collective opinion.

To start developing the counter-argument one can ask questions aimed at probing the applicability of the observation process and the concept of true observers on reference classes other than Homo sapiens.

Are human observers the only observers within the pool of terrestrial life forms? (Q1). •

Are there true observers within the pool of terrestrial life forms amongst the reference classes of observers that are not human? (Q3). •

Do human observers and other observers and true observers share common features? (Q4).

The Concept of Cognition from the Perspective of Four Disciplines of Biology

The entire process is tested by the filter of natural selection—the filter tests various structure–function forms that emerge from the process of biological abstraction leading to functions such as vision, flight, natural computation, etc. Thus, the empirical world is emerging from the epistemological-ontological unity implicit in Rosen's model.

All organisms, from bacteria to animals, acquire knowledge about their environments through the process of natural learning based on a universal algorithm.

The key feature of living systems is not their composition, the nature of chemical constituents that make them up, or matter, but rather the pattern in which the matter is organized to produce various organismal forms (Capra and Luisi [10]).

According to Maturana and Varela every organism, from single-cell microbes to complex multicellular animals, is an autopoietic unit—a system that sustains itself due to the network pattern of organization, which allows constant self-regeneration within the boundary that separates the autopoietic unit from its environment.

Answering Q1–Q4

The first planets were formed much earlier than Earth (Lineweaver [11]), it follows that life, in the form of first fully functioning observers such as bacteria-like organisms, might have emerged early in the cosmic history at some first-generation Earth-like planet, following the emergence of the Earth-equivalents of the 'RNA world' and viruses as proto-observers.

As currently understood, the planet Earth was formed 4.5 BYA (billion years ago), leading soon after that to the emergence of the Earth-bound 'RNA world' and viruses as proto-observers, 4.1 BYA—3.8 BYA, paving the way for the emergence of first living organisms or observers.

The H-W thesis predicts that biogenic particles, including viruses and bacteria, are formed in the interior of comets and that there is a constant flow of biogenic particles from the cosmos towards Earth (Hoyle and Wickramasinghe [12]; Wickramasinghe et al. [13]; Wickramasinghe [14]).

Discussion

If the capacity of cognition/semiosis/observation is a biological universal, several new elements should be added to the AP concept.

The key conundrum of AP, known as the fine-tuning principle, according to which all physical constants of the universe are fine-tuned to the extent that a small change in parameters would mean that life, as we know it, would not emerge at all, is resolved through the concept of physical multiverse (Garriga and Vilenkin [15]; Koonin [16]; Tegmark [17]).

The human body is a cognitive/semiotic mini-multiverse: we are conglomerates of viruses, bacteria, archaea, protists, eukaryotic cells housing former bacteria (mitochondria) (McFall-Ngai et al. [18]), culminating in the emergence of the corporate body dominated by the neural-type intelligence and the consciousness (Slijepcevic [19]).

Is our consciousness falsely projecting the existence of the physical multiverse as a result of the composite nature of our cognitive/observational faculties that (i) originated at the dawn of life with bacteria and archaea and (ii) reflect the entire cognitive/semiotic multiverse that developed in the last 3.8 BY?

Acknowledgments

A machine generated summary based on the work of Slijepcevic, Predrag (2020 in Biosemiotics).

Comment on 'The Aestivation Hypothesis for Resolving Fermi's Paradox'

https://doi.org/10.1007/s10701-019-00289-5

Abstract-Summary
Sandberg et al. implicitly assume, however, that computer-generated entropy can only be disposed of by transferring it to the cosmological background.

While this assumption may apply in the distant future, our universe today contains vast reservoirs and other physical systems in non-maximal entropy states, and computer-generated entropy can be transferred to them at the adiabatic conversion rate of one bit of negentropy to erase one bit of error.

Introduction

Our main point is related to these quotes: The argument is that the thermodynamics of computation make the cost of a certain amount of computation proportional to the temperature ...As the universe cools down, one Joule of energy is worth proportionally more.

For the purposes of this paper we will separate the resources into energy resources that can power computations and matter resources that can be used to store information, process it ... We adopt their terminology, using "civilization" to refer to an arbitrarily technologically powerful agent in the universe, "reservoir" to refer to a bounded thermodynamic system (e.g., a battery) that can be manipulated by a civilization, and "bath" to refer to an effectively infinite thermalized system whose temperature is exogenously determined (e.g., the cosmic microwave background) [20].

Our Model

If work is supplied to the bit eraser, it pumps entropy from the memory tape to the bath.

Following Sandberg et al., we have assumed the key thermodynamic features of an expanding spacetime are fully captured by the thermal-bath temperature schedule $T(t)$ as a model for the CMB, and so are ignoring potentially important features like the increasing volume, decreasing pressure, and decreasing particle density of the real CMB as it evolves into the future.

By the second law, the entropy of the bath must increase by at least this amount because, by assumption, neither the reservoir nor the bit eraser have an appreciable number of internal states.

The conclusions of Sandberg et al. then follow: if $T(t)$ is decreasing with time, the civilization performs more total bit erasures before the work reservoir is exhausted (and so, by assumption, more total computations) if it waits until a later time when T is lower.

Critique of Assumption

The key feature necessary for the above conclusions is that the reservoir cannot accept an appreciable amount of entropy despite being able to do prodigious amounts of work.

This appears in Ref. [21] as an assumption that all of the reservoir's internal energy is available to do useful work, a conflation of the energy with the free energy.

The maximum number of bits erased is set by the difference between the total initial entropy of the two parts and their total final entropy when they have thermalized at the joint temperature set by conservation of energy.

As more random bits—generated by external computations—are swapped into the memory, one can keep erasing bits in the memory until all parts of the reservoir have equilibrated to the same temperature, i.e., the reservoir is at maximal entropy for its energy.

Reservoir-Bath Coupling

Let us assume that the reservoir has been exhausted (internally thermalized) as described in the previous section. (This situation differs from our initial toy model in that the reservoir, on its own, can do no useful work.) We suppose civilization desires to power further bit erasures by exploiting the temperature difference between the bath and the reservoir that is induced by the changing bath temperature $T(t)$, i.e., powering the bit eraser with heat flow between the reservoir and the bath.

This process—converting some of the bath into a new reservoir and then mining the temperature difference between reservoirs until they equilibrate—can be repeated for as long as there is any matter in the universe accessible to the civilization.

Only once the civilization has commandeered all accessible matter in the universe will they face the incentive to aestivate since they now must push bit-erasure entropy into the uncontrolled bath.

Final Comments

We have concluded that a civilizations capable of reversible computing has no incentive to aestivate until after it has taken control of, and fully exploited, all accessible matter in the universe.

The incentive to wait, or lack thereof, applies just as well to a civilization whose terminal desires involve expending work to move matter into particular configurations (e.g., galactic-scale art projects) whose limiting cost is residual frictions, analogous to residual computational faults necessitating error correction.

Acknowledgments

A machine generated summary based on the work of Bennett, Charles H.; Hanson, Robin; Riedel, C. Jess (2019 in Foundations of Physics).

The Sun and Exoplanets: The Solitude of Man

https://doi.org/10.3103/S0190271712010111

Abstract-Summary
Solar pulsations with a period of $P_0 = 9600.606(12)$ were discovered in 1974.

A more recent discovery is that planetary distances in the solar system are subject to spatial resonance with the parameter $L_0 \equiv cP_0 \approx 9600$ ls and that the P_0 pulsation itself has cosmological significance (coherent cosmic oscillation, or the pace of absolute time of the universe; c is the speed of light).

As of June 2011, 552 extrasolar planets have been discovered.

The scale L_0 indicates that the sun is a special quantum object, where L_0 is a wave function parameter that is not subject to the rational principles of the classical world, but rather follows a peculiar, quantum logic.

Acknowledgments
A machine generated summary based on the work of Kotov, V. A. (2012 in Bulletin of the Crimean Astrophysical Observatory).

Too Early? on the Apparent Conflict of Astrobiology and Cosmology

https://doi.org/10.1007/s10539-005-8305-2

Abstract-Summary
When coupled with naturalistic understanding of the origin of life and intelligence, which follows the basic tenets of astrobiology, and with some fairly incontroversial assumptions in the theory of observation selection effects, this infinity leads, as Ken Olum has recently shown, to a paradoxical conclusion.

Olum's paradox is related, to the famous Fermi's paradox in astrobiology and "SETI" studies.

This strategy has consequences of importance for both astrobiological studies and philosophy.

Acknowledgments
A machine generated summary based on the work of Ćirković, Milan M. (2006 in Biology & Philosophy).

Book Reading List

If the Universe Is Teeming with Aliens … Where Is Everybody?

by Webb, S. *(2015).*

Given the fact that there are perhaps 400 billion stars in our Galaxy alone, and perhaps 400 billion galaxies in the Universe, it stands to reason that somewhere out there, in the 14-billion-year-old cosmos, there is or once was a civilization at least as advanced as our own. The sheer enormity of the numbers almost demands that we accept the truth of this hypothesis. Why, then, have we encountered no evidence, no messages, no artifacts of these extraterrestrials?

Please see https://www.springer.com/gp/book/9783319132358 for original source.

Searching for Extraterrestrial Intelligence

by Shuch, H. P. *(2011)*.

This book is a collection of essays written by the very scientists and engineers who have led, and continue to lead, the scientific quest known as SETI, the search for extraterrestrial intelligence. Divided into three parts, the first section, 'The Spirit of SETI Past', written by the surviving pioneers of this then emerging discipline, reviews the major projects undertaken during the first 50 years of SETI science and the results of that research.

Please see https://www.springer.com/gp/book/9783642131950 for original source.

The Search for Extraterrestrial Intelligence

by Montebugnoli, S. (Ed), Melis, A. (Ed), Antonietti, N. (Ed) *(2021)*.

This book presents the latest knowledge of the newly discovered Earth-like exoplanets and reviews improvements in both radio and optical SETI. A key aim is to stimulate fresh discussion on algorithms that will be of high value in this extremely complicated search. Exoplanets resembling Earth could well be able to sustain life and support the evolution of technological civilizations, but to date, all searches for such life forms have proved fruitless.

Please see https://www.springer.com/gp/book/9783030638054 for original source.

The Search for Life Continued

by Jones, B. W. *(2008)*.

Barrie Jones addresses the question "are we alone?", which is one of the most frequently asked questions by scientists and non-scientists alike. In *The Search for Life Continued*, this question is addressed scientifically, and the author is not afraid to include speculation. Indeed, the author believes beyond reasonable doubt that we are not alone and this belief is based firmly on frontier science of the most imaginative kind.

Please see https://www.springer.com/gp/book/9780387765570 for original source.

The Cosmic Zoo

by Schulze-Makuch, D., Bains, W. *(2017)*.

Are humans a galactic oddity, or will complex life with human abilities develop on planets with environments that remain habitable for long enough? In a clear, jargon-free style, two leading researchers in the burgeoning field of astrobiology critically examine the major evolutionary steps that led us from the distant origins of life to the technologically advanced species we are today.

Please see https://www.springer.com/gp/book/9783319620442 for original source.

References

1. Ćirković MM (2006) Too early? On the apparent conflict of astrobiology and cosmology. Biol Philos 21:369–379
2. Lineweaver CH (2001) An estimate of the age distribution of terrestrial planets in the universe: quantifying metallicity as a selection effect. Icarus 151:307–313
3. Lineweaver CH, Fenner Y, Gibson BK (2004) The galactic habitable zone and the age distribution of complex life in the milky way. Science 303:59–62
4. Annis J (1999) An astrophysical explanation for the great silence. J Br Interplan Soc 52:19–22, (preprint astro-ph/9901322)
5. Freeman J, Lampton M (1975) Interstellar archaeology and the prevalence of intelligence. Icarus 25:368–369
6. Ćirković MM, Vukotić B, Dragićević I (2008) Galactic 'Punctuated Equilibrium': how to undermine carter's anthropic argument in astrobiology. Astrobiology (in press)
7. Carter, B. (1974). Large number coincidences and the anthropic principle in cosmology. In: Longair MS (ed) Confrontation of cosmological theories with observational data. International astronomical union/union Astronomique Internationale, vol 63. Springer, Dordrecht, pp 291–298
8. Barrow JD, Tipler FJ (1996) The anthropic cosmological principle. Oxford University Press, Oxford
9. Hoffmeyer J (2015) Introduction: semiotic scaffolding. Biosemiotics 8(2):153–158
10. Capra F, Luisi PL (2014) The systems view of life: a unifying vision. Cambridge University Press, New York
11. Lineweaver CH (2001) An estimate of the age distribution of terrestrial planets in the universe: quantifying metalicity as a selection effect. Icarus 151(2):307–313
12. Hoyle F, Wickramasinghe NC (1981) Evolution from space. Simon and Schuster, New York
13. Wickramasinghe J, Wickramasinghe NC, Napier B (2010) Comets and the origin of life. World Scientific, Singapore
14. Wickramasinghe NC (2017) Proofs that life is cosmic. World Scientific, Singapore
15. Garriga J, Vilenkin A (2001) Many worlds in one. Phys Rev D 64:43511
16. Koonin EV (2007) The cosmological model of eternal inflation and the transition from chance to biological evolution in the history of life. Biol Direct 2:15
17. Tegmark M (2009) The multiverse hierarchy. arXiv 0905.1283
18. McFall-Ngai M, Hadfield MG, Bosch TCG, Carey HV, Domazet-Lošo T, Douglas AE, Dubilier N, Eberl G, Fukami T, Gilbert SF, Hentschel U, King N, Kjelleberg S, Knoll AH, Kremer N, Mazmanian SK, Metcalf JL, Nealson K, Pierce NE, Rawls JF, Reid A, Ruby EG, Rumpho M, Sanders JG, Tautz D, Wernegreen JJ (2013) Animals in a bacterial world, a new imperative for the life sciences. Proc Natl Acad Sci U S A 110(9):3229–3236
19. Slijepcevic P (2018) Evolutionary epistemology: reviewing and reviving with new data the research programme for distributed biological intelligence. Biosystems 163:23–35
20. Bennett CH (1973) Logical reversibility of computation. IBM J Res Dev 17:525–532

21. Sandberg A, Armstrong S, Cirkovic M (2016) That is not dead which can eternal lie: the aestivation hypothesis for resolving Fermi's paradox. J Br Interplanet Soc 69:406–415. arXiv: 1705.03394

Chapter 11
The Gaia Hypothesis, Evolution and Ecology

Introduction by Guido Visconti

The Gaia hypothesis was introduced in the 1970s by James Lovelock and Lynn Margulis. The original idea proposed that near homeostatic conditions on Earth have been maintained "by and for the biosphere". A major justification for this approach was that the atmospheric composition for an anabiotic Earth would be quite different from the observed one. However, the authors did not provide details of how these calculations were made and on the basis of the biogeochemical cycles knowledge at that time (fifty years ago) it is quite dubious that those results represented reality.

Gaia hardly can be classified as a scientific theory. Following Karl Popper idea, a scientific theory is such that it can be disproved (falsified). Now one of the problems with Gaia is that it pretends to explain how some equilibrium is maintained (atmospheric composition, salinity, cloudiness, etc.) but does not offer predictions that can be falsified. It is more like the old fashioned inductive reasoning (observation, pattern, hypothesis, theory) rather than the present accepted view (theory, hypothesis, observation, confirmation). It is to note however that in Russia Vladimir Vernadsky in the 1930s had imagined a similar strong role (as Gaia) of the biosphere on the atmospheric composition.

One of the consequences of Gaia hypothesis was the origin of the Earth System Science approach that was initiated by NASA in 1983. In this case the Earth was considered as an integrated system where interactions between the different components were considered. Actually the program did not produce a series of dedicated satellites and was nothing more than a cultural approach.

Gaia raised a series of criticisms mainly on two different points. The first one was the lack of any detailed mechanism for the homeostatic processes and the other was related to the connection between the regulation mechanism and natural selection. As a matter of fact, the Gaia hypothesis suggests that the biological processes could evolve by natural selection although there is no evidence of that. Gaian processes are found at microbial scale but it is hard to imagine how these could be transferred to a global scale.

G. Visconti (ed.), *Climate, Planetary and Evolutionary Sciences*,
https://doi.org/10.1007/978-3-030-74713-8_11

To respond to such criticism very simple toy models were developed like the notorious Daisyworld. It is astonishing how so many papers were written on such a subject which had very few connections with reality. There was also some attempt to estimate quantitatively the role of the biosphere in the environmental regulation using the Maximum Entropy Production approach. The debate that followed was another typical academic endeavor to discuss nothing but the conclusions were again that Gaia did not satisfy the criteria of testable scientific theory. In a review paper by Free et al., the conclusion was "…. If many issues raised by the Gaia hypothesis are also considered by conventional evolutionary and ecological science, do we need the hypothesis at all?".

Nevertheless, some of the predictions of Gaia have been subjected to experimental tests and have failed. The most evident is the so-called CLAW (Charlson, Love-lock, Andreae, Warren) hypothesis. According to CLAW dimethyl sulfide (DMS) produced by marine organisms could be converted into cloud condensation nuclei (CCN) that could facilitate the clouds formation and then change the reflective prop-erties of the atmosphere. Because the DMS productivity increases with temperature and light, this would constitute a strong negative feedback mechanism. The original claims by CLAW reported that a 50% increase in DMS would compensate for the warming of a CO_2 doubling while the real data showed that a 1% increase in DMS would lead to 0.1% increase in CCN. Also over the past thirty years, observations in the marine boundary layer, laboratory studies, have been conducted seeking evidence for the CLAW hypothesis. The results indicate that a DMS biological control over cloud condensation nuclei probably does not exist and that sources of these nuclei in the marine boundary layer and the response of clouds to changes in aerosol are much more complex than was supposed by CLAW. As an example in regions which show some correlation between DMS and CCN the formation of the latter is contributed by many other processes.

In general, if we look at the geological record we see very tiny evidence of the presence of Gaia. In the most recent history ice ages (driven probably by the astronomical modulation of solar radiation) have determined a large impact of the biospheric activity and the climate changes have not been mitigated. On a longer time scale the snow ball Earth episode, half a billion years ago, has almost destroyed life on Earth and destroyed also the claim that Gaia maintains stable conditions for the biosphere. In the last century global warming has shown a constant increase of temperature due to anthropic perturbation of the carbon dioxide. We now understand better the role of microorganisms in the biogeochemical cycles but it is very hard to see that homeostatic processes are at work.

Machine-Generated Summaries

Keywords: *Earth, life, feedback, environment, system, earth system, gaian, organism, process, make, theory, benefit, biological, condition, one.*

The GAIA Theory: From Lovelock to Margulis. From a Homeostatic to a Cognitive Autopoietic Worldview

https://doi.org/10.1007/s12210-012-0187-z

Abstract-Summary

Since the end of 1980s, Lynn Margulis, Lovelock's longstanding co-author, proposed replacing Gaia's homeostatic nature with an autopoietic and evolutionary one that is connected to second-order cybernetic processes.

This included symbiogenetic processes concerning the birth and evolution of microbiotic organisms at the planetary level, which led to the construction of macroorganisms and their properties that stabilize the environment.

A close relationship between symbiogenetic and autopoietic theory (the latter proposed by Maturana and Varela in Autopoiesis and Cognition: the Realization of the Living, D. Reidel Publishing Co., Dordecht [1]) is represented by the fact that both theories refer primarily to the epigenetic–cytoplasmatic mechanisms in cellular constitution and evolution, and only secondarily to the established, DNA-mediated genetic code.

The second part of this work consists of more speculative comments about some important articulations of the Gaian construct, in particular: (1) The apparent lack of information on the chemical–physical nature of living and inert matter and on their possible interaction in the construction of organisms and environment. (2) The substantial weakness in the descriptive processes leading to the auto-organization of the two terrestrial matrices (organisms and environment) that is Lovelock's engineeristic and physiological automatisms without consciousness and Margulis' cognitive symbiogenetic processes operating at elementary matter.

The latter appears to privilege the theory of spontaneous and istantaneous cooperative phenomenon between elementary particles, at the base of the change from chaos to order and from one ordered state to another, both in physical and living realm. (3) Finally, the substantial underevaluation, operated by Lovelock in his holistic approach to the study of planet Earth, of the role played by the physical phenomenon of the distance interaction between quantum objects, leading to their entanglement.

Introduction and Gaia Scientific Background

The catastrophe theory (Thom [2]), applied in biology and geophysics, will allow Lovelock to further hypothesize a synchronic form of evolution "by jumps", and not necessarily a "diachronic" or continuous kind, which in the long term was suggested by Darwinian theory.

General systems theory (Boulding [3]) is a term that has come into use to describe a level of theoretical mathematical models linking specific theories of several different disciplines, such as cybernetics, information theory, organization theory, autopoietic theory, and so on, with applications to several fields.

As we will see, however, Lovelock has chosen an approach that requires minimal internal organization, postulating a low level of conflict between the biosphere and in general living components and environmental elements.

The most general principles of the scientific explanation of the natural world are: (1) quantum conception, (2) the second principle of thermodynamics and (3) self-organization in physical and nonphysical dynamical systems (Azzone [4]; Haken [5]).

The Birth of Gaia

In those years, James Lovelock was appointed by NASA to investigate the chemical composition and physical characteristics of the atmosphere surrounding Mars, in order to determine the planet's suitability for life.

Lovelock accepted the assignment, even though he was sure that there was no life on Mars.

In the mid-1960s, even before NASA sent the Mariner probe to observe the desolate surface of the planet from orbit, observations had been carried out from Earth with infrared telescopes, showing that the atmosphere on Mars, as on Venus, was dominated by carbon dioxide and only a small percentage of oxygen and nitrogen.

Life could have contributed to creating an atmosphere that is completely different and almost complementary to the ones on Mars and Venus, and predictably, with chemical, physical and thermal disequilibrium.

Lovelock does not seem to indicate whether the creation of a particular kind of atmosphere by life is due to chance or necessity.

The Formal Theory and Its Critics

The Gaia hypothesis presents a new view of the atmosphere, which is seen as a component part of the biosphere rather than a mere environment for life.

Gaia theory has been considered ill defined and difficult to test, mainly regarding the need for optimization of the environment by biota for its own benefit (Kirchner [6]; Kleidon [7]).

Kirchner [6] starts by defining a weak form of Gaia, which holds that life collectively has a significant effect on the Earth's environment (influential Gaia), and that therefore the evolution of life and of its environment are intertwined, with each affecting the other (co-evolutionary Gaia).

The stronger forms of Gaia depart from these traditions, claiming that the biosphere can be modeled as a single giant organism (geophysiological Gaia), or that life optimizes the physical and chemical environment to best meet the biosphere's needs (optimizing Gaia).

Daisyworld: Can a Mathematical Model Support a Theory?

Daisyworld is a model planet with black and white daisies with separate fitness curves, in which the black ones like it cooler and the white ones like it warmer.

As the sun heats up over 3.5 billion years, black daisies approach their optimum temperatures, become more fit and thereby increase their population, causing the albedo (reflectivity) of the planet's surface to drop.

Black daisies increase until the temperature surpasses their fitness peak and moves into the fitness range for white daisies.

In the same direction, Kirchner [6] notes that vegetation's response to temperature is weaker on the real Earth than on Daisyworld, where a temperature shift of only 1 °C can expand daisy cover from 0 to 45% over the planet's surface.

On the real Earth, vegetation would actually respond to temperatures in the opposite direction from what Daisyworld predicts; all else equal, warmer temperatures would expand forests poleward, making the surface darker and thus amplifying the warming (Kirchner [6]).

From Lovelock to Margulis, from First to Second Cybernetics

Margulis' biological/evolutionary approach is based on the fundamental role of primitive germs that appeared in the Archean and provoked life on Earth, which largely contributed to maintaining the physical–chemical atmospheric conditions that were suitable for life on the planet.

It is Margulis who introduced autopoietic theory most directly into Gaia science, to the extent of presenting Gaia as "the autopoietic planet where the biosphere as a whole is autopoietic, in the sense that it maintains itself.

In her enthusiastic openness to an autopoietic theory of life, Margulis [8] accepted an even more impressive notion of cognition as necessarily involved in the event of life, at the lowest level of life, such as isolated cells.

It is hard to affirm that Gaia theory has obtained substantial advantages from the opening to the autopoiesis hypothesis, beyond a few gains regarding the property of autonomy and perhaps evolution.

Final Considerations

Lovelock, mentor of the living Earth, did not pose, on the one hand, the problem of the emergence of life and what relationship this process could have had with inert nature, and on the other, if it could be possible to propose a global approach to study the entire planet on a physical basis today.

Lovelock knew that the major atomic physicists, Schrödinger [9], Bohr [10] and Heisenberg [11], agree that quantum principles have given the basis for the knowledge of the inert and living matter and precisely located analogies and differences between the emergence of the quantum event and of the phenomena of life.

Lovelock (and the same can be said for Margulis) does not seem to have found it necessary to introduce quantum theory in his description of the structure/function of the living planet, nor did he find possible analogies between the quantum world and life.

Acknowledgement

A machine generated summary based on the work of Onori, Luciano; Visconti, Guido (2012 in Rendiconti Lincei. Scienze Fisiche e Naturali).

Gaia and Natural Selection

https://doi.org/10.1038/28792

Abstract-Summary

Evidence indicates that the Earth self-regulates at a state that is tolerated by life, but why should the organisms that leave the most descendants be the ones that contribute to regulating their planetary environment?

Main

Organisms alter their material environment and their environment constrains and naturally selects organisms.

The Gaia theory [12] proposes that organisms contribute to self-regulating feedback mechanisms that have kept the Earth's surface environment stable and habitable for life.

Natural selection and Gaia pose a puzzle: how can self-regulation at the planetary level emerge from natural selection at the individual level [13]?

I attempt to address this question by focusing on the feedbacks to biospheric growth and selective pressures that can arise from environment-altering traits of the biota.

To illustrate how biospheric feedbacks affect natural selection, and to explore the impact of random mutation, I extend the Daisyworld model that has underpinned much of the early modelling work on the Gaia hypothesis.

The traits selected at the individual level are ones that change the global environment in a manner favourable to growth.

Origin of the Gaia Hypothesis

The atmosphere of a planet without life (forced only by solar ultraviolet radiation) should show less disequilibrium (attributable to photochemical processes).

The presence of abundant life on a planet may be detectable by atmospheric analysis.

This perturbed state is remarkable in that the atmospheric composition is fairly stable over periods of time that are much longer than the residence times of the constituent gases, indicating that life may regulate the composition of the Earth's atmosphere [14].

Of the major atmospheric gases, nitrogen maintains much of the atmospheric pressure and dilutes oxygen, which constitutes 21% of the atmosphere, just below the fraction at which fires would disrupt land life [15].

Lovelock and Margulis [16–18] therefore proposed the Gaia hypothesis of "atmospheric homeostasis by and for the biosphere", adding that both the redox potential and the acidity of the Earth's surface are anomalous, compared with our planetary neighbours, and can be tolerated by life.

Development of the Gaia Theory

A long history of life on Earth may not be evidence in itself for self-regulation: it is merely a prerequisite for conscious observers to have evolved [19].

Given the large perturbations and changes in forcing of the Earth's surface, being 'just lucky' is a less probable explanation for the persistence of life than the existence of some form of planetary self-regulation.

The silicate-weathering negative feedback could not prevent runaway of the ice–albedo positive feedback as the latter operates much faster [20], indicating that abiotic feedbacks may not be sensitive enough to maintain a habitable climate on Earth for 3.8 billion years.

This increases the growth rate of all daisies, an environmental positive feedback that reinforces the spread of life.

The self-regulation of Daisyworld is impressive: although the solar input changes over a range equivalent to 45 °C the surface of the planet is maintained within a few degrees of the optimum temperature for daisy growth.

The Basis of Environmental Regulation

The first step is to add organisms that alter their environment in a manner that affects their growth, without altering the forces of selection on the responsible trait.

Selective feedback occurs whenever the spread of a trait critically alters the environmental variable that determines the benefit of that trait.

The alteration in the variable can maintain, or even promote, conditions in which the trait is advantageous, generating positive selective feedback.

The environmental alteration can start to reduce the advantage of the trait, generating negative selective feedback.

The daisy traits of 'darkness' and 'paleness' change the world in a way that alters the forces of selection on them, generating selective feedback.

Daisyworld shows that selective feedback is the result of a trait altering the same environmental variable at the level of selection and at a large scale.

Evolution on Daisyworld

Introducing selective herbivory on the daisies slightly impairs regulation but does not destroy it, because selective pressures from the environment dominate.

Adding the unselective herbivores results in smaller populations of daisies at any given time and, therefore, a small decrease in the range of temperature regulation.

A variation of this model is to compare the effects of an unselective herbivore (type 0) and three different types of selective herbivore (types 1–3) that favour more abundant over less abundant daisies to varying degrees [21].

If selection by the herbivores dominates the system, we expect there to be small populations of many daisy colours right up to the extents of regulation, which would therefore be reduced.

At extreme solar luminosities there is an innate tendency for a reduction in daisy biodiversity, because only one or two daisy shades (dark ones at the beginning and light ones at the end) are selected by the environment and can provide regulation.

From Models to Reality

The robust self-regulation of Daisyworld is an outcome of the direct and strong coupling of plant growth to planetary temperature, but the real world is far more complex.

If the daisies were real, their individual temperatures would depend somewhat on their neighbours and their region.

In a two-dimensional extension of Daisyworld [22], albedo mutation generates a heterogeneous distribution of daisy colours and the inclusion of spatial heat transport extends the range of climate regulation.

In the two-dimensional Daisyworld [22], destructive habitat fragmentation impairs temperature regulation once a threshold is reached at which the diminishing areas of daisies become disconnected.

By modelling separate land and ocean biotas, it can be shown that the demise of regulation in one region may not spell disaster for the entire system, as long as enough organisms in other areas are contributing to regulation [23].

Land Ecosystems

Ecosystem-level environmental feedbacks must be understandable in terms of natural selection.

The trees of the Amazon rainforest, through generating a high level of water cycling, maintain the moist environmental conditions in which they can persist [24] (a positive feedback on growth and selection).

Shifts in the balance between boreal forest and tundra amplify external forcing: 115,000 years ago, orbital forcing reduced summer temperatures and seems to have triggered the spreading of the arctic tundra southwards to replace boreal forest [25].

The positive feedback has probably also operated in the opposite direction: 6000 years ago, orbital forcing warmed the high latitudes, which would have triggered boreal forests to spread northwards and amplify the initial warming [26].

Internal changes in ecosystems, involving feedbacks on growth and natural selection, may drive changes in climate [12, 27].

Through promoting soil acidification, iron-capping, water storage and the build up of peat, peat bog plants exclude trees and other plants [27, 28], generating positive selective feedback.

Marine Phytoplankton

Different species of marine phytoplankton produce varying amounts of dimethylsulphoniopropionate (DMSP), the precursor of DMS [29].

The main pathway generates sulphur dioxide which is further oxidized to sulphate, and which can ultimately contribute to sulphate aerosol formation. (A secondary pathway generates methanesulphonate which can be measured in ice cores and used as a proxy for DMS production in the past [30, 31].) Sulphate aerosol is a major source of cloud condensation nuclei [32, 33], which can form cloud droplets that are important scatterers of solar radiation.

A (non-selective) negative feedback on growth of DMS-emitting phytoplankton and climate was first proposed [34] whereby a reduction in temperature and light beneath clouds reduces photosynthesis and restricts the spread of DMS producers.

It, however, an increase in temperature may be amplified by a decrease in photosynthetic production [35], DMS production and cloud reflectivity, generating positive feedback.

Evidence that DMS production in the Southern Hemisphere was enhanced during the last ice age [30, 31] indicated that the feedback may then have been negative, but switched to become positive as temperatures rose at glacial termination [23, 31, 36].

Testing Gaia

One approach would be to test whether adding a realistic environment and random generation of environment-altering traits would enhance the stability of a community ecology model (W. D. Hamilton, personal communication), thus allowing ecosystem-level predictions to be made.

Modelling of interacting ecosystems competing for space in a shared planetary environment might test whether those with stabilizing feedbacks come to predominate.

A dynamic vegetation component for future global circulation models (GCMs), being developed at present, could be put to this task and offer comparison with the real world (P. M. Cox, personal communication).

A simpler (energy-balance) model of feedbacks between the biosphere and climate would provide a framework with which to explore effects over longer timescales (for example, those involved in glacial–interglacial transitions).

Feedbacks that have yet to be incorporated in GCM simulations (such as those involving DMS production) could be quantitatively evaluated and predictions made.

Conclusions

Darwin focused on the exponential growth of organisms, the constraints imposed on them by their environment and the resulting natural selection.

The fact that organisms also alter their environment means there is an inevitable feedback connection between the living and non-living.

Gaian models suggest that we must consider the totality of organisms and their material environment to fully understand which traits come to persist and dominate.

Acknowledgement

A machine generated summary based on the work of Lenton, Timothy M. (1998 in Nature).

Seven Misconceptions Regarding the Gaia Hypothesis

https://doi.org/10.1007/s10584-011-0382-4

Abstract-Summary

Gaia is not a species—she is the summation of all species—and natural selection is a mechanism designed to account for the origin of species, not another level of biological organization altogether.

Gaia theory violates natural selection because it requires that individual organisms or species behave in an altruistic manner or sacrifice individual fitness on behalf of the welfare of the whole.

The single most pervasive misconception associated with the Gaia hypothesis is that global regulation can arise only by means of the ongoing processes of natural selection.

The global envelope of conditions conducive to life is maintained in the face of a universe exceedingly hostile to life; and the integrity of Gaia theory cannot be overturned by a single instance of partial or temporary failure of regulation.

Gaia theory fails to specify on behalf of which life forms Gaia regulates.

Acknowledgement

A machine generated summary based on the work of Moody, David E. (2011 in Climatic Change).

The Emergence and Evolution of Earth System Science

https://doi.org/10.1038/s43017-019-0005-6

Abstract-Summary

Earth System Science (ESS) is a rapidly emerging transdisciplinary endeavour aimed at understanding the structure and functioning of the Earth as a complex, adaptive system.

We discuss the emergence and evolution of ESS, outlining the importance of these developments in advancing our understanding of global change.

Inspired by early work on biosphere–geosphere interactions and by novel perspectives such as the Gaia hypothesis, ESS emerged in the 1980s following demands for a new 'science of the Earth'.

Introduction

It was only in the early twentieth century that contemporary systems thinking was applied to the Earth, initiating the emergence of Earth System Science (ESS).

Building on the recognition that life exerts a strong influence on the Earth's chemical and physical environment, ESS originated in a Cold War context with the rise of environmental and complex system sciences [37–39].

The ESS framework has since become a powerful tool for understanding how Earth operates as a single, complex, adaptive system, driven by the diverse interactions between energy, matter and organisms.

Although other definitions of ESS include the whole planetary interior [40, 41], the processes of which become increasingly important as the timescale of consideration increases [42], we focus on the surface, where the majority of materials are cycled within the Earth System.

The Emergence of ESS

Triggered by the growing recognition of global changes such as human-driven ozone depletion and climatic change, a series of workshop and conference reports in the 1980s called for a new 'science of the Earth' [43, 44].

The challenge of international commitment and disciplinary integration was addressed in 1986 by the International Council for Science (ICSU) with the formation of the International Geosphere-Biosphere Programme (IGBP) [41, 45–47], which joined the World Climate Research Programme (WCRP), formed in 1980 to study the physical-climate component of the Earth System.

The International Human Dimensions Programme (IHDP) on Global Environmental Change was founded, providing a global platform for social science research that explored the human drivers of change to the Earth System and the consequences to human and societal well-being.

The Amsterdam Declaration [48], signed by the Chairs of the International Geosphere-Biosphere Programme (IGBP), International Human Dimensions Programme (IHDP), World Climate Research Programme (WCRP) and DIVERSITAS at the 2001 'Challenges of a Changing Earth' conference, described the key findings of a decade of Earth System Science (ESS).

ESS Tools and Approaches
Supporting the evolutionary development of ESS are three interrelated foci that drive science forwards: observations of a changing Earth System, computer simulations of system dynamics into the future and high-level assessments and syntheses that initiate the development of new concepts.

Such process-level studies complement remote-sensing observations by providing critical insights into the underlying dynamics that generate the patterns of a changing Earth System observed from space.

Although best known for their capability to simulate potential future trajectories of the Earth System, models are probably most valuable as knowledge-integration tools: they bring our rapidly growing understanding of individual processes into an internally consistent framework, they generate new ideas and hypotheses, and, most importantly, the model–observation interface is the ultimate test of our understanding of how the Earth System works.

New Concepts Arising from ESS
Research on tipping elements and cascades highlights the ultimate risks of not only climate change but also of biosphere degradation and the destabilization of the Earth System as a whole [49].

A final example is the planetary boundaries framework, which links biophysical understanding of the Earth (states, fluxes, nonlinearities, tipping elements) [49] to the policy and governance communities at the global level [50].

Built around nine processes that collectively describe the state of the Earth System (including climate change, biodiversity loss, ocean acidification and land-use change), the planetary boundaries framework guides the levels of human perturbations that can be absorbed by the Earth System whilst maintaining a stable, Holocene-like state—a 'safe operating space' for humanity—the only state that we know for certain can support agriculture, settlements and cities, and complex human societies.

In an Earth-System context, the Anthropocene was proposed as a very rapid trajectory away from the 11,700-year, relatively stable conditions of the Holocene [48].

Future Directions

ESS then developed rapidly, from the 'new science of the Earth' movement in the 1980s to the global research efforts of international programmes such as the IGBP.

The big challenge is to fully integrate human dynamics, as embodied in the social sciences and humanities, with biophysical dynamics to build a truly unified ESS effort.

The human dimensions of ESS must, therefore, go well beyond economic models (IAMs) and incorporate the deeper human characteristics that capture our core values and how we view our relationship to the rest of the Earth System.

Although long-ignored by the physical perspectives that have dominated ESS, understanding these human dynamics is essential for the effective guidance systems required for steering the future trajectory of the system [51–53].

Projections of the trajectory of the Earth System—ranging from the biophysical dimensions (for example, climate) to the social sciences and humanities—provide a very wide range of perspectives on the future [54–56].

Acknowledgement

A machine generated summary based on the work of Steffen, Will; Richardson, Katherine; Rockström, Johan; Schellnhuber, Hans Joachim; Dube, Opha Pauline; Dutreuil, Sébastien; Lenton, Timothy M.; Lubchenco, Jane (2020 in Nature Reviews Earth & Environment).

The Gaia Hypothesis: Conjectures and Refutations

https://doi.org/10.1023/A:1023494111532

Abstract-Summary

Gaia theory predicts that the composition of the atmosphere should be tightly regulated by biological processes, but rates of carbon uptake into the biosphere have accelerated by only about 2% in response to the 35% rise in atmospheric CO_2 since pre-industrial times.

Gaia theory would predict that atmospheric CO_2 should be more sensitively regulated by terrestrial ecosystem uptake (which is biologically mediated) than by ocean uptake (which is primarily abiotic), but both processes are about equally insensitive to atmospheric CO_2 levels.

Gaia theory predicts that biological feedbacks should make the Earth system less sensitive to perturbation, but the best available data suggest that the net effect of biologically mediated feedbacks will be to amplify, not reduce, the Earth system's sensitivity to anthropogenic climate change.

Gaia theory predicts that biological by-products in the atmosphere should act as planetary climate regulators, but the Vostok ice core indicates that CO_2, CH_4, and dimethyl sulfide—all biological by-products—function to make the Earth warmer when it is warm, and colder when it is cold.

Gaia theory predicts that biological feedbacks should regulate Earth's climate over the long term, but peaks in paleotemperature correspond to peaks in paleo-CO_2 in records stretching back to the Permian; thus if CO_2 is biologically regulated as part of a global thermostat, that thermostat has been hooked up backwards for at least the past 300 million years.

Gaia theory predicts that organisms alter their environment to their own benefit, but throughout most of the surface ocean (comprising more than half of the globe), nutrient depletion by plankton has almost created a biological desert, and is kept in check only by the nutrient starvation of the plankton themselves.

Acknowledgement

A machine generated summary based on the work of Kirchner, James W. (2003 in Climatic Change).

Long Neglected Damper in the El Niño—Typhoon Relationship: A 'Gaia-Like' Process

https://doi.org/10.1038/srep11103

Abstract-Summary

Tropical cyclone (TC) is one of the earth's most hazardous disasters; it is intriguing to explore whether 'Gaia-like' processes may exist in nature to regulate TC activities.

El Niño can shift the forming position of the Western Pacific typhoons away from land.

This shift enables typhoons to travel longer distances over ocean and is known to be a positive process to promote TCs to achieve higher intensity.

We show that during El Niño, typhoons intensify over region undergoing strong ocean subsurface shoaling where upper ocean heat content can drop by 20–50%.

We find this an elegant, 'Gaia-like' process demonstrating nature's self-regulating ability.

During El Niño, typhoons can take advantage of the longer travelling distance over ocean to achieve higher intensity, nature is also providing a damper to partially cancel this positive impact.

Extended

We show that the linkage is more complex and there is a co-existing negative linkage from subsurface shoaling (blue linkages).

Introduction

The relationship between ocean's subsurface thermal condition, El Niño and typhoon intensification.

For long the above framework neglects the fact that ocean subsurface condition is also different under El Niño.

Reversely, the colder the pre-TC SST, the lower the pre-TC UOHC, the shallower the pre-TC ocean subsurface warm layer, the stronger the TC-induced cooling effect during intensification, the less fluxes will be for intensification.

Given the strong pre-existing subsurface shoaling over the western North Pacific Ocean during the El Niño TC season, we conducted numerical experiments to examine the possible consequential impact.

The above results suggest that during El Niño, the shoaling effect creates a different initial ocean setting for TCs, thus can induce stronger cooling effect during TC's intensification over ocean.

If there was no pre-existing ocean subsurface shoaling, then given the long TC travelling distance (3066 km) during El Niño, the compounded distance-integrated flux supply would be 2.182×10^9 Wm^{-1} (712 Wm$^{-2} \times 3066$ km).

Methods

Ocean Reanalysis System 4 (ORAS4) data of the European Centre for Medium-Range Weather Forecasts (ECMWF) was used.

The atmospheric data was based on the monthly data from the ECMWF (European Centre for Medium-Range Weather Forecasts)'s Interim Reanalysis database at each $1.5°$ grid.

TC track and intensity data was from the US Joint Typhoon Warning Center (JTWC)'s Best Track data base (obtained via the web site of Professor Kerry Emanuel, Massachusetts Institute of Technology, USA).

The long-term climatological mean genesis and intensity peak (i.e. life-time maximum of a TC) locations were calculated based on the average of all western North Pacific TCs (746 cases), from August to November of each year, between 1958 and 2010.

The mean genesis location, mean peak location and mean intensification track was obtained for the El Niño years.

Additional Information

Acknowledgement

A machine generated summary based on the work of Zheng, Zhe-Wen; Lin, I.-I.; Wang, Bin; Huang, Hsiao-Ching; Chen, Chi-Hong (2015 in Scientific Reports).

The Gaia Hypothesis: Fact, Theory, and Wishful Thinking

https://doi.org/10.1023/A:1014237331082

Abstract-Summary

Organisms can greatly affect their environments, and the feedback coupling between organisms and their environments can shape the evolution of both.

Accepted facts, the Gaia hypothesis advances three central propositions: (1) that biologically mediated feedbacks contribute to environmental homeostasis, (2) that they make the environment more suitable for life, and (3) that such feedbacks should arise by Darwinian natural selection.

Many of the biological mechanisms that affect global climate are destabilizing, and it is likely that the net effect of biological feedbacks will be to amplify, not dampen, global warming. (2) Nor do biologically mediated feedbacks necessarily enhance the environment, although it will often appear as if this were the case, simply because natural selection will favor organisms that do well in their environments – which means doing well under the conditions that they and their co-occurring species have created. (3) Finally, Gaian feedbacks can evolve by natural selection, but so can anti-Gaian feedbacks.

Daisyworld models evolve Gaian feedback because they assume that any trait that improves the environment will also give a reproductive advantage to its carriers (over other organisms that share the same environment).

Acknowledgement

A machine generated summary based on the work of Kirchner, James W. (2002 in Climatic Change).

Testing the Effect of Life on Earth's Functioning: How Gaian is the Earth System?

https://doi.org/10.1023/A:1014213811518

Abstract-Summary

The Gaia hypothesis of Lovelock states that life regulates Earth's functioning for its own benefit, maintaining habitable, or even optimum conditions for life.

Based on these definitions, I put forward four null hypotheses, describing increasing beneficial effects of life on the conditions of Earth, ranging from an 'Antigaian' to an 'optimising Gaian' null hypothesis.

I list some indications for rejection of all but one hypothesis, and conclude that life has indeed a strong tendency to affect Earth in a way which enhances the overall benefit (that is, carbon uptake).

This does not imply that the biota regulates Earth's environment for its own benefit.

Acknowledgement
A machine generated summary based on the work of Kleidon, Axel (2002 in Climatic Change).

Beyond Gaia: Thermodynamics of Life and Earth System Functioning

https://doi.org/10.1023/B:CLIM.0000044616.34867.ec

Abstract-Summary
Most prominently, the Gaia hypothesis addresses this question by proposing that near-homeostatic conditions on Earth have been maintained "by and for the biosphere."

The role of the biota in the Earth system is described from a viewpoint of nonequilibrium thermodynamics, particularly with respect to the hypothesis of maximum entropy production (MEP).

We should expect biotic activity, and Earth system processes affected by the biota, to evolve to states of MEP.

Potential implications of the MEP hypothesis for global change research are also discussed.

It is concluded that the resulting behavior of a biotic Earth at a state of MEP may well lead to near-homeostatic behavior of the Earth system on long time scales, as stated by the Gaia hypothesis.

Homeostasis is a result of the application of the MEP hypothesis to biotically influenced processes rather than a postulate.

Besides providing a fundamental perspective on homeostasis, the MEP hypothesis also provides a framework to understand why photosynthetic life would be a highly probable emergent characteristic of the Earth system and why the diversity of life is an important characteristic of Earth system functioning.

Acknowledgement
A machine generated summary based on the work of Kleidon, Axel (2004 in Climatic Change).

Book Reading List

Ecosystem Services
By Muddiman, S. *(2019).*

This book bridges the gap between economic and ecological theory and practice. Its main focus is on how the principles of the Austrian School of economics could improve the validity of Ecosystem Services.

Please see https://www.springer.com/gp/book/9783030138189 for original source.

Earth System Analysis
By Schellnhuber, H. (Ed), Wenzel, V. (Ed) *(1998)*.

As humanity approaches the 3rd millennium, the sustainability of our present way of life becomes more and more questionable. New paradigms for the long-term coevolution of nature and civilization are urgently needed in order to avoid intolerable and irreversible modifications of our planetary environment. **Earth System Analysis** is a new scientific enterprise that tries to perceive the earth as a whole, a unique system which is to be analyzed with methods ranging from nonlinear dynamics to macroeconomic modelling.

Please see https://www.springer.com/gp/book/9783642523564 for original source.

Gaia's Body
By Volk, T. *(1998)*.

Is Earth alive? Put more rigorously, is the biosphere a self- sustaining meta-organism? This is the essence of Gaia theory: if the biosphere really is a single coherent system, then it must have something like a physiology. It must have systems and processes that perform living functions. OK, then, what systems, what processes, what functions? Gaia's Body is Tyler Volk's answer to this question. In this book, he describes the environment that enables the biosphere to exist; various ways of looking at its "anatomy" and "physiology," the major biogeographical regions such as rainforests, deserts, and tundra; the major substances the biosphere is made of; and the chemical cycles that keep it in balance.

Please see https://www.springer.com/gp/book/9781461274520 for original source.

The Biosphere
By Vernadsky, V. I. *(1998)*.

Long unknown in the West, The Biosphere established the field of biogeochem-istry and is one of the classic founding documents of what later became known as Gaia theory. It is the first sustained expression of the idea that life is a geological force that can change Earth's landforms, its climate, and even the contents of its atmosphere. A complete, unabridged translation has never before been available in English. This edition - complete with extensive annotations, an introductory essay placing the work in its historical context and explaining its relevance to readers today, and a foreword cosigned by a stellar group of international experts - will be the definitive edition of this classic work.

Please see https://www.springer.com/gp/book/9780387982687 for original source.

The Dynamics of Small Solar System Bodies
By Wood, J. *(2019)*.

This SpringerBrief summarizes the latest relevant research and discoveries that have been made in the area of ringed small bodies and small body taxonomy, including those that lay the groundwork for future discoveries.

Please see https://www.springer.com/gp/book/9783030281083 for original source.

The Earth System and Evolution of Life
By Maruyama, S., Santosh, M. W. *(2021).*

During the last 200 years since Geology has been established as an integrated science, nearly the same duration as modern Biology, our understanding of the Earth has taken great leaps forward through the works of several experts, and by contributions from a large number of scientific community. In the 21st Century, however, we face a massive challenge to understand and integrate the voluminous data and break-through made in several fields of Genome-Biology, Astronomy, Climate in the near future, fast depleting resources and the fate of human beings in this Planet. The well illustrated chapters in this book provide a succinent summary of the multidisciplinary nature of science and attempts to bridge genome-level biology through astronomy and earth history.

Please see https://www.springer.com/gp/book/9789048190478 for original source.

Understanding the Earth System
By Ehlers, E. (Ed), Krafft, T. (Ed) *(2001).*

This volume includes revised versions of most of the presentations made at the International Conference «Understanding the Earth Sys tem: Compartments, Processes and Interactions" held on November 24–26, 1999 in Bonn. The Conference was organized by the German National Committee on Global Change Research as part of the Bonn Science Festival 1999–2000.

Please see https://www.springer.com/gp/book/9783540675150 for original source.

References

1. Maturana HF, Varela FJ (1980) Autopoiesis and cognition: the realization of the living. In: Cohen RS, Wartofsky MW (eds) Boston studies in the philosophy of science 42. D. Reidel Publishing Co., Dordecht
2. Thom R (1989) Structural stability and morphogenesis: an outline of a general theory of models. Addison-Wesley, Reading
3. Boulding KE (1965) The image. University of Michigan Press, Ann Arbor
4. Azzone GF (2010) The second evolution: after the genes the cultural birth. Rend Lincei 21:283–299
5. Haken H (1975) Cooperative phenomena in systems far from thermal equilibrium and in non physical systems. Rev Mod Phys 47:91–121
6. Kirchner JW (1989) The Gaia: can it be tested. Rev Geophys 27:223–235

7. Kleidon A (2002) Testing the effect of life on Earth's functioning: how Gaian is the Earth system? Clim Change 52:383–389
8. Margulis L (2001) The conscious cell. Ann NY Acad Sci 929:1–16
9. Schrödinger E (1944) What's life. Cambridge University Press, USA
10. Bohr N (1965) I quanti e la vita. Bollati Boringhieri, Torino
11. Heisenberg W (1958) Physics and philosophy. Allen and Unwin, London
12. Lovelock JE (1995) The ages of Gaia, 2nd edn. Oxford Univiersity Press, Oxford
13. Doolittle, W. F. Is Nature really motherly? CoEvol Quart Spring 58–63 (1981).
14. Lovelock JE (1972) Gaia as seen through the atmosphere. Atmos Environ. 6:579–580
15. Watson AJ, Lovelock JE, Margulis L (1978) Methanogenesis, fires and the regulation of atmospheric oxygen. Biosystems 10:293–298
16. Lovelock JE, Margulis L (1974) Atmospheric homeostasis by and for the biosphere: the gaia hypothesis. Tellus 26:2–10
17. Margulis L, Lovelock JE (1974) Biological modulation of the Earth's atmosphere. Icarus 21:471–489
18. Lovelock JE, Margulis L (1974) Homeostatic tendencies of the Earth's atmosphere. Origins Life 5:93–103
19. Watson AJ (1991) Gaia. New Sci Inside Sci 48:1–4
20. Caldeira K, Kasting JF (1992) Susceptibility of the early Earth to irreversible glaciations caused by carbon dioxide clouds. Nature 359:226–228
21. Harding SP, Lovelock JE (1996) Exploiter-mediated coexistence and frequency-dependent selection in a numerical model of biodiversity. J Theor Biol 182:109–116
22. Von Bloh W, Block A, Schellnhuberr HJ (1997) Self-stabilization of the biosphere under global change: a tutorial geophysiological approach. Tellus 49B:249–262
23. Lovelock JE, Kump LR (1994) Failure of climate regulation in a geophysiological model. Nature 369:732–734
24. Salati E (1987) In: Dickinson RE (ed) The geophysiology of Amazonia: vegetation and climate interactions . Wiley, New York, pp 273–296.
25. Gallimore RG, Kutzbach JE (1996) Role of orbitally induced changes in tundra area in the onset of glaciation. Nature 381:503–505
26. Foley JA, Kutzbach JE, Coe MT, Levis S (1994) Feedbacks between climate and boreal forests during the Holocene epoch. Nature 371:52–54
27. Klinger LF (1991) In: Schneider SH, Boston PJ (eds) Scientists on Gaia. MIT, London, pp 247–255
28. Hamilton WD (1996) Gaia's benefits. New Sci 151:62–63
29. Liss PS, Hatton AD, Malin G, Nightingale PD, Turner SM (1997) Marine sulphur emissions. Phil Trans R Soc Lond B 352:159–169
30. Legrand M et al (1991) Ice-core record of oceanic emissions of dimethylsulphide during the last climate cycle. Nature 350:144–146
31. Legrand M (1997) Ice-core records of atmospheric sulphur. Phil Trans R Soc Lond B 352:241–250
32. Andreae MO, Crutzen PJ (1997) Atmospheric aerosols: biogeochemcial sources and role in atmospheric chemistry. Science 276:1052–1058
33. Ayers GP, Gras JL (1991) Seasonal relationship between cloud condensation nuclei and aerosol methanesulphonate in marine air. Nature 353:834–835
34. Charlson RJ, Lovelock JE, Andreae MO, Warren SG (1987) Oceanic phytoplankton, atmospheric sulphur, cloud albedo and climate. Nature 326:655–661
35. Roemmich D, McGowan J (1995) Climatic warming and the decline of zooplankton in the California current. Science 267:1324–1326
36. Lovelock JE (1997) Ageohysiologist's thoughts on the natural sulphur cycle. Phil Trans R Soc Lond B 352:143–147
37. Vernadsky VI (1924) La Géochimie. Librairie Félix Acan
38. Vernadsky VI (1998) The biosphere (complete annotated edition: Foreword by Margulis L et al. Introduction by Grinevald J. translated by Langmuir DB. Rrevised and annotated by McMenamin MAS). Springer

39. Lovelock J (1979) Gaia: A New Look at Life on Earth. Oxford University, Oxford
40. National Research Council (1986) Earth system science. A Program for Global Change (National Academies Press, Overview
41. Dutreuil S (2016) Gaïa: Hypothèse, Programme de Recherche pour le Système Terre, ou Philosophie de la Nature? Thesis, University Paris 1 Panthéon-Sorbonne
42. Lenton TM (2016) Earth system science. A very short introduction. Oxford University Press
43. Waldrop MM (1986) (1986) Washington embraces global earth sciences. Science 233:1040–1042
44. Edelson E (1988) Laying the foundation. MOSAIC 19:4–11
45. Kwa C (2005) Local ecologies and global science: discourses and strategies of the International Geosphere-Biosphere Programme. Soc Stud Sci 35:923–950
46. Kwa C (2006) The programming of interdisciplinary research through informal science-policy interactions. Sci Public Policy 33:457–467
47. Uhrqvist O (2014) Seeing and knowing the earth as a system: an effective history of global environmental change research as scientific and political practice. Thesis, Linköping University
48. Steffen W et al (2004) Global change and the earth system: a planet under pressure. Springer, 2004.
49. Aykut S (2015) Les, "limites" du changement climatique. Cités 63:193–236
50. Rockström J et al (2009) A safe operating space for humanity. Nature 461:472–475
51. Cai Y, Lenton TM, Lontzek TS (2016) Risk of multiple interacting tipping points should encourage rapid CO_2 emission reduction. Nat Climate Change 6:520–525
52. Folke C, Biggs R, Norström AV, Reyers B, Rockström J (2016) Social-ecological resilience and biosphere-based sustainability science. Ecol Soc 21:41
53. Carpenter SR, Folke C, Scheffer M, Westley FR (2019) Dancing on the volcano: social exploration in times of discontent. Ecol Soc 24:23
54. Heymann M, Dahan Dalmedico A (2019) Epistemology and politics in Earth system modelling: historical perspectives. J Adv Model Earth Syst 11:1139–1152
55. Malm A, Hornborg A (2014) The geology of mankind? A critique of the Anthropocene narrative. Anthrop Rev 1:62–69
56. Picketty T (2014) Capital in the twenty-first century. Harvard University Press

Printed by Printforce, the Netherlands